THE HENRY A. WALLACE SERIES
ON AGRICULTURAL HISTORY AND RURAL STUDIES

IOWA STATE UNIVERSITY PRESS • AMES

THE HENRY A. WALLACE SERIES
ON AGRICULTURAL HISTORY AND RURAL STUDIES

*Richard S. Kirkendall,* SERIES EDITOR

The American Farmer and the New Deal
*by Theodore Saloutos*

Roswell Garst: A Biography
*by Harold Lee*

Railroad Development Programs in the Twentieth Century
*by Roy V. Scott*

Agricultural Science and the Quest for Legitimacy: Farmers, Agricultural Colleges, and Experiment Stations, 1870–1890
*by Alan I Marcus*

To Their Own Soil: Agriculture in the Antebellum North
*by Jeremy Atack and Fred Bateman*

Corn and Its Early Fathers
*by Henry A. Wallace and William L. Brown*

Toward a Well-Fed World
*by Don Paarlberg*

Prairie Grass Roots: An Iowa Small Town in the Early Twentieth Century
*by Thomas J. Morain*

The Farm Policy Game: Play by Play
*by Lauren Soth*

Agriculture and National Development: Views on the Nineteenth Century
*edited by Lou Ferleger*

# Agriculture and National Development: Views on the Nineteenth Century

# Agriculture and National Development

*Views on the Nineteenth Century*

*Edited by* LOU FERLEGER

THE HENRY A. WALLACE SERIES
ON AGRICULTURAL HISTORY AND RURAL STUDIES

IOWA STATE UNIVERSITY PRESS • AMES

To my parents
Samuel and Jean Ferleger

LOU FERLEGER, editor, is Associate Dean of the College of Arts and Sciences and Associate Professor of Economics at the University of Massachusetts/Boston. He has written articles on the postbellum southern economy, statistics, and the contemporary macroeconomy. He is coauthor of *Statistics for Social Change* and a recent recipient of a fellowship from the National Endowment for the Humanities. He is finishing a monograph tentatively entitled *Tools and Time: Southern Farmers after Reconstruction.*



Manufactured in the United States of America
♾ This book is printed on acid-free paper



First edition, 1990

**Library of Congress Cataloging-in-Publication Data**

Agriculture and national development : views on the nineteenth century / edited by Lou Ferleger. — 1st ed.

p. cm. — (The Henry A. Wallace series on agricultural history and rural studies)

Includes bibliographical references.

**ISBN 0-8138-0314-4. — ISBN 0-8138-0313-6** (pbk.)

1. Agriculture — United States — History — 19th century. 2. Agricultural innovations — United States — History — 19th century.

I. Ferleger, Lou. II. Series.

S441.A357 1990

338.1′0973 — dc20 89-24736

# CONTENTS

# SERIES EDITOR'S INTRODUCTION

THE HENRY A. WALLACE SERIES on Agricultural History and Rural Studies is designed to enlarge publishing opportunities in agricultural history and thereby to expand public understanding of the development of agriculture and rural society. The Series will be composed of volumes that explore the many aspects of agriculture and rural life within historical perspectives. It will evolve as the field evolves. The Press and the editor will solicit and welcome the submission of manuscripts that illustrate, in good and fresh ways, that evolution. Our interests are broad. They do not stop with Iowa and U.S. agriculture but extend to all other parts of the world. They encompass the social, intellectual, scientific, and technological aspects of the subject as well as the economic and political. The emphasis of the series is on the scholarly monograph, but historically significant memoirs of people involved in and with agriculture and rural life and major sources for research in the field will also be included.

Most appropriately, this Iowa-based Series is dedicated to a highly significant agriculturist who began in Iowa, developed a large, well-informed interest in its rural life, and expanded the scope of his interests beyond the state to the nation and the world. An Iowa native and son of an agricultural scientist, journalist, and secretary of agriculture, Henry A. Wallace was a 1910 graduate of Iowa State College, a frequent participant in its scientific activities, editor of *Wallaces' Farmer* from 1921 to 1933, founder in 1926 of the Hi-Bred Corn Company (now Pioneer Hi-Bred International, Inc.), secretary of agriculture from 1933 to 1940, and vice president of the United States from 1941 to 1945. In the agricultural phases of his wide-ranging career, he was both a person of large importance in the development of America's agriculture and the leading policymaker during the most creative period in the history of American farm policy.

A man of several parts, Henry A. Wallace had an interest in history as well as in the other disciplines with which his name is more often associated, especially genetics, statistics, and agricultural economics. His knowledge of history, his sense of it, and his belief that there were lessons to be learned from it appeared often in his public statements. Many of his writings in *Wallaces' Farmer* were essentially historical essays, often with much statistical data, for he was fascinated by numbers, regarding them as essential for the definition and understanding of change. Although he had spent little time in history classrooms, he had some understanding of agricultural history in the nineteenth century, the subject of this book. He knew that, as Professor Ferleger writes, "the agricultural sector vastly increased its capacity to produce" during that century, and "the changes in agriculture supported and facilitated the industrial revolution of the late nineteenth century." His most illustrious ancestor, his paternal grandfather "Uncle Henry" Wallace, had participated in, observed, and interpreted those developments and had passed his observations and interpretations on to "Young Henry."

This volume of essays provides, in a sense, a justification for the Wallace Series. The essays illustrate how lively, provocative, and challenging the field of agricultural and rural history has become. The volume presents evidence of the large importance of agriculture and rural life in the history of the United States, and its authors offer help for others who are ready to do work that needs to be done. The book both shows where we are in our understanding of a major part of American history and makes recommendations that could enrich even more a major field of exploration.

RICHARD S. KIRKENDALL
The Bullitt Professor of American History
The University of Washington

# PREFACE

THIS BOOK FOCUSES on the transformation of American agriculture in the nineteenth century—a century that marked a great change. First, the agricultural sector vastly increased its capacity to produce both old and new products. Second, the changes in agriculture supported and facilitated the industrial revolution of the late nineteenth century. Yet far too few analysts of American agriculture note and understand the many facets of this transformation as it took place in different regions under diverse conditions.

Understanding the transformation is not just a question of a familiarity with a breadth of trends and experiences, of historical or economic data. It is also a question of interpretation. To date, very little attention has been directed to the varying and at times conflicting interpretations by scholars of the pace and impact of regional changes in the agricultural sector. Such differences in interpretation have been addressed only modestly (most notably in studies on the slave south) and usually in specialist journals.

It is the intent of this book, therefore, to offer a wide range of views on the changing patterns of nineteenth-century American agriculture. I suggested that authors review the work that has been done in their specific fields and identify areas of serious debate. However, these essays were not intended to be merely reviews of the literature. The authors individually determined the mix of factors, theories, themes, and interpretations they wanted to include. That is, the theoretical and historical approach adopted in each essay conforms to the author's own outlook, the author's own sense of what facts and issues must be discussed to convey "what happened" in agriculture in a given region and time. I did not require an author to argue against every other scholar's view of the issues; rather I let each essayist determine whose view was worthy of note.

This book shows that there are many interpretations of nineteenth-century American agriculture. Undoubtedly some readers will argue that the essayists and I have left out still other approaches, have ignored issues of equal importance to the ones discussed. They may be right. I look forward to additional books that reassert in other ways what Wayne Rasmussen proposes in the Introduction: "The changes that took place in American agriculture during the nineteenth century were of vital importance to the development of the nation."

This is not a compilation of papers presented at a conference. Essays were individually solicited, reviewed, and accepted or rejected by myself in consultation with the editorial staff at the Iowa State University Press. Many people were helpful in suggesting contributors to this volume, including Stanley Engerman, Elizabeth Fox-Genovese, Eugene Genovese, William Parker, Morton Rothstein, Harold Woodman, and Gavin Wright. I am grateful for their advice and support. Dick Cluster provided valuable editorial suggestions on the style and content of the essays for which I am thankful. Robert Schultz, acquisitions editor at the Iowa State University Press, shepherded me through the editorial process, which took longer than I (though not he) expected. Once all the papers arrived I was assisted in the final preparation of the manuscript by Suzanne Lowitt. Finally, Carol Ivan has supported the project since its inception with her usual grace, wit, and advice.

LOU FERLEGER

# INTRODUCTION

*Wayne D. Rasmussen*

The Changes That Took Place in American agriculture during the nineteenth century were of vital importance to the development of the nation. As the century began, farmers made up some 90 percent of the nation's working force. The real farm gross product was $333 million, with each worker producing food and fiber valued at $292. The nation's exports, valued at $71 million, were almost entirely farm and forest products. As the century ended, farmers were 38 percent of the working force. The real farm gross product was $5,740 million, with each farm worker producing products valued at $526. Farm products were some 62 percent of exports. Agriculture had supplied much of the capital and working force that transformed the nation, whether for good or for ill, from a rural to an industrial nation.

During the century farmers moved from being somewhat self-sufficient to producing for the market place, although southern cotton and tobacco growers had been producing for the market before the Revolution. Farmers in the North and West adopted horsepowered machinery, while southern farmers saw slavery come to an end, to be replaced mainly by sharecropping. Continuing changes in federal land laws, culminating in the Homestead Act of 1862, made it easier for individuals to acquire title to farmland. The number of farms increased from 335,000 in 1800 to 5,727,000 in 1900 even though farmers were decreasing steadily as a percentage of the population.

Many of these new farms were established by immigrants who came to the United States to secure free land under the Homestead Act or to settle on farms offered at low cost by the railroads. These newcomers influenced the shape of American agriculture and of rural life, but in ways, as Kathleen Neils Conzen points out in her chapter, "Immigrants in Nineteenth-Century Agricultural History," not yet fully understood. The transformation of rural America is a familiar story, but we are only now beginning to understand "the transformation of European peasants into American farmers and the transformation of American farming by them. . . ."

At about the same time the Homestead Act was passed, Congress, in the Morrill Land-Grant College Act, gave public land to the states for colleges that would teach agriculture, mechanical engineering, and military science, and, by another act, established the U.S. Department of Agriculture. This greatly strengthened the move towards improving agricultural production, which had been started by agricultural societies and farm journals and had been carried forward in the 1830s and 1840s by inventors of laborsaving horsepowered machines. The demand for farm products during the Civil War led farmers to adopt new methods and new machines, basically replacing hand power by horsepower. This has been called the first American agricultural revolution. The application of science and technology to agriculture was further strengthened by the Hatch Agricultural Experiment Station Act of 1887, which provided federal funding for establishing one or more experiment stations in each state.

As production increased both from improved practices and from more land being brought under cultivation, surpluses developed and farmers went through successive periods of economic difficulty. Some farm organizations such as the National Grange and the Farmers Alliances organized cooperatives and called for government regulation of railroads, elevators, and big business monopolies. While some successes were achieved, failures were more frequent.

Rural life also was changing, though perhaps not as rapidly as farming. Many of the farm journals carried "women's" or "family" pages, featuring such items as recipes, sewing hints, and health-care techniques. The National Grange, from its founding in 1867, stressed family rather than individual memberships and had programs for both women and youth. The Department of Agriculture's annual reports or yearbooks carried articles on health, labor saving in the farm home, and more effective planning of the farmstead. During the nineteenth century, according to Elizabeth Fox-Genovese in her chapter "Women in Agriculture during the Nineteenth Century," women contributed to agricultural households in three primary ways. They produced essential goods and services for members of farm households, they worked on their own to produce goods directly for the market, and they worked outside the household to earn money to contribute to its support. A fuller understanding of these roles depends upon the accumulation of a host of microcosmic studies.

While some generalizations may be made about American agriculture in the nineteenth century, it has been studied mainly by regions, which is the basic organization followed in this volume. During the 1920s and 1930s, as major regional studies were underway, a number of studies of individual farms, localities, and rural communities were published. These were dismissed by many in the history profession as "local studies," interesting only to those in the immediate area. Today, however, many working in what has been called the "new rural history" say that we cannot truly understand the changes that have taken place in the structure of American agriculture without studying the individual farm and the rural community. The authors of several chapters in this book are leaders of this new group of historians. Drawing upon the local studies available, while pointing out the need for more such studies, they have written chapters centered on changes within particular agricultural regions. Incidentally, many of the scholars in the field of "new rural history" write of the yeomen or yeomen farmers, a term virtually never used by American farmers about themselves or their fellow farmers.

The most important development faced by northern farm communities during the nineteenth century, according to Hal S. Barron in his chapter "Listening to the Silent Majority: Change and Continuity in the Nineteenth-Century Rural North," was the expansion of the market economy and its social costs to rural communities. Most farmers participated in the market, but they were not quintessential economic men. Their participation in the market economy was mediated by noneconomic concerns relating to such concepts as independence and family security. In my opinion, this is true of the great majority of American farmers today, most of whom are "small" or "midlevel" farmers. Barron concludes that the rural North experienced a great deal of continuity during the transformation of nineteenth-century America into an urban-industrial society.

A study of 21,118 rural households in the Northeast and the Midwest, based upon the 1860 federal census manuscripts, demonstrates that "a substantial proportion of the population, including natives and immigrants, was eventually financially able to acquire land to start a farm and to make it eventually produce not only enough for one's family but for market sale as well." However, as Jeremy Atack and Fred Bateman further conclude in "Yeoman Farming: Antebellum America's

Other 'Peculiar Institution,'" problems were beginning to appear even though the northern rural economy was succeeding as an integrated social, familial, and economic institution. There was a tension between the national pursuit of a new social order of individualism and independence based upon a family farm system, on one side, and the drive toward an interdependent capitalistic market economy in which farming functioned largely as a business enterprise, on the other.

Agricultural historians, as R. Douglas Hurt points out in "Northern Agriculture after the Civil War, 1865–1910," have often emphasized the westward movement of both farm people and types of farming in their studies of northern agriculture. The spread of the production of particular commodities and of agricultural institutions was traced in a series of descriptive articles, mainly by agricultural economists and economic historians, in the *Yearbooks* of the U.S. Department of Agriculture published from 1921 through 1925. It is only in recent years that historians have paid attention to the human implications involved and to the economic and cultural adjustments made in the older regions of the North. More specific studies in the subjects outlined by Hurt would be helpful, but, as he suggests, the time may be nearing for a synthesis of northern agricultural history since the Civil War.

During the nineteenth century the Midwest became the agricultural heartland of the nation. This is one reason that historians have given it close attention in the past and continue to do so in the present. As Donald Winters points out in "The Economics of Midwestern Agriculture, 1865–1900," infrastructure—including railroads, banks, marketplaces, and roads—affected the development of midwestern agriculture. Without this infrastructure, commercial farming would have developed much more slowly and individual farmers would have found their opportunities to acquire and develop farms definitely limited. However, the infrastructure was developed and farms became very productive. During the last thirty years of the century farmers produced more than the domestic and export markets could absorb at prices that would give farmers a reasonable return on labor and investment. This, however, was not clear either to farmers or to the nation as a whole. Farmers blamed the railroads, the banks, and manipulation in the marketplace for their problems, while some economists and others wrote that the problem could be solved by better farm management, including more attention to marketing. Winters urges economic historians to deal with this broader theoretical question. Future research, he states, must address the dispute over which conceptual framework, the one that assumes manipulation by special interests or the one that assumes predominance of market forces, offers the most appropriate model of the nineteenth-century economy.

Farmers, aided by such new institutions as the land-grant universities and the agricultural experiment stations, sought to solve their economic problems through more efficient farming and marketing and through such organizations as the National Grange and the various Alliances. It was difficult, Dorothy Schwieder shows in "Agricultural Issues in the Middle West, 1865–1910," to deal with such problems as the exodus of farm youth to the cities and the declining political, economic, and social influence of the farm population. A good corn crop would not solve many of the problems faced by farm women, such as inadequate education for one's children, lack of conveniences in the farm home, lack of opportunities for socializing, and the unrelenting routine of hard physical work from sunup to sundown. From about 1900 until the end of World War I, farmers shared in the general national prosperity, but many social problems remained. This was pointed out in the 1909 report of the Country Life Commission.

The South has been the most studied of all agricultural regions and continues to attract research. Southern agriculture was different. It was the first to become highly dependent upon the marketplace and its basic commodities were different from those of the North and Midwest, but, most of all, its commercial agriculture was based upon human slavery until after the Civil War and then became dependent upon an exploitive labor system known as sharecropping.

There was one important similarity between southern and northern agriculture, as David F. Weiman points out in "Staple Crops and Slave Plantations: Alternative Perspectives on Regional Development in the Antebellum Cotton South." Both were expanding rapidly from the eastern seaboard to the west. He concludes that the viability of the plantation system as a social order depended upon the extension of the legal boundaries of slavery, but that the drive to expand the slave regime heightened sectional conflicts and the external threat to slavery. The plantation, with its slaves, was singularly efficient in settling sparsely populated frontier regions and producing cotton for export, a fact usually overlooked in discussions of frontier settlement. Overall, according to Weiman, the plantation was a more effective producer than the small farmer. By the early 1850s, the widening economic gap between the plantation owner and the free farmers and wage laborers was becoming more apparent and was threatening to shatter the organic unity that preserved the social order.

While many economic historians like Weiman believe that the slave-worked plantation was an efficient unit of production, there are some, as pointed out by Richard H. Steckel in "Growth and Development in the Antebellum South: Old Debates and New Directions," who disagree. Data is available and could be brought together in a useful form to permit careful econometric analyses of production efficiency. Other areas of antebellum southern agriculture that should be given such analysis include health, fertility, and migration. Further research is also warranted in the place of the southern free, nonslaveholding population in a plantation-dominated economy and social structure.

There is even more controversy over what took place in agriculture in the postbellum South than in the pre-Civil War period. The most important factor, according to Jay R. Mandle in "The Role of Markets and Institutions in Postbellum Southern Development," is race. With freedom, the former slaves possessed neither formal education nor substantial material assets and the economy in which they were living was severely damaged. For some years the ex-slaves and some reformers believed that the federal government could and would provide each ex-slave with "forty acres and a mule." This never happened and it was left to the marketplace, influenced at least in part by the question of race, to find a solution. Black labor, however, was compelled to work in the South until World War I because northern employers preferred European immigrants to southern blacks, perhaps in part because the blacks were almost totally without education and because of northern racial prejudice. Immobility and racism were characteristic aspects of the plantation system.

The situation was complex. It is necessary to understand what crops were grown and where, the types of labor contracts that evolved, and credit arrangements made as well as the existence of racial discrimination. These are the matters discussed by Lee J. Alston in "Issues in Postbellum Southern Agriculture." That the South devoted even more of its resources to the production of cotton after the war than it did before shaped the entire structure of southern agriculture. Cotton could be grown by uneducated workers even without close supervision, yet it could not be consumed by the worker and there was always a market even though at times the price might be low. This was reflected in the labor contracts and the types of credit available. One reason why there was little possibility for change lay in the fact that the plantation owners, bankers, and merchants completely dominated the political process. Thus Alston observes, "As a result of racism, low educational expenditures, inappropriate technology and little out-migration, the South at the end of the postbellum period remained an isolated low-wage region within the U.S. economy."

During the second half of the nineteenth century, according to Joseph P. Reidy in "Slavery, Emancipation, and the Capitalist Transformation of Southern Agriculture, 1850–1910," capitalistic social relations spread in the South, bringing about a social revolution. The revolution was not completed during this period, but forces had been put in place that left the outcome virtually inevitable. This was true even though the rate of change differed among commodities and slowed when workers and croppers combined, always in the long run unsuccessfully, to resist the demands of the planters. After 1900 the South continued internal capital accumulation, outside capital investment, mechanization, and the qualitative exploitation of labor. Not until after World War II did a combination of insect pests, natural disasters, New Deal programs, racist violence in the South, and employment opportunities in the North bring about a transformation of the labor force from an agricultural to an industrial one.

Western agriculture also underwent major changes in the nineteenth century. What little agriculture there was in the region at the beginning of the period was found along the Pacific Coast in regions not yet part of the United States. Over the century the United States expanded to the Pacific Ocean and the West was tied to the rest of the nation by railroads. Farming moved from open-range livestock grazing and large-scale wheat production to livestock ranching and the beginnings of an irrigation farm economy that was to transform agriculture in California, the Columbia Basin, Utah, and other smaller areas, even though large-scale wheat farming continued to dominate large parts of the region. Much of the Great Plains was marginal so far as farming was concerned and a number of the more aggressive farm protest movements originated or were strongly pressed there.

The essays in this volume trace in some detail the changes taking place in agriculture during the nineteenth century that transformed farmers' goals from independence and self-sufficiency to profit from the marketplace and improved standards of living. The long-term results of these changes are less apparent than the changes themselves, even though the authors give such problems careful attention. For example, by the end of the century most farmers did not make decisions or build their lives entirely around the dictates of the marketplace. Indeed, most who did left farming. Their resources, whether capital or labor, brought greater returns in industry than in farming. Those who remained in farming did so from a mixture of motives and of alternative opportunities. As several authors point out, this is an area worthy of more detailed studies and analyses based upon a rigorous use of the data available. At the same time, as is obvious from these studies, the analyst must develop or adopt a theoretical framework within which to conduct research, but even this is not simple. One of the most difficult problems and yet one that is basic to understanding the changes taking place is determining whether the farmer was most influenced by manipulation by special interests or was driven by the impersonal forces of the marketplace. Or is it possible that the farmer was caught somewhere in-between? The studies in this volume help us answer that question.

# CONTRIBUTORS

LEE J. ALSTON is Associate Professor of Economics at the University of Illinois at Urbana/Champaign. His specialty is the economic history of the southern United States after the Civil War. He has published many articles and is at present working on a book manuscript with Joseph P. Ferrie entitled "Paternalism in the Southern United States: A Study of Institutional Decay."

JEREMY ATACK is Professor of Economics at the University of Illinois at Urbana/Champaign. He is author and coauthor of many articles on agricultural and industrial development in the United States during the nineteenth century. Most recently, he is coauthor with Fred Bateman of *To Their Own Soil* which won the 1987 Theodore Saloutos Prize awarded by the Agricultural History Society.

HAL S. BARRON is an Associate Professor of History at Harvey Mudd College and a member of the History Faculty at the Claremont Graduate School. For several years he has served as the Convener of the Rural History Network of the Social Science History Association. Besides several articles on rural social history, he is the author of *Those Who Stayed Behind: Rural Society in Nineteenth-Century New England.* He is at present writing a book on the rural North between 1880 and 1930, which focuses on the effects of urban culture and an emerging organizational society on the countryside.

FRED BATEMAN is Professor of Business Economics in the Indiana University School of Business. He has published extensively in the areas of U.S. industrial history and agricultural development. Among his publications are *A Deplorable Scarcity* and *To Their Own Soil* (with Jeremy Atack), which won the 1987 Theodore Saloutos Prize awarded by the Agricultural History Society. He is now working with Jeremy Atack on a monograph on American industrialization and papers on the emergence of large-scale enterprise under support from the National Science Foundation.

KATHLEEN NEILS CONZEN is Associate Professor of History at the University of Chicago. She has written many articles on rural immigrants including "Peasant Pioneers: Generational Succession among German Farmers in Frontier Minnesota," "German-Americans and the Invention of Ethnicity," and "Historical Approaches to the Study of Rural Ethnic Communities."

ELIZABETH FOX-GENOVESE is the Eleonore Raoul Professor of the Humanities and Director of Women's Studies at Emory University. Her publications include *The Origins of Physiocracy: Economic Revolution and Social Order in Eighteenth-Century France; Within the Plantation Household: Black and White Women of the Old South;* and with Eugene D. Genovese, *Fruits of Merchant Capital: Slavery and Bourgeois Property in the Rise and Expansion of Capitalism.*

R. DOUGLAS HURT is Director of the Center for the Study of Agriculture and Rural Life, Iowa State University. He is the author of *The Dust Bowl: An Agricultural and Social History; American Farm Tools: From Hand Power to Steam Power; Indian Agriculture in America: Prehistory to the Present;* and *The Department of Agriculture,* and many articles in American agricultural history.

JAY R. MANDLE is Professor of Economics and History at Temple University. He has published extensively on the postemancipation experience of both the United States South and the Caribbean. His recent book, coauthored with Joan D. Mandle, is *Grass Roots Commitment: Basketball and Society in Trinidad and Tobago.*

WAYNE D. RASMUSSEN has been historian of the United States Department of Agriculture until 1986 (when he retired), including being chief of the Agricultural History Branch. He has been president and founder of the Association for Living Historical Farms and Agricultural Museums and president of the Agricultural History Society. He is at present the society's Executive Secretary-Treasurer. He has published dozens of articles and many monographs including the classic reference documenting the history of American agriculture, *Agriculture in the United States: A Documentary History* (4 vols.). He has recently completed a book on the United States extension service, *Taking the University to the People: Seventy-five Years of Cooperative Extension.*

JOSEPH P. REIDY is an Associate Professor of United States History at Howard University. He was formerly a member of the Freedmen and Southern Society Project at the University of Maryland and is a coeditor of its first three published volumes: *The Black Military Experience, The Destruction of Slavery,* and *The Wartime Genesis of Free Labor.* He has authored several articles on the nineteenth-century South and is now at work on a study of slavery and emancipation in nineteenth-century Georgia.

RICHARD H. STECKEL is Associate Professor of Economics at Ohio State University and a Research Associate at the National Bureau of Economic Research. With research interests in historical demography, he has published numerous articles on migration, fertility, mortality, and health and a book on *The Economics of U.S. Slave and Southern White Fertility.*

DOROTHY SCHWIEDER is a Professor of History at Iowa State University. Her specialties are women and nineteenth-century social history in Iowa. She has recently co-authored *Buxton: Work and Racial Equality in a Coal Mining Community.* She is at present working on the history of the Iowa State Extension Service.

DAVID F. WEIMAN is Associate Professor of Economics at Yale University. He has written many articles on the Upcountry and regional development in the nineteenth-century South and is at present working on a monograph, *Uneven Regional Development and the Origins of the New South: The Formation and Development of the Southern Upcountry, 1820 to 1920.*

DONALD L. WINTERS is Professor of History at Vanderbilt University. He is author of *Henry Cantwell Wallace as Secretary of Agriculture 1921–1924* and *Farmers Without Farms: Agricultural Tenancy in Nineteenth-Century Iowa.* More recently, he has published articles on southern agriculture in the nineteenth century.

# PART I

# The North and the Midwest

CHAPTER 1

# Listening to the Silent Majority: Change and Continuity in the Nineteenth-Century Rural North

HAL S. BARRON

THE RELATIONSHIP of the rural North to the larger thrust of American history during the nineteenth century is a time-honored theme in American historiography.[1] It was the North's shifting western frontier, according to Frederick Jackson Turner, that provided the crucible for basic American values and institutions. From an economic point of view, the increasing productivity of northern agriculture played a significant role in the economic development of the nation. With the expansion of Afro-American history and the rise of the new social history during the 1960s and 1970s, however, historians studying the nineteenth century focused overwhelmingly on the South and on urban and industrial society. Consequently, rural northerners became a kind of historiographical "silent majority."

Recently this majority has been "rediscovered" as a growing number of historians, economists, geographers, anthropologists, and others have turned their attentions away from the cities and toward the countryside.[2] Following the theoretical and methodological leads developed in the many studies of nineteenth-century urban communities, and informed by their significant findings, these scholars are exploring previously unexamined aspects of northern rural society, posing new questions, and reformulating older ones in different ways: What were the patterns of social structure and mobility in the countryside? The dimensions of family life? The roles of class, race, gender, and ethnicity? The influences of ideology and *mentalité?*

Although still incomplete, the answers to these questions allow a fresh perspective on northern rural history—one that not only departs from our previous understanding of the countryside, but provides important revisions of the prevailing view of nineteenth-century American society more generally. From this perspective, the crucial experiences in the agrarian past are not just the processes of settlement and the economic and technological development of agriculture. Rather, it is equally important to look at the social and political relationships that tied the countryside to the great transformation of nineteenth-century America into an urban-industrial society. In contrast to the dramatic changes that characterized urban and industrial society in the nineteenth-century North, however, the countryside experienced much greater continuity.

Perhaps the most important development confronted by northern agrarian communities during the nineteenth century was the expansion of the market economy. Breaking from the almost universal celebration of economic modernization that typified historical writing a generation ago, recent studies of industrialization consider the social costs of these changes and point to the various patterns of resistance and accommodation to capitalism that grew out of and, in turn shaped, working-class identity and culture.[3] But how disruptive was the growth of the market in the countryside? What was the extent of market involvement, and, more importantly, what did it mean to the historical actors involved?

Such concerns have fueled a heated debate between scholars who regard northern farmers in every period essentially as capitalists motivated by a spirit of liberal individualism and historians who argue that preindustrial farmers, at least, had values different from and in opposition to those of the ideal-typical economic man. In the leading statement of the former view, James T. Lemon (1972) argues that most colonial Pennsylvania husbandmen produced some cash crops in order to make a profit and were prevented from becoming completely commercial only by the lack of adequate transportation and sufficient markets. In contrast, James A. Henretta (1978) in his pathbreaking essay, "Families and Farms," delineates an alternative *mentalité* that limited participation in the market. Preindustrial yeomen subordinated individualism and acquisitive behavior to concerns for family security and the welfare of others in the community, especially the desire to continue the family line on the family land. A New Hampshire man in 1770, for example, found no corn for sale at any price because local farmers honored their promises to friends, neighbors, and kinfolk instead of maximizing profits.[4] Using the concept of a "household mode of production," Marxist historians have similarly described northern agrarian society during the early nineteenth century. Household production, in the words of Michael Merrill (1977, 61, 63), "is regulated by neither profit nor individual need alone, but by social need. Production is planned with one eye on the needs of one's own household, and the other on the needs of neighboring households." Like their urban artisan counterparts, then, preindustrial farmers were not predisposed to accede to the demands of the market at the expense of the bonds of kin and community.

At times, these farmers' attitudes went beyond simply inhibiting production for the market and formed the basis for active resistance to the incursions of capitalism in the countryside. In particular, conflicts came to a head as yeomen confronted the rise of industry in rural areas. In agrarian New England, for instance, the development of textile mills and blast furnaces created much friction between mill owners and farm communities over water and fishing rights and taxation (Prude 1983; Kulik 1985). At other times, however, yeomen embraced the market, particularly industrial outwork, albeit for traditional reasons. In his analysis of the development of capitalism in the Connecticut river valley between 1800 and 1860, Christopher Clark (1979) shows how rural families turned to putting-out work in an effort to maintain family security and continuity—a point reiterated by Thomas Dublin's (1985) more recent analysis of palm-leaf hatmaking in rural New Hampshire. Although, as Clark (1979, 183) argues, the market eventually undermined the social bases of the preindustrial *mentalité,* "the values of the household economy, while doing much to ease the transition to capitalist institutions, also provided a basis for objection to them."

The most comprehensive analysis of the struggles between yeomen and the dictates of the market, however, is not set in the North, but in the South in Steven Hahn's (1983) history of the Georgia upcountry before and after the Civil War. Here, the spread of cotton as a cash crop diminished the traditional independence of Georgia yeomen (who shared the same *mentalité* as their northern counterparts), eventually causing them to lose their land to an emerging merchant class. The altercations over stock laws and grazing rights during the 1870s were class-based attempts to preserve older bonds of communality in the face of capitalist imperatives and served as preludes to the more widespread battles during the Populist revolt.[5]

One problem with the new studies of the preindustrial rural North is that most of them do not deal with farmers and their crops, but focus instead on situations where the growth of capitalist institutions was particularly unsettling and coercive. Factories in rural communities by their very nature mark a decisive break with the agrarian past. Was commercialized agriculture seen by yeomen in the same way, or might it instead have been accepted as readily as putting-out work because of its congruity with traditional family goals? If so, did it necessarily have the same consequences? Hahn's study of Georgia, while a model to be emulated, is not representative of northern rural society because the rapid spread of the cotton economy and the monopolistic power of merchants in upcountry Georgia did not have widespread analogues in the North. Or did they? What is missing, unfortunately, is a close-grained analysis of the trials and tribulations of ordinary northern farmers as they grew more and more of their crops for the market—the thousands of little transformations that added up to the great transformation of the American agricultural revolution.

In one sense, albeit a limited one, critics on the other side of the debate have tried to do this by focusing on the extent of market participation. Winifred Rothenberg (1979, 1981, 1984a, 1985, and 1988), for example, has used Massachusetts farmers' account books and other quantitative sources from the late eighteenth and early nineteenth centuries to construct careful neoclassical economic analyses that demonstrate a substantial degree of market involvement. According to Rothenberg, Massachusetts farmers were not content with casual local marketing opportunities and instead traveled surprisingly long distances to sell their crops for better prices. The prices they received in both local and distant markets conformed over time to levels established in the major centers, and they fluctuated in synchronous cycles with more general price indices in Philadelphia and New York City, indicating the importance of those markets in the countryside. Similarly, more and more farmers paid their laborers a wage that was specified by the market rather than by custom.

The social meanings of participation in the market, however, remain the subject of more heated speculation than empirical investigation. Although she attempts to distance herself from questions of moral economy and *mentalité,* Rothenberg ultimately rejects that portrayal as a kind of pastoral Marxist celebration of the world we have lost or a New Left version of the agrarian myth. Yet, as many have pointed out, there are substantial problems in moving from purely structural data about the extent and role of markets to assumptions about the worldviews of the participants. The issue of social significance, in the end, derives less from the simple existence of market interactions than from the character of those economic relationships and the social meanings embedded within them.[6]

Like Rothenberg, Bettye Hobbs Pruitt (1984) also focuses on the structure of the agrarian economy in late eighteenth-century Massachusetts. Her evocative analysis of the Massachusetts Tax Valuation List of 1771 marks the beginning of a more satisfactory resolution to the debate. Very few eighteenth-century farmers, it seems, actually achieved the yeoman ideal of self-sufficiency. Instead, they were structurally interdependent on their neighbors to make up the produce, labor, or machinery that they lacked for subsistence. Moreover, patterns of dependence did not flow only from rich to poor, but cut across all strata of the community, lessening inequality and resulting in communities that were self-sufficient even though individual farmers were not. Yet these communities were also affected by market factors, and Massachusetts farmers relied on both local exchange and the outside market to determine the value of their produce. Thus, writes Pruitt (1984, 356, 364), "eighteenth-century farmers and planters perceived self-sufficiency and commercialization as complementary, not mutually exclusive, modes of agriculture," and "although many factors quite unrelated to cost and profit may have influenced the terms of trade among neighbors, the market value of the commodity in question was surely an important part of the negotiation, and that value was governed by forces originating beyond town and provincial boundaries."

During the first part of the nineteenth century, then, production for the market was beyond neither the imagination nor the experiences of most northern farm families, but that is not to say that the northern farmer was the quintessential economic man. Instead, participation in the market continued to be mediated by noneconomic concerns, and the persistence of these concerns underlies much of the confusion about the attitudes of antebellum farmers and their relationship to the market economy. Agriculture had a dual nature. In the words of economic historians Jeremy Atack and Fred Bateman (1987, 12, 272–73), it was "simultaneously a complex, successful economic activity as well as an engine of family and social organization with strong noneconomic motivations." In their view, northern farmers before the Civil War were "satisficers" who pursued the psychic satisfactions of independence and family security rather than more economically optimal goals. Farmers cashed in on commercial production when times were good, but relied on their farm's ability to feed the family when the markets turned sour, according to Atack and Bateman (1987, 273), "successfully straddling the fence between agriculture as a way of life and as a business enterprise like all others."

Other work also underscores the dual nature of agriculture embodied in the family farm and the enigmatic combination of attributes that it engendered. My own study of nineteenth-century Chelsea, Vermont, indicates that the continued importance of the family farm facilitated the preservation and adaptation of older values and attitudes even though local farmers were enmeshed in the market economy (Barron 1984). In addition to their traditional concern for the family line, these Vermont farmers were also curiously ambivalent about the imperatives of the market. On the one hand, they sought to maximize agricultural efficiency to get the most out of their farms, but they also looked down on those who made money speculatively as betrayers of a producer's ethic.

Taken together, these studies suggest that specializing in cash crops for the market during the first half of the nineteenth century entailed significant changes in farm production, but not, perhaps, the dramatic challenges to rural sensibilities that industrialization posed in urban society. Although most scholars, including Atack and Bateman, assume the ultimate incompatibility or inevitable conflict between familistic goals and complete profit maximization in a commercialized agricultural economy, the family farm remained the dominant form of economic organization in the North until the aftermath of World War II. Indeed, a recent theoretical essay by two Marxist anthropologists suggests that it may not make sense to think of the agrarian North as a capitalist society until the past half-century when such family concerns finally succumbed to the logic of agribusiness (Mann and Dickinson 1978). This view begs other questions: To what extent, and in what ways, did so-called noneconomic priorities continue to influence farm life and rural communities; and what did those continuities imply for the relationship of agrarian society to the larger whole? Perhaps the great transformation in the countryside was not that great.

The dual nature of agriculture did more than influence individual decisions on the family farm; it also affected the nature of the rural community as a whole. From settlement on, family and kinship considerations shaped rural communities. More than we had previously thought, frontier settlers in the Midwest were not lone pioneers, but arrived at a given location as part of a larger family group or a chain of migration—either responding to calls from kinfolk who had gone before, or making the way for those still to follow. Once in place, kinship networks were the most significant building blocks of community, as John M. Faragher's (1986) recent work on the Sugar Creek area in Illinois illustrates. Europeans immigrating directly to the Midwest, like those immigrating to eastern cities, also chose their destinations and structured their communities on the basis of kinship ties and village or regional associations from the Old Country (Ostergren 1980, 1983; Gjerde 1985). Nor did such ties necessarily evaporate after the settlement period or in response to market pressures and commercial crop production. Instead, throughout the rural North family strategies for passing the farm to the next generation and for settling their children nearby created communities that grew more interconnected and homogeneous, often serving as buffers against the more disruptive effects of the larger society. Thus life in the rural North continued to be informed by values of independence, localism, and familism well after extensive involvement in the market economy.

In particular, access to land and the connections between land and family shaped much of northern rural society before the Civil War and continued to play a central role throughout the century. In the preindustrial period land ownership was the measure of success for farm families, less because of the market value of the land than for what it meant in terms of guaranteeing family security in the present generation as well as the next (Henretta 1978; Bushman 1981). These concerns were so basic that they influenced the size of farm families, and using data from the middle of the century, Richard Easterlin and others have shown that decisions about family size were determined by the availability of land and the farmer's assessment of his ability to provide a start in life for the next generation (Easterlin 1976a, 1976b; Easterlin et al. 1978).[7]

This concern for the family line continued long into the twentieth century, according to Mark Friedberger's (1983, 1984a, 1984b) study of farm inheritance patterns in Iowa. Through their use of wills and predeath, or *inter vivos* transfers, Iowa farmers tried to ensure a continued family presence on the land as well as support for themselves in their old age, at the same time treating all of their heirs equitably. *Inter vivos* transfers in particular, which farmers from all different ethnic groups used, reflected the farm parents' desires to help their children get established and become independent at the outset of their careers rather than wait for the death of the father. In addition to the transfer of their estates, Iowa farm families also assisted the next generation and kept them close to home by providing credit and selling land to family members at substantial discounts.

The practices and concerns that I found in Chelsea, Vermont, during the second half of the nineteenth century parallel those that Friedberger delineates for Iowa, and they had a marked effect on the nature of that community. Partly as a result of family strategies, outmigration from Chelsea was selective and in-migration diminished substantially over time, so that after several generations, the local population consisted primarily of those who had or could expect an economic stake in the community. Demographic stability and homogeneity, in turn, led to a broad-based political and ideological consensus, which supplanted the earlier factions that had divided the community (Barron 1984).

Patterns of land transmission, however, varied considerably from place to place and according to ethnicity, with differing effects on the nature of the rural community. In her study of German immigrants in nineteenth-century rural Minnesota, Kathleen N. Conzen (1985) analyzes the effects of land transfers on the development of a stable and cohesive ethnic community in the countryside. The Catholic peasants who settled in St. Martin township in Stearns County brought with them the traditional view of the land as a family trust, which they managed to adapt to frontier conditions. Aided by federal land policies that made it relatively easy to acquire large tracts of land, these German pioneers attempted to provide nearby farms for all of their children, which they transferred to them at the beginning of their careers. In the process, they bought out other ethnic groups (especially Yankees, who in Minnesota were less committed to the cultural precept of familial continuity) and transformed St. Martin into an extremely homogeneous community. Paradoxically, the commercial economy aided rather than hindered the realization of traditional goals. Because land was bought and sold so readily in Minnesota, it was easier to provide for the next generation in the New World than it had been in the Old where family fortunes were inexorably tied to a particular piece of land that could not be liquidated and had to be divided and redivided among the heirs.

Just as Yankees were more stable in rural Vermont than on the Minnesota frontier, so, too, were there variations within immigrant groups. Robert Ostergren (1981) has studied the continuing importance of Old World attitudes in his analysis of inheritance among Swedish immigrants in rural Minnesota and found that Swedish settlers from different regions brought markedly different patterns of land transfer. Those from Rattvik parish in Dalarna came from a tradition that regarded land as an ancestral resource and from a region that was still relatively unaffected by modernization in the late nineteenth century. Consequently, Rattvik settlers in Athens, Minnesota, transferred property to their heirs while they were still alive and rarely had their estates liquidated after death. In contrast, Swedish immigrants from the Attmar parish of Medelpad province, a more commercialized region, saw land as a capital resource that could be sold in order to take advantage of local opportunities outside of agriculture or to pursue other opportunities away from the community. Attmar Swedes settled in Isanti township, which had one of the highest liquidation rates of the eight Swedish communities that Ostergren studied.

In spite of these variations, family strategies and patterns of land transfer often led to a degree of demographic stability in the rural North unmatched in urban-industrial society. This conclusion contrasts with early studies of the frontier, which typically found a great deal of population turnover—a pattern similar to that uncovered in the cities by the more recent work of Stephan Thernstrom, Peter Knights, and other urban historians. Indeed, the notion of a people in motion is one of the most compelling and widespread images to emerge from the "new" social history of the nineteenth century (Thernstrom and Knights 1971). More recently, however, scholars studying older, commercialized rural areas have uncovered a pattern of population stability rather than mobility (Katz et al. 1978; Davenport 1983; Barron 1984). Persistence rates for rural Erie and Schoharie counties in New York, and Chelsea, Vermont, were much higher than comparable rates measured in other nineteenth-century communities, either urban or frontier, echoing James C. Malin's (1935) classic study of Kansas, which also shows a trend toward decreasing population turnover after the settlement period.[8]

Those who did move, it seems, often did not move far. Geographer David P. Davenport (1983, 1984) managed to trace 92 percent of his sample of over 2,000 male household heads in rural Schoharie County, New York, from the 1855 state census to their whereabouts five years later. Almost three-fourths did not move, but of the one-quarter who did, the majority moved only short distances to neighboring or nearby rural communities; and migrants to the cities were typically not ex-farmers, but professionals, businessmen, and unskilled laborers. "This finding," states Davenport (1983, 190), "challenges the traditional interpretation of a close association between rural emigration and depopulation and migration to cities and the western frontier." Unfortunately, Davenport's sample does not include the numerous young men and women who were not yet the heads of families in 1855 (and who probably did migrate to the frontier and the cities in more substantial numbers), but his study is still very suggestive because it points out the degree to which rural northerners stayed close to home during the nineteenth century.[9]

The influence of family strategies and the greater degree of social homogeneity in the rural North is also indicated in an aggregate way by Lee Soltow's studies of wealth distribution. About 60 percent of American farmers (including adult farm laborers) owned real estate between 1850 and 1870 in contrast to only 30 percent of the free adult males who were not farmers; and almost three-fourths of the farmers owned either real or personal property, whereas half of the nonfarmers were propertyless (Soltow 1975, 33, 38). In his earlier study of wealth holding in Wisconsin between 1850 and 1870, Soltow (1971, 3–14) also finds a more equal distribution of wealth among farmers and farm laborers than among those who did not work the land. Both of these findings have been reconfirmed more recently by the work of Atack and Bateman (1987, 86–101, 100) who write, "wealth was clearly not perfectly equal, but when compared with the wealth inequality in urban areas, the South, or the entire United States in 1860 or at other times, one is struck by the much more equal distribution prevailing in the rural North during the late antebellum period."

Of course, the fact that farmers' family strategies tried to ensure security and continuity does not mean these strategies always succeeded. From Joseph Smith, Sr., in Vermont to Omar Morse in Minnesota, the history of the northern countryside is filled with families who, in spite of their best efforts, either lost their land because of debt and bad luck, or never managed to own any to begin with (Bushman 1981; Marshall 1986). And as commercialization and mechanization intensified, especially after the Civil War, both the risks associated with market fluctuations and the costs of getting into agriculture and starting a farm became higher.

We know little about the losers in the history of the rural North because there are few studies of northern tenant farmers and farm laborers. In general, questions of class in northern agrarian history are problematic. Much of our thinking about the relations between owners, tenant farmers, and agricultural laborers in the North has been shaped by the histories of the South and California, where distinctions between owners and nonowners were the sharpest and were reinforced by patterns of racial and ethnic discrimination and oppression. Yet the rural North cannot be understood as a class society simply in terms of who owned land and who did not, because the lines between these groups were very fluid. Likewise, economic power was not concentrated, and there were comparatively few landholding elites who could dominate society to the same degree as southern planters or northern industrialists. Landownership in the rural North was more universal than in the South, and the status of laborer or tenant may have only been a temporary one in the climb towards owning a farm.

This, obviously, is the famous agricultural ladder—from laborer to renter to owner—that agricultural historians and economists have debated for some time. The most recent entry to that debate, Donald L. Winters, traced the fortunes of tenant farmers in several Iowa counties during the second half of the nineteenth century and found that almost one-fifth stayed in the sample counties and eventually became farm owners. According to Winters (1978, 83, 19), "Whether these mobility rates confirm or invalidate the agricultural ladder thesis depends, of course, upon one's conception of how the system should have worked." But in his estimation, "a significant number of nineteenth-century Iowa farmers used the institution of tenancy as a stepping stone to landownership," and "the agricultural ladder thesis provides a useful framework for investigation and experience of Iowa farmers in the postbellum period." It will take more comprehensive study to go beyond this preliminary consideration of rural social mobility in order to understand the northern tenant farmer's place in the rural community.[10]

At the lowest rung of the ladder, the northern farm laborer is also an obscure historical figure. The principal study of farm labor in the antebellum Midwest concentrates primarily on the work performed by hired hands and says very little about patterns of mobility or the relations between farmers and their workers.[11] Nonetheless, David Schob (1975, 228, 233) asserts, "farm hands in pre–Civil War America were usually considered members of the family, or at least treated on a fairly equal basis," and "the relationship between the farmer and his hired hand was more personal than that of an employer and common laborer in the town or city."

In contrast, other studies chronicle growing tensions between farmers and farm laborers in the wake of market involvement and technological change. Looking at the initial transition from family labor to hired help in Concord, Massachusetts, during the first half of the nineteenth century, Robert Gross (1982, 51) documents the disruption of the older bonds of kinship and community as farmers capitalized on the heightened demand for hay created by the Boston market. In his words, "now it was money—and money alone—that kept help working on the farm. The social relations of production were imbued with the ethos of agricultural capitalism."

In the Midwest traditional relationships between farmers and some of their hired hands also changed after the Civil War as the adoption of the reaper and other farm machinery created a new group of seasonal laborers who worked at harvest time. During the depression years of the 1870s, farmers began using self-binding reapers, which displaced these laborers, and tensions between the migrant workers and farmers came to a head. In several midwestern states, according to Peter and JoAnn Argersinger (1984), farm workers threatened strikes and destroyed crops and machinery, while farmers retaliated by forming protective associations and began shooting at unwanted trespassers. The return of economic prosperity eventually defused this particular situation, but a marked separation between farmer and laborer remained. Clearly, for farm laborers as for tenant farmers, more research needs to be done to clarify the full range and significance of class relations in northern agrarian society.

Like agricultural laborers, women's roles in the rural North were also affected by the great transformations of the nineteenth century. Perhaps the greatest change for nineteenth-century rural women was what Bengt Ankarloo (1979) has termed the "defeminization" of farm work. Preindustrial farm women were helpmates to their husbands and were essential contributors to the economic well-being of their families. In addition to caring for the children and the usual range of housework, rural women were also responsible for the dairy, the garden, and the chicken coop as well as a variety of home manufactures including spinning, weaving, and processing agricultural produce. In the words of John M. Faragher (1979; 1981, 547), "the family was the basic unit of production and its success depended upon the full participation of both partners . . . certainly, the multiplicity and complexity of women's work signals their central position in the family economy."

The rise of the market, however, made women's work seem less valuable. In the Midwest the expansion of cash grain crops and the use of machinery after the Civil War emphasized traditionally male work and undermined the economic significance of women's contributions (Ankarloo 1979). When Norwegian farmers in Minnesota specialized in wheat and relied on farm machinery, for example, the traditional sexual division of labor within the family altered dramatically. As their husbands assumed sole responsibility for agricultural production, women concentrated exclusively on domestic work, especially child rearing and housekeeping. Even when these farms diversified into hog production and dairying after the wheat boom, Norwegian women did not return to their traditional tasks of milking and caring for the animals. Instead, greater prosperity and the availability of hired labor allowed them to remain in a separate sphere of household work and to subscribe to a new "bourgeois" ethic that looked down on women's work out-of-doors (Gjerde 1985). In the East, where farming remained more diversified, the centralization of dairy production off the farm and the expansion of commercial truck farming and fruit growing displaced women's traditional tasks in the family economy and shifted women's work away from the farm and into the house (McMurry 1984a, 1984b).

This portrayal of the defeminization of agriculture, though, underestimates significant continuities. As Joan Jensen (1982, 1986) has shown, Delaware Valley farm women quit spinning and weaving when commercial cloth came on the market during the early nineteenth century. Instead, they began making larger quantities of butter, which they sold in the market themselves, providing an important source of income and economic stability for their families until the second half of the century. In a more general sense, the ideal of the "profitable farm wife," which Sally McMurry finds expressed by women and men in the agricultural press during the 1840s and 1850s, was actually a continuation of the older helpmate ideal, and was touted precisely because she could produce marketable commodities. During this period, McMurry (1984a, 86) writes, "women continued to perform traditional tasks, but their products more often went to market."

Studies set after the Civil War also indicate the importance of women's work on the farm long after the commercialization of agriculture. Women in the Nanticoke Valley of New York were active in all phases of commercial dairy operations during the second half of the nineteenth century, and according to Nancy Grey Osterud (1984; 1985, 5), "the dairy process was the least gender-marked of any farm and household operations." Similarly, poorer Norwegian women, in contrast to Gjerde's more successful immigrants, shouldered much of the work on the farm, especially milking, up until the middle of the twentieth century (Pederson 1985). Finally, in her analysis of surveys of farm families from the 1920s, Joann Vanek (1980) stresses the continued high regard for women's work and its value to the farm household, even though that work was not directly related to farm production.

Within this mix of continuity and change in women's work, to what extent did new cultural definitions make inroads in the northern countryside? In antebellum urban America and the larger towns of New England, an emerging ethic of domesticity had enabled middle-class women to adopt new roles for themselves based on their duties as mothers, educators, and guardians of morality, as their economic contributions were altered by industrialization. Similarly, in her study of "progressive," or innovative, farm families and their houses in the settled areas of the North, Sally McMurry (1984a) documents the shift away from the "profitable farm wife" ideal and the embrace of the prevailing image of domestic womanhood during the second half of the nineteenth century. Even among the innovative farm families that McMurry studies, however, this adoption of urban middle-class sensibilities was blunted and modified by the realities of farm life. Whether the cult of domesticity and other aspects of bourgeois culture made similar inroads among the less affluent and less "enlightened" strata of rural society is open to question and investigation.

In contrast to urban and industrial society during the nineteenth century, then, the rural North experienced a great deal of continuity over the divide of the great transformation. Although increased specialization in cash crops often redefined relationships between farmers and their laborers, and between rural men and women, nonetheless the persistence of the family farm as the basic unit of production and building block of the rural community ensured the continuation of "noneconomic" considerations in agrarian life. Values of familism, localism, and independence continued to hold sway long after the transition to commercial agriculture. Life in the rural North changed over the course of the nineteenth century, but it was not transformed, and the so-called yeoman and the so-called business farmer still had much in common.

More and more, however, that continuity became harder to maintain in the face of new and more substantial pressures from outside the rural community. Agrarian values came into conflict with the larger thrust of American society. Initially, those conflicts focused on economic issues as commercial farmers confronted centralized and powerful economic forces, especially the railroads and the "moneyed" interests. Although angry northern Populist farmers were definitely making an ideological statement, given the simultaneous importance of both commercial and noneconomic attitudes in northern rural society, it is difficult to see them either as full-fledged critics of capitalism or as refugees bemoaning the loss of a Jeffersonian never-never land of self-sufficiency. Instead, farmers wanted to defend the status quo—the autonomy of their communities, their families, and their own independence—against new outside threats beyond their control. In this vein, part of the Populist appeal in the North can be understood as a power struggle between townspeople and farmers, as Stanley Parsons (1973) has argued, brought on by growing tensions as the larger agencies of change were funnelled through rural villages and service centers.

Beyond questions of economics, beginning in the late nineteenth century the emergence of large-scale organizational society spawned new bureaucratic institutions and cultural forms that also threatened the localism and familism central to country life, as well as the political power of the rural electorate. To what extent and to what effect farmers at the beginning of the twentieth century resisted or accommodated to these changes, are, like so many others, questions remaining to be answered.

## NOTES

1. In this essay, the rural North includes New England, New York, New Jersey, Pennsylvania, Ohio, Indiana, Michigan, Illinois, Wisconsin, Minnesota, Iowa, Kansas, Nebraska, and North and South Dakota. The discussion of developments before the Civil War focuses mainly on the Northeast and the Old Northwest and, like the original settlers, moves west over the course of the nineteenth century.

2. See an earlier version of this essay (Barron 1986). For other discussions of the "new" rural history, see Robert P. Swierenga's writings (1972, 1981, 1982, and 1983); as well as Argersinger (1977), Winkler (1985), Hahn and Prude (1985), and Frederickson (1987). For a good introduction to recent work on southern rural history, see Burton and McMath (1982).

3. For the two most prominent examples of this literature, see Dawley (1976) and Wilentz (1984).

4. See also the exchange between Lemon and Henretta generated by this article (Lemon 1980).

5. For an analysis of the tensions created by the market in Virginia, see Schlotterbeck (1982).

6. Rothenberg's argument has touched off two lively exchanges in the *Journal of Economic History* (Weiss 1983; Bernstein and Wilentz 1984; and Rothenberg 1983, 1984b). In the same vein see James Oakes's (1984) discussion of Steven Hahn's work; and two essays informed by cultural anthropological theory (Merrill 1985; Vickers 1986).

7. Easterlin's study of farm family fertility is part of a large literature on the subject (Leet 1975, 1976; Laidig et al. 1981).

8. Richard G. Bremer (1974) has also analyzed persistence in the Midwest. Bremer, however, regards the glass as half empty rather than as half full and emphasizes the continued high-mobility rates of Nebraska farmers even though their persistence in the late nineteenth and twentieth centuries was much higher than earlier frontier levels. Canadian scholars have also studied population persistence in rural areas (Gagan 1981; Akenson 1984). Recently, Donald H. Parkerson (1982) has questioned the validity of the high degree of population turnover in nineteenth-century cities, arguing instead that low-persistence rates are artifacts of the record-linking process. However, Michael Katz's study (Katz et al. 1978) does not rely on linked data and still finds more population stability in rural Erie County than in the city. See also David P. Davenport's (1985a) discussion of this issue.

9. Other studies have also examined internal migration from the countryside (Knights 1983, 1985; Davenport 1985b; Barron 1984; Adams and Kasakoff 1984; M. Conzen 1974).

10. LaWanda Cox (1944, 1948) has written the classic critique of the agricultural ladder hypothesis. See also Cogswell (1975).

11. For a study of the mobility of farm laborers, see Shepherd (1980).

## REFERENCES AND FURTHER READINGS

Adams, John W., and Alice Bee Kasakoff. 1984. "Migration and the Family in Colonial New England: The View from Genealogies." *Journal of Family History* 9:24–43.

Akenson, Donald, ed. 1984. *Canadian Papers on Rural History.* Vol. 4. Gananoque, Ontario: Langdale Press.

Ankarloo, Bengt. 1979. "Agriculture and Women's Work: Directions of Change in the West, 1700–1900." *Journal of Family History* 4:111–20.

Argersinger, Peter H. 1977. "The People's Past: Teaching American Rural History." *History Teacher* 10:403–24.

Argersinger, Peter H. and JoAnne E. Argersinger. 1984. "The Machine Breakers: Farmworkers and Social Change in the Rural Midwest of the 1870s." *Agricultural History* 58:393–410.

Atack, Jeremy, and Fred Bateman. 1987. *To Their Own Soil: Agriculture in the Antebellum North.* Ames: Iowa State Univ. Press.

Barron, Hal S. 1984. *Those Who Stayed Behind: Rural Society in Nineteenth-Century New England.* New York: Cambridge Univ. Press.

______. 1986. "Rediscovering the Majority: The New Rural History of the Nineteenth-Century North." *Historical Methods* 19:141–52.

Bernstein, Michael A., and Sean Wilentz. 1984. "Marketing, Commerce, and Capitalism in Rural Massachusetts." *Journal of Economic History* 44:171–73.

Bidwell, Percy W., and John I. Falconer. 1925. *History of Agriculture in the Northern United States, 1620–1860.* Washington: Carnegie Institute.

Bogue, Allan G. 1963. *From Prairie to Corn Belt: Farming on the Illinois and Iowa Prairies in the Nineteenth Century.* Chicago: Univ. of Chicago Press.

Bremer, Richard G. 1974. "Patterns of Spatial Mobility: A Case Study of Nebraska Farmers, 1890–1970." *Agricultural History* 48:529–42.

Burton, O. Vernon, and Robert C. McMath, Jr., eds. 1982. *Class, Conflict, and Consensus: Antebellum Southern Community Studies.* Westport, Conn: Greenwood Press.

Bushman, Richard L. 1981. "Family Security in the Transition from Farm to City, 1750–1850." *Journal of Family History* 6:238–56.

Clark, Christopher. 1979. "Household Economy, Market Exchange, and the Rise of Capitalism in the Connecticut Valley, 1800–1860" *Journal of Social History* 13:169–89.

Cogswell, Jr., Seddie C. 1975. *Tenure, Nativity and Age as Factors in Iowa Agriculture, 1850–1880.* Ames: Iowa State Univ. Press.

Conzen, Kathleen N. 1985. "Peasants into Pioneers: Patterns of Intra-Family Land Transmission among German Immigrants in Rural Minnesota, 1856–1905." In *The Countryside in the Age of Capitalist Transformation,* ed. Hahn and Prude, 259–92. Chapel Hill: Univ. of North Carolina Press.

Conzen, Michael P. 1974. "Local Migration Systems in Nineteenth-Century Iowa." *Geographical Review* 64:339–61.

Cox, LaWanda. 1944. "Tenancy in the United States, 1865–1900." *Agricultural History* 18:97–105.

______. 1948. "The American Agricultural Wage Earner, 1965–1900" *Agricultural History* 22:95–114.

Curti, Merle. 1959. *The Making of an American Community: A Case Study of Democracy in a Frontier County.* Stanford: Stanford Univ. Press.

Danhoff, Clarence. 1969. *Change in Agriculture: The Northern States, 1820–1870.* Cambridge: Harvard Univ. Press.

Davenport, David P. 1983. "Population Persistence and Migration in Rural New York, 1855–1860." Ph.D. diss., University of Illinois, Urbana.

______. 1984. "Tracing Rural New York's Out-Migrants, 1855–1860." *Historical Methods* 17:59-67.

______. 1985a. "Duration of Residence in the 1855 Census of New York State." *Historical Methods* 18:5–12.

______. 1985b. "Stepwise Migration and the Growth of Albany, New York, to 1855." Paper presented at the Annual Meeting of the Social Science History Association, November, Chicago.

Dawley, Alan. 1976. *Class and Community: The Industrial Revolution in Lynn.* Cambridge: Harvard Univ. Press.

Dublin, Thomas. 1985. "Women and Outwork in a Nineteenth-Century New England Town: Fitzwilliam, New Hampshire, 1830–1850." In *The Countryside in the Age of Capitalist Transformation,* ed. Steven Hahn and Jonathan Prude, 51–70. Chapel Hill: Univ. of North Carolina Press.

Easterlin, Richard A. 1976a. "Factors in the Decline of Farm Fertility in the United States: Some Preliminary Research Results." *Journal of American History* 63:600–14.

______. 1976b. "Population Change and Farm Settlement in the Northern United States." *Journal of Economic History* 36:45–75.

Easterlin, Richard A., George Alter, and Gretchen Condran. 1978. "Farms and Farm Families in Old and New Areas: The Northern United States in 1860." In *Family and Population in Nineteenth-Century America,* ed. T. K. Hareven and M. A. Vinovskis. Princeton, N.J.: Princeton Univ. Press, 28–84.

Faragher, John M. 1979. *Women and Men on the Overland Trail.* New Haven: Yale Univ. Press.

———. 1981. "History from the Inside-Out: Writing the History of Women in Rural America." *American Quarterly* 33:537–57.

———. 1986. *Sugar Creek: Life on the Illinois Prairie*. New Haven: Yale Univ. Press.

Fite, Gilbert C. 1966. *The Farmers' Frontier, 1865–1900*. New York: Holt, Rinehart, and Winston.

Frederickson, George M. 1987. "Down on the Farm." *New York Review of Books*, Apr. 23, 1987, 37–39.

Friedberger, Mark. 1983. "The Farm Family and the Inheritance Process: Evidence from the Corn Belt, 1870–1950." *Agricultural History* 57:1–13.

———. 1984a. "Probate and Land Records in Rural Areas." *Agricultural History* 58:123–26.

———. 1984b. "Handing Down the Home Place: Farm Inheritance Strategies in Iowa, 1870–1945." *Annals of Iowa* 47:518–36.

Gagan, David. 1981. *Hopefull Travellers: Families, Land, and Social Change in Mid-Victorian Peel County, Canada West*. Toronto: Univ. of Toronto Press.

Gates, Paul W. 1960. *The Farmer's Age: Agriculture, 1815–1860*. New York: Holt, Rinehart, and Winston.

Gjerde, Jon. 1985. *From Peasants to Farmers: The Migration from Balestrand, Norway, to the Upper Middle West*. New York: Cambridge Univ. Press.

Gross, Robert A. 1982. "Culture and Cultivation: Agriculture and Society in Thoreau's Concord." *Journal of American History* 69:42–59.

Hahn, Steven. 1983. *The Roots of Southern Populism: Yeoman Farmers and the Transformation of the Georgia Upcountry, 1850–1890*. New York: Oxford Univ. Press.

Hahn, Steven and Jonathan Prude. 1985. Introduction. In *The Countryside in the Age of Capitalist Transformation*, ed. by Hahn and Prude. Chapel Hill: Univ. of North Carolina Press, 3–21.

Henretta, James A. 1978. "Families and Farms: *Mentalité* in Pre-Industrial America." *William and Mary Quarterly*, ed. ser., 35:3–32.

Jensen, Joan M. 1982. "Churns and Buttermaking in the Mid-Atlantic Farm Economy." *Working Papers from the Regional Economic History Research Center* 5.

———. 1986. *Loosening the Bonds: Mid-Atlantic Farm Women, 1750–1850*. New Haven: Yale Univ. Press.

Katz, Michael B., Michael J. Doucet, and Mark J. Stern. 1978. "Migration and the Social Order in Erie County, New York: 1855." *Journal of Interdisciplinary History* 8:669–702.

Knights, Peter R. 1983. "The Facts of Lives; or Whatever Happened to 2,808 Nineteenth-Century Bostonians." Paper presented at the Annual Meeting of the Organization of American Historians, April, Los Angeles.

———. 1985. "How 'Rural' were Rural Migrants to Boston? A Case Study in 1860 and 1870." Paper presented at the Annual Meeting of the Social Science History Association, November, Chicago.

Kulik, Gary. 1985. "Dams, Fish, and Farmers: Defense of Public Rights in Eighteenth-Century Rhode Island." In *The Countryside in the Age of Capitalist Transformation*, ed. Steven Hahn and Jonathan Prude, 25–50. Chapel Hill: Univ. of North Carolina Press.

Laidig, Gary L., Wayne A. Schutjer, and C. Shannon Stokes. 1981. "Agricultural Variation and Human Fertility in Antebellum Pennsylvania." *Journal of Family History* 6:195–204.

Leet, Don R. 1975. "Human Fertility and Agricultural Opportunities in Ohio Counties:

From Frontier to Maturity, 1810–1860." In *Essays in Nineteenth-Century Economic History: The Old Northwest,* ed. D. C. Klingaman and R. K. Vedder. Athens, Ohio: Ohio Univ. Press, 138–58.

______. 1976. "The Determinants of the Fertility Transition in Antebellum Ohio." *Journal of Economic History* 36:359–78.

Lemon, James T. 1972. *The Best Poor Man's Country: A Geographical Study of Early Southeastern Pennsylvania.* Baltimore: Johns Hopkins Univ. Press.

______. 1980. "Comment on James A. Henretta's 'Family and Farms: *Mentalité* in Pre-Industrial America.' " *William and Mary Quarterly,* 3d. ser., 37:688–700.

McMurry, Sally, 1984a. "American Farm Families and Their Houses: Vernacular Design and Social Change in the Rural North, 1830–1900." Ph.D. diss., Cornell University.

______. 1984b. "Progressive Farm Families and Their Houses, 1830–1855: A Study in Independent Design." *Agricultural History* 58:330–46.

Malin, James C. 1935. "The Turnover of Farm Population in Kansas." *Kansas Historical Quarterly* 4:339–72.

Mann, S. A., and J. M. Dickinson. 1978. "Obstacles to the Development of a Capitalist Agriculture." *Journal of Peasant Studies* 5:466–81.

Marshall, James M. 1986. *Land Fever: Dispossession and the Frontier Myth.* Lexington: Univ. of Kentucky Press.

Merrill, Michael. 1977. "Cash is Good to Eat: Self-Sufficiency and Exchange in the Rural Economy of the United States." *Radical History Review* 3:42–71.

______. 1985. "Gifts, Barter, and Commerce in Early America: An Ethnography of Exchange." Paper presented at the Annual Meeting of the Organization of American Historians, April, Minneapolis.

Oakes, James. "The Politics of Economic Development in the Antebellum South." *Journal of Interdisciplinary History* 15:305–16.

Ostergren, Robert C. 1980. "Patterns of Migration to a Swedish Settlement on the Dakota Frontier." In *Ethnicity on the Great Plains,* ed. F. Luebke, 73–91. Lincoln: Univ. of Nebraska Press.

______. 1981. "Land and Family in Immigrant Communities." *Annals of the Association of American Geographers* 71:400–411.

______. 1983. "Settlement and Ethnicity Patterns on the Agricultural Frontiers of South Dakota." *South Dakota History* 13:49–82.

Osterud, Nancy Grey. 1984. "Strategies of Mutuality: Relations Among Women and Men in an Agricultural Community." Ph.D. diss., Brown University.

______. 1985. "The Valuation of Women's and Men's Work: Gender, Kinship and the Market in an Upstate New York Dairy Farming Community During the Late Nineteenth Century." Paper presented at the Annual Meeting of the Social Science History Association, November, Chicago.

Parkerson, Donald H. 1982. "How Mobile Were Nineteenth-Century Americans?" *Historical Methods* 15:99–109.

Parsons, Stanley. 1973. *The Populist Context: Rural Versus Urban Power on a Great Plains Frontier.* Westport, Conn.: Greenwood Press.

Pederson, Jane M. 1985. "Woman's Work, Woman's Place in Rural Wisconsin, 1870–1970." Paper presented at the Annual Meeting of the Social Science History Association, November, Chicago.

Prude, Jonathan. 1983. *The Coming of the Industrial Order: Town and Factory Life in Rural Massachusetts, 1810–1860.* New York: Cambridge Univ. Press.

Pruitt, Bettye Hobbs. 1984. "Self-Sufficiency and the Agricultural Economy of Eighteenth-

Century Massachusetts." *William and Mary Quarterly,* 3d ser., 41:333–64.

Rothenberg, Winifred. 1979. "A Price Index for Rural Massachusetts, 1750–1855." *Journal of Economic History* 39:975–1001.

______. 1981. "The Market and Massachusetts Farmers, 1750–1855." *Journal of Economic History* 41:283–314.

______. 1983. "Reply to Rona S. Weiss." *Journal of Economic History* 43:479–80.

______. 1984a. "Farm Account Books: Problems and Possibilities." *Agricultural History* 58:106–12.

______. 1984b. "Markets, Values, and Capitalism: A Discourse on Method." *Journal of Economic History* 44:174–78.

______. 1985. "The Emergence of a Capital Market in Rural Massachusetts, 1730–1838." *Journal of Economic History* 44:781–808.

______. 1988. "The Emergence of Labor Markets and the Transformation of the Rural Economy: Massachusetts, 1750–1855." *Journal of Economic History.* 48:537–566.

Schlotterbeck, John T. 1982. "The 'Social Economy' of an Upper South Community: Orange and Green Counties, Virginia, 1815–1860." In *Class, Conflict, and Consensus,* ed. O. Vernon Burton and Robert C. McMath, Jr., 3–28. Westport, Conn.: Greenwood Press.

Schob, David E. 1975. *Hired Hands and Plowboys: Farm Labor in the Nineteenth-Century Midwest, 1815–1860.* Urbana: Univ. of Illinois Press.

Shepherd, Rebecca A. 1980. "Restless Americans: The Geographic Mobility of Farm Laborers in the Old Midwest, 1850–1870." *Ohio History* 89:25–46.

Soltow, Lee. 1971. *Patterns of Wealthholding in Wisconsin Since 1850.* Madison: Univ. of Wisconsin Press.

______. 1975. *Men and Wealth in the United States, 1850–1870.* New Haven: Yale Univ. Press.

Swierenga, Robert P. 1972. "Towards the 'New Rural History': A Review Essay." *Historical Methods Newsletter* 6:111–22.

______. 1981. "The New Rural History: Defining the Parameters." *Great Plains Quarterly* 1:211–23.

______. 1982. "Theoretical Perspectives on the New Rural History: From Environmentalism to Modernization." *Agricultural History* 56:495–502.

______. 1983. "Agriculture and Rural Life: The New Rural History." In *Ordinary People and Everyday Life: Perspectives on the New Social History,* ed. J. B. Gardner and G. R. Adams, 90–113. Nashville: American Association for State and Local History.

Thernstrom, Stephan, and Peter Knights. 1971. "Men in Motion: Some Data and Speculations about Urban Population Mobility in Nineteenth-Century America." In *Anonymous Americans: Explorations in Nineteenth-Century Social History,* ed. T. Hareven, 17–47. Englewood Cliffs, N.J.: Prentice-Hall.

United States Department of Agriculture. 1940. *Farmers in a Changing World.* Washington: USDA Yearbook.

Vanek, Joann. 1980. "Work, Leisure, and Family Roles: Farm Households in the United States, 1920–1955." *Journal of Family History* 5:422–31.

Vickers, Daniel. 1986. " 'Cherries . . . are Ready Cash': A Case Study in Early American Economic Values." Paper presented at the Annual Meeting of the Social Science History Association, October, St. Louis.

Weiss, Rona S. 1983. "The Market and Massachusetts Farmers, 1750–1850: Comment." *Journal of Economic History* 43:475–78.

Wilentz, Sean. 1984. *Chants Democratic: New York City and the Rise of the American Working Class, 1789–1850.* New York: Oxford Univ. Press.

Wilson, Harold F. 1936. *The Hill Country of Northern New England: Its Social and Economic History, 1790–1930.* New York: Columbia Univ. Press.

Winkler, Karen. 1985. "The Greening of American Social History: Scholars Rediscover the Countryside." *Chronicle of Higher Education,* Mar. 20, 1956.

Winters, Donald L. 1978. *Farmers Without Farms: Agricultural Tenancy in Nineteenth-Century Iowa.* Westport, Conn.: Greenwood Press.

CHAPTER 2

# Yeoman Farming: Antebellum America's Other "Peculiar Institution"

JEREMY ATACK and FRED BATEMAN

A DOMINANT AMBITION among nineteenth-century Americans was the establishment of economic, political, and social systems founded on the institution of independent landowning farmers. In a sense, this represents the North's "peculiar institution." Indeed, on the stage of world history the dedication to the small farmer as self-reliant landowner in a universe of economic equality was rarer than agricultural slavery. Those who espoused this vision, among them Thomas Jefferson, thought it essential to the realization of political democracy. Yet a tension existed between this and other national goals, most notably those of industrialization, economic diversity, and nationalism. The concept is that of an American yeomanry abjuring urbanism, manufacturing, big business, and centralization, while celebrating rural life, self-sufficiency, stability, individualism, and localism. It was not "one nation indivisible," but a collection of self-sustaining individuals, each one dependent on his own resources and responsible for his own destiny. In the economic world, as Jefferson stated it, no one would depend entirely upon "the caprice of customers." In the political environment, power would be decentralized.

For all its poetical attractiveness, and despite the imprint that it has left on the American mentality, this Yeoman Dream was never uncontested. Although the vast majority of farmers in the North seems to have owned the land being farmed, perhaps 16 percent of the region's farmers were tenants in 1860, and tenancy became even more common and widespread later in the century (Atack 1988). Contending political views, and alternative socioeconomic ones, were also present. Even among the yeomanry, individual commitment to this ideal weakened as market opportunities developed. To a large extent the cash-seeking yeomen themselves undercut the very system within which they would be kings. The onrush of European industrialism into the New World during the second half of the century, and the national infrastructure created to facilitate it, sealed the agrarian dream's fate before the century's end.

Although we know the outcome of this contest for the national destiny, we have never fully understood whether yeoman farming could have succeeded as the foundation for a stable social order even if unchallenged. In this chapter we evaluate the attainments of the yeoman agricultural system at the peak of its prominence on the eve of the Civil War. If ever it were to succeed unfettered by the complications of industrialism, large-scale business, economic interdependence, urbanization, and centralization, it could have done so under no better circumstances. Ours are tests of the attainments of northern yeoman agriculture at midcentury and its viability as the foundation for the American nation.

The data underlying this study derive from a sample of 21,118 rural households selected from the 1860 federal census manuscripts by Fred Bateman and James D. Foust.[1] These households, encompassing more than 100,000 people, lived in 102 randomly selected rural townships scattered across the northern portion of the United States and have proved to be statistically representative of that parent population.[2] Over 11,000 of the households also operated farms, typically small ones (80–160 acres) that they usually owned. They are thus the very essence of the yeoman farmer economy—the North's "peculiar institution." For these household-operated farms, the household data from the manuscript census population schedules have been linked to the agricultural schedule data.[3]

The information covers a variety of explicit individual and household characteristics such as age, sex, occupation, literacy, birthplace, and wealth as well as implicit information about household structure and personal relationships. In addition, the agricultural data describe the farm and its operations—value, acreage, livestock, and crops—during the census year from June 1, 1859, to May 31, 1860.[4] The sample has also been divided into two subregions, the Northeast and the Midwest, each of which is representative of its statistical population. As a result we can build a detailed quantitative picture of rural life and agricultural activity in the North and its two subregions at this time, some aspects of which we deal with here insofar as they pertain to the vitality of yeoman farming in the North.

## YEOMAN AGRICULTURE AND FARM FORMATION

Federal policy toward land disposal clearly reflected national aspirations for the Yeoman Dream by offering plots in small sizes and at low prices. Theoretically such policy should have played a role in America's agricultural development and certainly we would argue that its impact is apparent in the manuscript data from the U.S. Census of Agriculture (Fogel and Rutner 1972; Dennen 1977). Examination of the size distribution of farms, for example, indicates that from Ohio westward disproportionately large numbers of farms consisted of 40, 80, 120, 160, and 200 acres, with 40, 80, and 160 acres being the most common. These acreages are fractions (one-sixteenth, one-eighth, and one-quarter) of the Land Act section of 1 square mile and were the sizes offered for sale by the federal government at different times.[5] Farms that were multiples of 40 acres are thus likely to be those established on land purchased from the federal government. Land secured from other sources, such as through inheritance or from speculators or other private vendors, was less likely to be in lots of such multiples of 40 acres. Speculators in particular should have been willing to sell land in whatever quantity the purchaser desired. Indeed this was one commonly cited advantage of purchasing land from them. In the Northeast, on the other hand, where settlement predated the Land Ordinance, farms were disproportionately concentrated on 50, 100, 150, and 200 acres (Fig. 2.1).

Originally, the 1785 Land Ordinance had established 640 acres—one square mile—as the minimum purchase, but by the time Ohio was being opened for settlement in the early 1800s the minimum had been reduced to 160 acres at $2 per acre. Sales, however, were to remain sluggish until 1820 when prices were cut to $1.25 and the minimum lot reduced to 80 acres with a further cut to 40 acres in 1832. The impact of these revisions is apparent in the size distribution of farms in the various midwestern states as late as 1860. For example, the Revision Act of 1820, which set minimum purchase requirements from the federal government at 80 acres (a size typically larger than could be efficiently and fully cultivated using only farm family labor and available technology), was passed just when the pace of settlement in Indiana and Illinois was accelerating. In those states we find that 80-acre farms were the most common. Extensive settlement in Michigan, on the other hand, took place only after the reduction of the minimum purchase to 40 acres and there 40-acre farms were most commonly found. Moreover, since the initial cost of a 40-acre farm was lower and its size more closely attuned to the needs and capabilities of a family, we find tenancy less prevalent in that state than elsewhere in the Midwest. Closer to the frontier in 1860, especially in Kansas and Minnesota, farmers seem to have taken advantage of the preemption provisions of the 1841 Act, as unreasonably large numbers

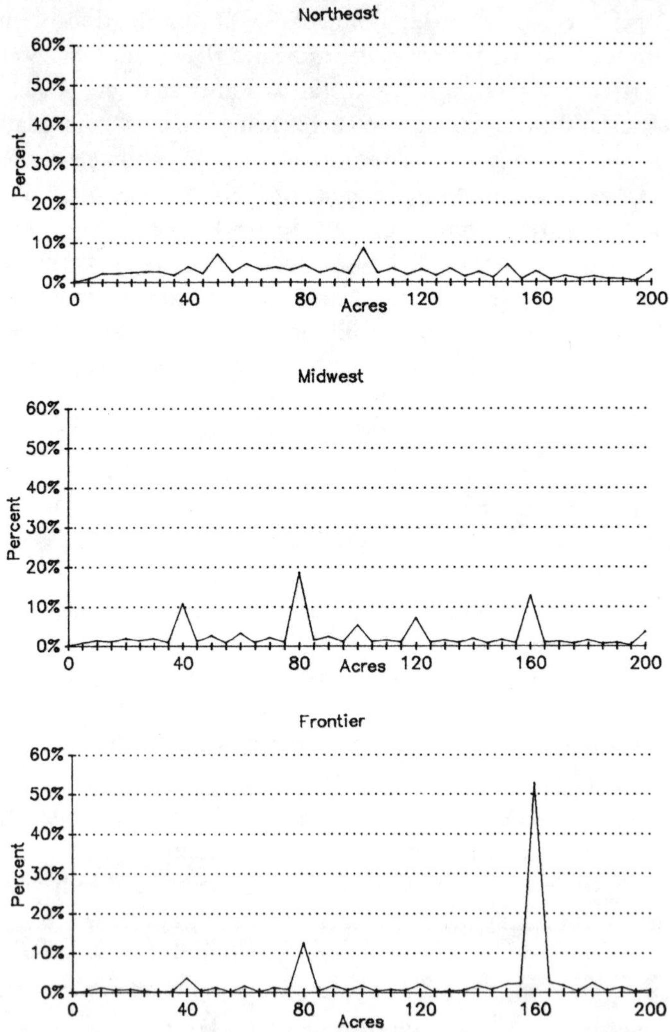


Fig. 2.1 The size distribution of farms, 200 acres and less, in the Northeast, the Midwest, and on the Frontier in 1860.

of farmers claimed no real wealth despite the land they farmed—which was usually 160 acres, the maximum for preemption.

## THE FAMILY ON THE FAMILY FARM

The process of farm creation in the Midwest placed heavy demands upon the supply of labor. However, hired workers were scarce thanks to the abundant opportunities for farm ownership. Besides, "manual labor . . . [of an] agricultural character, when followed as a necessary means of obtaining a livelihood, is considered by some as a degrading employment throughout our land" (U.S. Patent Office 1852, 323) and was therefore shunned. The new yeoman farmer who sought independence from product markets thus also needed independence from the tyranny of labor markets. The South had its own captive labor force. Northerners tried to create their own by having a large family "bound to the entrepreneur by ties of custom, law, fear, and affection" to use William Parker's vivid description (Lance Davis et al. 1972, 395). The implicit wage for this labor lay below that for hired workers due to the element of exploitation and the receipt of nonwage benefits, reinforced by the farmers' ability to trade to their favor on the discounted present value of the potential bequest to heirs.

Land availability figured heavily in the decision regarding family size.[6] Land created employment and, with land, labor could create capital. Consequently, unused land, which represented economic opportunity, probably influenced the age of marriage as well as the proportion of women who ever married in an age when couples waited upon a "nest egg" before wedding. It would also affect the incentives for married couples to regulate their fertility and influence their choice of completed family size. On a new midwestern farm, work could always be found for family labor; there were fences to build, ditches to dig, and land to clear in addition to the everyday routine of farm life. In these circumstances, there were few, if any, incentives to check fertility.

As a result, midwestern women tended to marry sooner and in greater proportions, and they bore children later into life and with shorter spacing between births than women in the Northeast. Using a synthetic measure of fertility developed by Steckel (1980), over 40 percent of the difference in fertility rates between farm families in the two regions is attributable to the higher proportion of midwestern women who married, and a quarter is explained by the earlier age of marriage (Fig. 2.2). In the East the demand for farm labor for breaking ground and creating farm capital was less and the opportunity cost of work on the farm higher. As a result, northeasterners produced smaller families, and family members, particularly women, found alternative sources of employment.

| | Northeast | Midwest |
|---|---|---|
| Total Fertility Rate | 3.49 | 5.12 |
| Proportion of Women who Marry | 0.69 | 0.81 |
| Age at Marriage | 26.9 | 25.3 |
| Interval between Children | 3.1 | 2.8 |
| Age of Mother at Last Child | 39.4 | 40.3 |

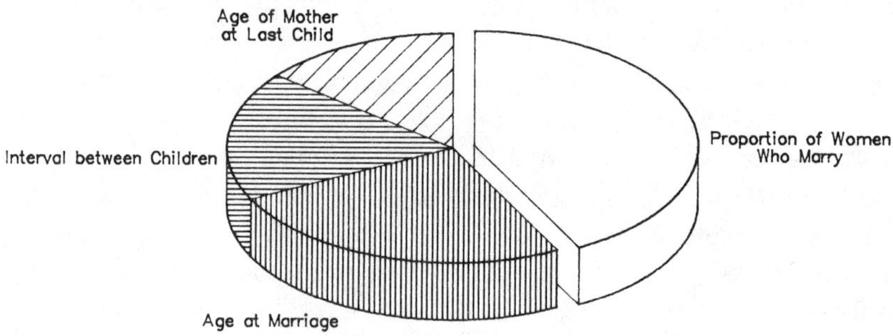


Fig. 2.2. Explaining East-West population fertility differences for farm families in 1860.

The observation that midwesterners tended to have larger families than northeasterners is not new. It was noted as long ago as 1843 and it shows up clearly in our data. Only one of our sample townships in the Northeast had a child-woman ratio (children under 10 per thousand women aged 16–44) greater than 1750 and in most the ratios were under 1250. On the other hand, in only a few midwestern townships were ratios that low and in many they exceeded 1750. Indeed, our data show one new wrinkle to the story: eastern and western nonfarm fertility rates were approximately the same orders of magnitude (1300–1500) so that the East-West difference seems to be generated by the radically different farm fertility rates that were almost twice as high in the Midwest as in the Northeast.

Regression analysis indicates that land availability was a significant explanatory variable in midwestern farm-family fertility but not among northeastern farm families. Other variables were also found to exert strong influences upon fertility. Wealth, for example, was associated with lower fertility. Ethnic and cultural differences were also associated with different fertility patterns. Communities with a large fraction of New Englanders, for example, showed significantly lower fertility levels than otherwise similar communities. Sometimes the same group responded differently in the two regions. Thus, in the Midwest the percentage of foreign-born tended to decrease fertility but in the Northeast this group tended to have greater fertility. And while the proportion of Irish was important in explaining fertility among farm households, they were associated with lower fertility in the land-scarce East and higher fertility in the Midwest.

## LAND POLICY AND FARM FORMATION

Public land policy, especially after 1832 when the minimum purchase was reduced to 40 acres at a minimum price of $1.25 per acre, seemed to insure almost universal access to sufficient land to start a yeoman farm in the Midwest. Nevertheless, even though an outlay of $50 for 40 acres of new land fell within the reach of 90 percent of the population, only 55 percent of the sampled rural households availed themselves of the opportunity to operate farms. In light of the pervasive agrarian ideal, this initially seems puzzling.

But easy access to cheap land was the least of a would-be farmer's financial concerns. Clarence Danhof, using contemporary literary accounts, suggested approximately $1,000 as the sum needed to establish a fully functional 40-acre farm on the midwestern frontier (Danhof 1941). Estimates based upon the census manuscript data are in broad agreement with this figure (Atack 1982). In Table 2.1 we show the sample estimates of the average cost of buying 40-, 80-, and 160-acre farms (the typical sizes) in the midwestern states, together with the cost of the average number of head of livestock for such a farm and the mean value of machinery. For 40-acre farms these outlays range from as little as $435 for a farm in Minnesota to $1,549 for the same size holding in Ohio. On the frontier, these costs were generally lower. The typical midwestern 40-acre farm, however, cost only half as much as a comparably

TABLE 2.1. Average Farm and Farm-making Costs for Owner-occupied Farms of Different Sizes in the Midwest in 1860

| State | Number of Observations | Farm ($) | Implements ($) | Livestock ($) | Total Cost ($) |
|---|---|---|---|---|---|
| 40-Acre Farms | | | | | |
| Illinois | 54 | 788 | 78 | 237 | 1,103 |
| Indiana | 215 | 769 | 43 | 206 | 1,018 |
| Iowa | 17 | 590 | 26 | 164 | 780 |
| Kansas | 3 | 267 | 37 | 153 | 457 |
| Michigan | 109 | 754 | 47 | 171 | 972 |
| Minnesota | 6 | 317 | 15 | 125 | 457 |
| Missouri | 44 | 329 | 34 | 219 | 582 |
| Ohio | 20 | 1,238 | 47 | 245 | 1,530 |
| Wisconsin | 26 | 858 | 44 | 113 | 1,015 |
| 80-Acre Farms | | | | | |
| Illinois | 128 | 1,345 | 105 | 297 | 1,747 |
| Indiana | 399 | 1,541 | 66 | 304 | 1,911 |
| Iowa | 33 | 1,060 | 45 | 221 | 1,326 |
| Kansas | 21 | 856 | 72 | 333 | 1,261 |
| Michigan | 124 | 1,282 | 52 | 222 | 1,556 |
| Minnesota | 19 | 630 | 34 | 141 | 805 |
| Missouri | 75 | 730 | 49 | 341 | 1,120 |
| Ohio | 40 | 2,323 | 73 | 388 | 2,784 |
| Wisconsin | 46 | 1,049 | 63 | 166 | 1,278 |
| 160-Acre Farms | | | | | |
| Illinois | 80 | 2,648 | 129 | 476 | 3,253 |
| Indiana | 195 | 3,457 | 113 | 509 | 4,073 |
| Iowa | 16 | 1,975 | 81 | 290 | 2,346 |
| Kansas | 74 | 1,744 | 92 | 354 | 2,190 |
| Michigan | 30 | 2,772 | 90 | 423 | 3,285 |
| Minnesota | 57 | 702 | 36 | 140 | 878 |
| Missouri | 61 | 1,434 | 73 | 522 | 2,029 |
| Ohio | 22 | 4,600 | 96 | 518 | 5,214 |
| Wisconsin | 30 | 1,633 | 84 | 294 | 2,011 |

sized northeastern unit. The northeastern farmer selling a 40-acre farm and moving to the Midwest could easily purchase there an 80- to 100-acre farm.

Eighty-acre units were the most common in the Midwest. Costs for such a functioning farm ranged from $805 in Minnesota to $2,784 in Ohio. The costs of an 80-acre farm in Ohio were comparable with a similar one in the Northeast. A northeastern farmer who sold and moved west to Ohio could purchase a farm perhaps one-eighth larger than the one he vacated. If, instead, he moved to the west or north of Indiana, he could purchase a farm at least twice as large. Although such a move could create possibilities of subdivision among heirs, there was little or no advantage to this in farm income.

Moving to the frontier was not, however, the only solution for those unable or unwilling to afford the cost of land purchase or the hardships of land breaking. Though contrary to the goal of agrarian democracy, tenancy offered an alternative entry choice. In the Midwest the average tenant of an 80-acre farm owned $57 worth of implements and $268 of livestock, and operated a farm valued at $1,460. However, the tenant's capital costs were limited to the $325 invested in implements and livestock. By contrast, the average owner-occupier of a similar unit held an investment of $1,715. Moreover, as farm size grew, the relative costs of tenancy fell so that the tenant of an average midwestern 160-acre farm had a capital outlay only one-seventh that of someone farming his own land.[7]

The census data do not account for all the costs included in Danhof's $1,000 figure. Missing are estimates of the cost of provisions for the period before the first crops could be raised (set at $100), seed, estimated at as much as $1–2 per acre or about $25–50 for the average 40-acre midwestern farm.[8] Additionally, an allowance might be needed for the hire of harvest labor. Neglecting the latter unspecified cost, the average 40-acre farm in the Midwest probably cost about $1200, but some substantial economies, totaling about one-third, could be realized, for example, by moving to the frontier, substituting oxen for horses, a cart for a wagon, or improving fewer acres in the early years.[9] For an 80-acre farm, the cost was likely to have exceeded $2,000; for a 160-acre holding, $3,000.

## THE DISTRIBUTION OF WEALTH IN A YEOMAN DEMOCRACY

Such costs put the Yeoman Dream out of the reach of many. An 80-acre farm was beyond the price range of 54 to 74 percent of the population unless they had an unusually understanding banker or were willing to settle for tenancy as a first step. Forty percent or more were too poor to participate at the 40-acre level, which we regard as the minimum viable size for a family farm that provided the sole source of support. Even tenancy on an 80-acre farm lay beyond the means of 5 to 25 percent of the population, and tenants had fewer borrowing opportunities than owner-occupiers.

With these levels of expenditure, it is not surprising that farmers were near the top end of the wealth distribution. Only those with median or higher wealth levels could afford the entry costs. There is strong evidence of a life cycle in wealth accumulation within the data (Atack and Bateman 1981a). Wealth increased and became more equally distributed as people grew older (Atack and Bateman 1981b). The association of higher wealth with age reflects, among other influences, land appreciation and the long-term effects of production greater than family consumption levels; the increasing equality partly reflects a smoothing out over time of random events, most notably inheritance. Inadequate wealth accumulation for farm ownership, resulting in tenancy, seems to be largely accounted for by age (Atack 1988). Using his own resources, a farmer could have secured a 40-acre farm by his midtwenties, but not one of 160 acres until he neared fifty (Fig. 2.3). Many could hasten their entry, of course, through access to credit markets or migration to the frontier.

In a life-cycle pattern of accumulation about a third of the observed inequality could have resulted from age (Atack and Bateman 1979). This was aggravated by differential rates of population growth. Inheritance would also lead to a concentration of wealth similar to—indeed possibly greater—than we observe and in that case inequality would have been

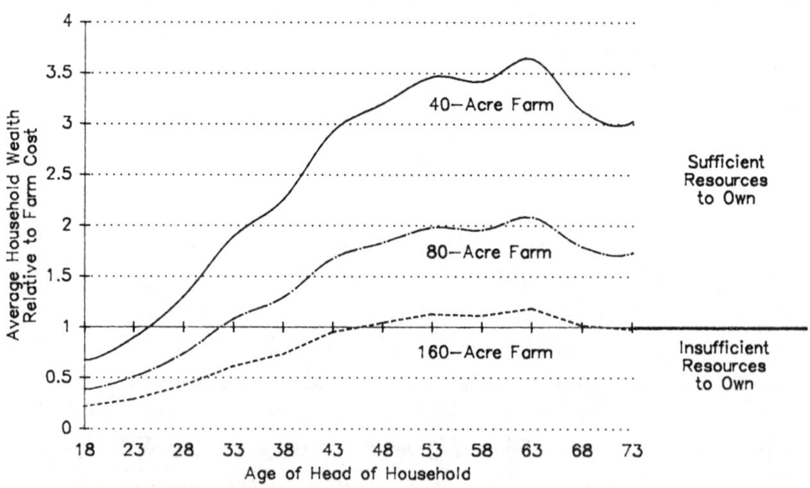


Fig. 2.3. Midwestern wealth accumulation by age relative to the cost of the typical 40-, 80-, and 160-acre farm in 1860.

determined by forces no more sinister and conspiratorial than decisions regarding family size (Soltow 1982, 1983).

In 1860 the wealthiest 1 percent of the rural midwestern population owned about 12 percent of the total wealth; the wealthiest 5 percent held 31 percent. This is slightly more egalitarian than in the Northeast.

Clearly, however, wealth was not equally distributed among the population in this supposed yeoman democracy. Nonetheless, wealth was much more equally distributed among farmers than among other groups. By definition, all farmers had at least some wealth, but large segments of other populations reported none. The Gini index of inequality for midwestern farmers was thus only 0.57; among the nonfarmers it was 0.72 (Atack and Bateman 1981b).

Other factors besides age, among them race, nativity, literacy, sex, and occupational status, contributed to rank in the wealth distribution. To establish the individual importance of these in determining the level of family wealth and to measure their relative importance, we ran a regression estimate of the form:

Family Wealth = g(Age, $Age^2$, Race, Sex, Literacy, Occupation, Birth) for the Northeast and the Midwest. The variables, Race, Sex, Literacy, Occupation, and Birth were all 0–1 dummy variables where:

Race = 1, if white
0, otherwise
Sex = 1, if male
0, if female
Literacy = 1, if literate
0, otherwise
Occupation = 1, if farmer
0, if otherwise
and Birth = 1, if native born
0, otherwise.

The variables are represented by the abbreviations: AGE, AGESQ, RACE, SEX, LIT, OCC, and BIRTH.

The raw regression coefficients from these estimates have an immediate and appealing interpretation. They represent the dollar increments (or decrements) to family wealth for each of the social and demographic attributes. The results, together with an indication of their statistical significance and the level of family wealth estimated with each of these attributes set at their mean values, are given in Table 2.2.

TABLE 2.2. The Change in Family Wealth in Response to Social and Demographic Attributes

| | REGRESSION COEFFICIENT OF | | | | | | | | Wealth Level at the Mean Attribute Values |
|---|---|---|---|---|---|---|---|---|---|
| Region | Constant | AGE | AGESQ | RACE | SEX | LIT | OCC | BIRTH | |
| Midwest | −7951.82 | 213.87[a] | −1.69[a] | 1578.84[a] | 474.00[a] | 1046.24[a] | 985.58[a] | 1228.46[a] | 2274.23 |
| Northeast | −8000.78[a] | 220.54[a] | −1.81[a] | 1623.11[a] | 318.18 | 1025.04[a] | 2144.69[a] | 1222.16[a] | 2916.15 |
| Rural North | −8350.68[a] | 220.59[a] | −1.75[a] | 1585.61[a] | 448.42[a] | 1134.18[a] | 1388.97[a] | 1355.06[a] | 2538.36 |

[a]Significantly different from zero at better than the 5 percent level.

AGE was significantly positive while the coefficient of age squared (AGESQ), was significantly negative. Based upon our regression equations, wealth increased with age, reaching a maximum at age 63 in the Midwest and at age 61 in the Northeast and declined thereafter.[10] The coefficient of RACE was significantly positive. This provides strong evidence of discrimination in the distribution of wealth based on race, such that nonwhite families, either Indian or black, had an average of $1579 to $1623 less wealth than a demographically and socially identical white family. This, however, does not tell the full story of discrimination, given the possibility that the head of a nonwhite family would be either literate, or a farmer, or both, was also much smaller than for white families. Actually, the only discrimination with which blacks did not have to contend was that against the foreign born. The coefficient of SEX was significantly positive in the Midwest suggesting significant discrimination against female-headed households, a trend reinforced by the lower probability that women were also farmers. The coefficients of LIT, OCC, and BIRTH were significantly positive and generally were of comparable magnitudes. The results imply discrimination against the foreign born and in favor of the farmer and the educated.

Although the aspirations for equality that formed part of the Yeoman Dream were not perfectly realized, compared with more recent periods and with other regions during the same period, wealth was more evenly distributed among the population of the rural townships in 1860. To the extent that human realities always diverge from lofty abstract goals, the closeness between them in this case is remarkable.

## SELF-SUFFICIENCY AND THE FAMILY FARM

Jefferson had spoken passionately in favor of the self-sufficient smallholder, but the 80-acre-minimum farms sold by the government before 1832 could not be completely cultivated by family labor with the available technology. The reduction of the public land sale minimum to 40 acres brought a closer concordance between reality and the Jeffersonian ideal. Indeed, the average farmer with 40 acres was self-sufficient and may have produced a small surplus for sale in the market beyond the needs of his family and livestock. Self-sufficiency accorded well with the goal of independence. Market sales, of course, portended future exposure to "the caprice of customers" which confounded Jefferson's agrarian vision, but at the same time they generated cash for market necessities such as implements and luxuries such as refined sugar, coffee, and better clothing. Small surpluses too were a necessary evil if farmers were to avoid the greater evil of market involvement as debtors who might ultimately be forced from the land.

According to our calculations, three-quarters of the midwestern farms of 40 to 60 acres produced a dairy surplus, more than half had meat surpluses, and two-thirds had grain surpluses, so that overall fewer than one-third of the farms produced less food than they probably consumed.[11] These surpluses, of course, went to feed the growing nonfarm population, especially the urban and industrial centers of the East Coast. Furthermore, about 90 percent of farmers could have had positive cash flows from the sale of their crops after allowing for all consumption needs, human and livestock, and seed retained for the next year's planting. This cash in turn provided a source of effective demand for the products of the industrial sector.

The average marketable surplus among the sample of northern farms was valued at $360, about 70 percent more than has been found for farms in Canada West at about the same time (McInnis 1984). In part this is attributable to superior fertility and a better climate; in part it reflects higher prices from market proximity. Northeastern farms realized, on average, a marketable surplus of $451, while those in the

TABLE 2.3. The Value of the Net Marketable Surplus at the Farm Gate, 1859

| Location[a] | Average Net Surplus | Average Surplus By Size of Farm: 31–40 | 61–80 | 121–160 | 161–320 | Percent with Deficits |
|---|---|---|---|---|---|---|
| New England | $370 | $144 | $267 | $451 | $558 | 4 |
| Middle Atlantic | 485 | 239 | 405 | 621 | 860 | 8 |
| Northeast[b] | 451 | 222 | 365 | 551 | 765 | 8 |
| Old Northwest | 331 | 143 | 228 | 445 | 544 | 9 |
| Recent Settlement | 315 | 134 | 217 | 290 | 434 | 7 |
| Midwest[c] | 307 | 136 | 215 | 379 | 459 | 12 |
| Slave States | 138 | −2 | 70 | 178 | 195 | 33 |
| With good transport | 392 | 177 | 275 | 491 | 609 | 9 |
| With poor transport | 276 | 97 | 208 | 337 | 448 | 14 |
| Entire Sample | 360 | 156 | 257 | 438 | 566 | 10 |

[a]New England: Connecticut, New Hampshire, and Vermont; Middle Atlantic: New Jersey, New York, and Pennsylvania; Old Northwest: Illinois, Indiana, Ohio, plus southern Michigan and Wisconsin; Recent Settlement: Iowa, Kansas, Minnesota, plus northern Michigan, northern Wisconsin, and the nonslave townships in Missouri; Slave: Maryland and the slave-owning townships in Missouri.

[b]Includes Maryland.

[c]Includes Missouri.

Midwest earned $307; this provides one justification for the comparatively higher land prices in the former region. (See Table 2.3.)

The highest-valued surpluses were in Illinois, New Jersey, New York, and eastern Pennsylvania, where surpluses often exceeded $750 per farm. Surpluses on the northern and western frontiers were typically lower than those elsewhere. In only a few townships was there a deficit that would have required additional sources of income for farm families to feed the livestock and put sufficient food on the table.[12] These were all located on the margin of settlement. At the same time, farms in two Minnesota townships in Polk and Mahnomen counties stood out as exceptions to this generalization. On average they produced surpluses with cash values of $1,111 and $886 respectively, which were generated by surpluses of both food and nonfood items.

Surpluses, in general, remained relatively small because farm scale was small—especially so long as farmers relied upon family labor and the occasional hired hand, and so long as they chose to produce a diversified range of crops and livestock products that had much in common with their desired consumption basket of goods. In the aggregate, about 8 percent of northeastern farms had negative net surpluses; in the Midwest about 12 percent were in that situation. The median surplus in the Northeast ($305) was about 50 percent greater than that in the Midwest ($200). These are neither particularly large nor impressive surpluses, and the low slope of the net farm surplus lines in Fig. 2.4 indicate generally similar conditions on most farms. More than two-thirds produced net surpluses valued at less than $500. In both regions, however, a small fraction of farms produced substantial surpluses. Ten percent in the Northeast earned in excess of $1,000 beyond the consumption needs of the livestock and farm family. Only 4.4 percent of midwestern farms, however, boasted a surplus that large.

Subdivision of the regions into New England, the Middle Atlantic, the Old Northwest, states of recent settlement (Iowa, Kansas, Minnesota, and the northernmost townships in Michigan and Wisconsin), and the two slave states points to some differences and surprising similarities (see Table 2.3). The average surplus was greatest by a considerable margin in the Middle Atlantic states and very low in the slave-owning regions. Surpluses in New England and in the two subregions of the Midwest were comparable. The distribution was virtually identical among farms in the Old Northwest and in the areas of recent settlement, except that the top 5 percent in the former had somewhat higher values

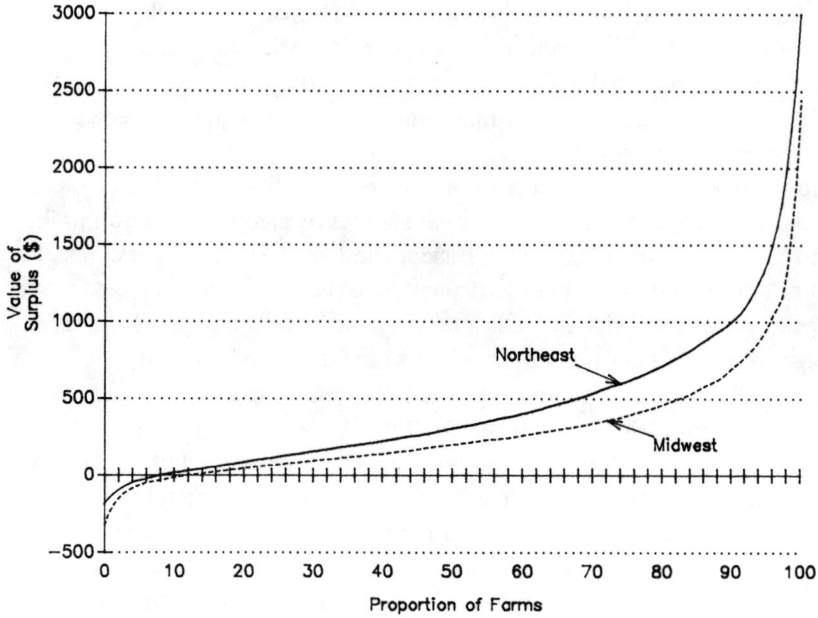


Fig. 2.4 The distribution of farm surpluses by region, 1859-1860. (Proportion of farms generating a surplus less than or equal to that shown.)

for their net surpluses. Surprisingly, New England farms were the least likely of all to have a net deficit. By size of farm, net surpluses in New England and the Old Northwest were also very similar but only about two-thirds of the levels realized from farms in the Middle Atlantic states. Surpluses in the slave-owning areas were half, or less, of those in the nonslave townships.

There were also important differences in the value of the marketable surplus depending upon whether a township was served by water or rail transportation, or was forced to depend upon wagon haulage. Surpluses where cheap transportation was available were larger, averaging $392 per farm, than those produced in townships dependent upon wagon haulage, which averaged only $276. The most pronounced differences were between the very small or the very large farms. Those with easy, inexpensive access to markets earned twice as much as comparably sized farms without that advantage. Cash from market sales must have been seductive for it seems to have caused farmers who had the chance to produce more (either by more extensive or intensive agriculture even on farms of the same nominal acreage) and perhaps specialize in more highly valued market crops. At the time, in 1860, this seems to have caused little conflict with the Yeoman Dream but it did eventually lead to what Jefferson had feared—dependency on markets—and the financial difficulties that farmers faced in the postbellum period, especially the 1890s (Saloutos 1948; Shannon 1948).

## PROFITABILITY AND AGRICULTURAL EXPANSION

The rapid expansion of the area under cultivation caused by sale of the public land drove down rents. We cannot measure the impact of this at the margin, but the effect is discernible in the average noncapitalized rents shown in farm profits. Our estimate of the average return per dollar invested in northern agriculture, 12 percent (Table 2.4), is quite low relative to that available in such alternative fields as manufacturing or transportation (Table 2.5). It is, however, quite close to that being earned by southern farmers at this time as well as to the traditionally assumed competitive rate. In the two sample slave areas the rate of return was lower.

While the estimated rate of return for the entire North is consistent with expectations, those for the separate Midwest and Northeast subregions differ from impressions gained in the historical literature. Scholars often explain the westward movement of northern agriculture, for example, in terms of relative profit opportunities, western rates being presumably higher than those in the East. Yet our figures indicate a marginally

TABLE 2.4. Northern Agricultural Profitability: Return per Dollar Invested, 1859

| State/Region | Rate of Return | State/Region | Rate of Return |
|---|---|---|---|
| Illinois | .191 | Connecticut | .106 |
| Indiana | .104 | Maryland | .076 |
| Iowa | .111 | New Hampshire | .068 |
| Kansas | .100 | New Jersey | .100 |
| Michigan | .057 | New York | .173 |
| Minnesota | .088 | Pennsylvania | .101 |
| Missouri | .060 | Vermont | .122 |
| Ohio | .133 | | |
| Wisconsin | .113 | Northeast[a] | .127 |
| | | | |
| Midwest[a] | .123 | The North[a] | .123 |

[a]Weighted by farm capital in the population.

TABLE 2.5. Mean Rates of Return by Sector and Region in the Antebellum Economy

| Economic Sector | Rate of Return (%) | Standard Deviation |
|---|---|---|
| *Agriculture:* | | |
| Northern (Atack-Bateman) | | |
| A Estimates[a] | | |
| Entire North | 7.1 | 23.0 |
| Northeast | 10.5 | 16.4 |
| Midwest | 5.0 | 26.2 |
| B Estimates[b] | | |
| Entire North | 12.1 | n.a. |
| Northeast | 12.6 | n.a. |
| Midwest | 12.1 | n.a. |
| C Estimates[c] | | |
| Entire North | 8.0 | n.a. |
| Northeast | 9.8 | n.a. |
| Midwest | 6.5 | n.a. |
| Southern | | |
| Cotton (Fogel-Engerman) | 10.0 | 7.9 |
| Cotton (Vedder-Stockdale) | 9.7 | n.a. |
| Land and Capital (Vedder-Stockdale) | 10.6 | n.a. |
| Rice (Swan) | −3.8 | n.a. |
| Slave Hiring (Evans) | 10.4 | n.a. |
| *Manufacturing:* (Bateman-Weiss) | | |
| East, 1860 | 21.8 | 45.0 |
| South, 1860 | 27.9 | 47.6 |
| West, 1860 | 26.1 | 48.6 |
| U.S., 1860 | 25.2 | 47.1 |
| *Transportation:* | | |
| Steamboats, 1850 (Atack et al.) | | |
| Trunk River | 8.5 | 5.5 |
| Tributary River | 24.1 | 13.7 |
| Central Pacific RR (Mercer) | | |
| Unaided Private Return | 13.4 | n.a. |
| Aided Private Return | 14.1 | n.a. |

Sources: Robert Fogel and Stanley Engerman, *Time on the Cross* (Boston, 1974); Richard Vedder and David Stockdale, "The Profitability of Slavery Revisited," *Agricultural History* 49(1975):392–404; Dale Swan, "The Structure and Profitability of the American Rice Industry, 1859" (Ph.D. diss., Univ. of North Carolina, Chapel Hill, 1972); R. J. Evans, "The Economics of American Negro Slavery," in *Aspects of Labor Economics* (Princeton: Princeton Univ. Press, 1962); Jeremy Atack et al., "The Profitability of Steamboating on the Western Rivers: 1850," *Business History Review* 49(1975):346–54; Lloyd Mercer, "Rates of Return for Land Grant Railroads: The Central Pacific System," *Journal of Economic History* 30(1970):602–26; Fred Bateman and Thomas Weiss, *A Deplorable Scarcity* (Chapel Hill: Univ. of North Carolina Press, 1981); Jeremy Atack and Fred Bateman, *To Their Own Soil* (Ames: Iowa State Univ. Press, 1987).

[a]Includes all townships in the sample. Owner-occupiers only; each farm weighted equally.

[b]Tenants and yeomen, weighted by capital. Frontier townships excluded.

[c]Tenants and yeomen, weighted by capital. Excludes capital gains on land and excludes frontier townships.

higher return in the latter. Nevertheless this is consistent with the impact of public land policy, which would have eroded any scarcity value to midwestern land as a whole. Despite this, the rate of return per dollar invested on the Kansas and Minnesota frontier, which might be thought of as the marginal midwestern investment at this time of between 8.8 percent and 10 percent is virtually identical to that being earned on bonds (Macaulay 1938). We have no comparable "frontier" figure for the Northeast. During the initial investment period in pioneer farming, the earned short-run rates of return (but not the expected long-term rates) can be quite low, but this signals not necessarily misallocation but rather the prevailing stage of investment. From this perspective, these rates of returns were certainly adequate to stimulate and maintain westward migration and interest in western land opportunities, particularly during periods of optimism regarding future land values.

The rates that we discuss above represent the average return per dollar invested in agriculture in a randomly selected township. Farmers, however, invested in farms, not in farming. Their return depended critically upon the size of farm they could afford to buy or rent. Similarly, average farm profit rate by state or region depended upon the size distribution of farms, a reality that worked against the interests of the small yeoman producer. Big farms were more profitable per dollar invested than small. This is consistent with our analysis of self-sufficiency, which demonstrated that on 43 percent of farms under 40 acres farmers could feed neither their families nor their livestock the recommended diet, whereas only 19 percent of those holding more than 80 acres fell short of this standard.

The rate of return per farm in many states averaged about 6 percent, though it ranged from as much as 17.9 percent in Illinois to −11.9 in Michigan. In some states, such as Minnesota, Missouri, and New Hampshire, where the average rate was very low but positive or in Michigan where it was negative, a person must have viewed starting a farm as an investment whose payoff lay well into the future.

Even where average returns were low, however, farmers who could afford a larger spread could still reap rates of return equal to or greater than those offered by such alternative investments as bonds. Moreover, they could live off their capital and invest time and energy in capital improvements that eventually would be rewarded either through higher productivity or enhanced land values. Thus, even in Michigan where the average "loss" for a farm under forty acres was 21.3 percent, the individual who could purchase a unit of 160 acres or larger could expect an average return of 12.5 percent.

Small midwestern farms were much less profitable than comparably sized northeastern ones, probably reflecting regional differences in agriculture. In the Midwest the principal cash crop was wheat, which was best grown extensively. For the Northeast, truck crops produced for urban markets and dairy goods were important income sources. These products could be successfully produced on a much smaller scale than grains. A most remarkable feature, however, is the marked relationship between farm size and profitability in each region (Fig. 2.5). A systematic relationship between rate of return and scale existed in farming in every state covered in the sample. A 40-acre family farm—the yeoman

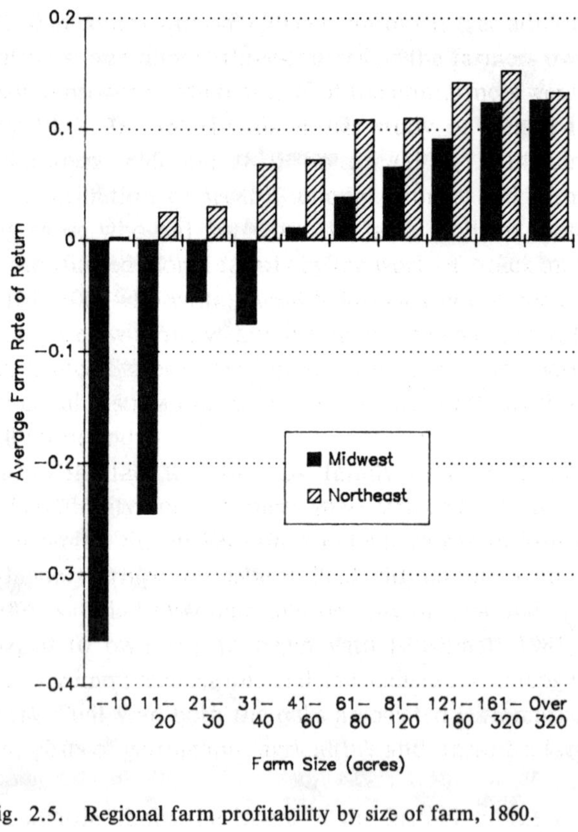


Fig. 2.5. Regional farm profitability by size of farm, 1860.

farm par excellence—was barely, if at all, profitable. To be secure a northeastern farmer probably had to work about 60 acres, while a midwesterner should have had at least 80 acres. Fortunately, in each region both the modal and median farm size equalled or exceeded these levels.

Northern agriculture would appear even less attractive as a pure economic investment if the basis for our profit estimates is adjusted for comparability with that used in other sectoral studies by excluding capital gains. By this measure, eastern producers were earning a current rate virtually identical (10.8 percent) to that of southern cotton growers. Western farmers were not doing so well, receiving only a 5.2 percent return. This suggests the importance that farmers, particularly those already in the West or considering migration there, attached to the future and to prospects for ultimately realizing capital gains on their land (Fig. 2.6). Agricultural historians long have stressed this perception. Paul Gates, for example, claims that, "The pioneer farmer was well aware that in the end his profits would come largely from rising land values"

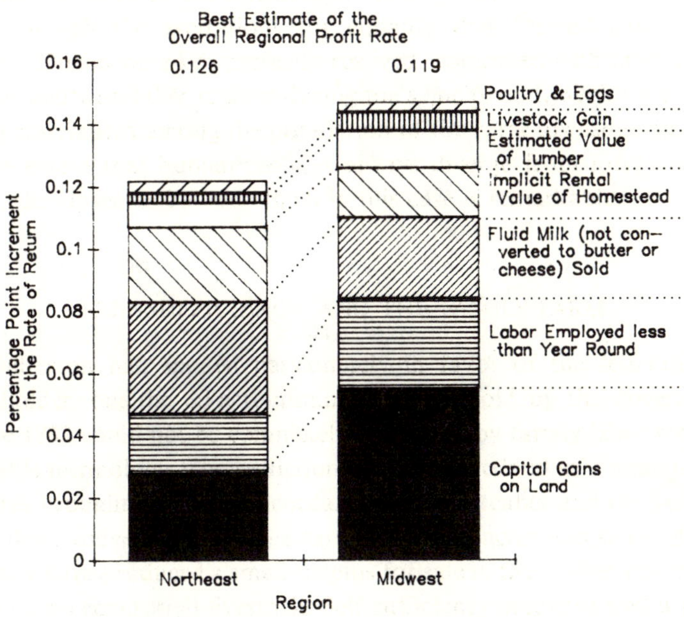


Fig. 2.6. The contribution of various adjustments to the estimated regional rate of return per dollar invested in agriculture in 1860.

(Gates 1960). By the 1850s and 1860s, Gates observes, this source of return had diminished considerably in the Northeast, but was still important in the newer western states. Our own analysis of the contributions to total profit from several components substantiates the large relative importance of capital gains to western farm profit. This was further reflected in land values. Between 1850 and 1860 the cash value per acre of farmland in the eastern states rose annually on average by 2.8 percent, while in the West it grew 7.3 percent.

Nevertheless, from the perspective of the national economy, the question remains why agricultural investors North and South would have been satisfied to earn a relatively low rate of return in farming when a considerably higher one was available, for example, in manufacturing (an activity open to even a small investor) or transportation. One must consider the farmers' economic preferences and behavior. Apparently not short-run profit maximizers, their behavior is more consistent with that of satisfiers. The yeomanry wanted the best of both worlds: independent self-reliance with a market-derived cash flow. They perceived family farming as a way of life and agriculture as a business enterprise; they wanted their farm to produce a "guaranteed" real income for their families as well as a cash income from surpluses sold to others. Their yeoman farm was a multiple-product, multiple-goal enterprise. To achieve their goals, they chose to satisfice so as to gain an acceptable return on investment in addition to producing most of their annual real income through the farm. Over a longer time horizon, they could also look forward to capital gains on their landholdings.

Nevertheless, even during the postbellum era, as the economy became increasingly industrialized and markets more formalized, this behavior continued. "Perhaps no development of the nineteenth century," said Theodore Saloutos, "brought greater disappointment to the American farmers than did their failure to realize the prosperity that they had expected from industrialism" (Saloutos 1948). Continued Saloutos, "Even before the wholesale transition from a subsistence to a commercialized status, the evidence was rather strong that the financial rewards of farming would be small." Farmers assumed that industrialization would increase the demand for food, especially in urban areas, and that in turn this would raise their total cash income from sales on the market. While at times and in some places it did, their expectations were never fully met.

Although low, our estimates for rate of return excluding capital gains are clearly in accord with nineteenth-century observations. If anything, they err on the high side. Contemporaries also recognized the wide variation between agricultural and industrial rates. The Secretary of the Treasury reported in 1845, for example, that "while the profit of agriculture varies from 1 to 8 percent, that of manufactures is more than double" (*Report from the Secretary of the Treasury* 1845). In this same document, reports of returns on agricultural investments of 3 to 6 percent were said to be typical for the 1830s and 1840s. As Saloutos (1948) observed:

> Certainly the period from the Civil War to 1897, was not profitable at least from the standpoint of remunerative farm operations. . . . Part of the answer for the unprofitableness of farming is to be found in the rapid territorial expansion—the land grant, immigration, and irrigation policies that encouraged people to take land without regard for the fact that they accelerated agricultural production beyond all reasonable market demands. . . . The capacity to produce foodstuffs, contrary to the predictions of Malthus, greatly overstripped the capacities to consume them.

Despite this, a substantial portion of the American population, native and immigrant alike, seems to have harbored a single-minded ambition to farm. For millions, family agriculture offered a way of life nonpareil. The conclusions from our analysis reflect the strength of the Yeoman Dream and the dogged persistence with which nineteenth-century Americans pursued it.

## CONCLUSION

Generations of scholars have debated whether slavery would have collapsed under its own weight even if the Civil War had never been fought. Would the North's equally peculiar institution have met a similar fate because of its own internal contradictions and its conflicts with the industrializing external economic environment? First, consider whether the yeoman agrarian economy had succeeded by midcentury. Limiting our perspective to the time preceding the Civil War, how would we evaluate the economic status of yeoman agriculture specifically and its relationship to American agrarian society generally? Clearly, if the economic underpinning provided by yeoman farming did not succeed, then neither could the entire Jeffersonian political vision. By the standards we applied above, the northern rural economy was succeeding as an integrated social, familial, and economic institution. In this chapter and elsewhere, we found the farmers of the North to be reasonably well-educated for their time, to be mobile geographically and socially, to have achieved a comparatively egalitarian economy, and to be well-fed and growing wealthier. The rural population, in which the yeoman's role was central, was surprisingly diverse and heterogeneous in 1860, encompassing small manufacturers and various service workers as well as those tied more closely to the soil.

A substantial proportion of the population, including natives and immigrants, was eventually financially able to acquire land to start a farm and in time to make it produce not only enough for a family but for market sale as well. And while the immediate return on their farm investment may have been meager, the "love of the land" as well as the promise of capital gains and generally of better things to come was sufficient to keep people establishing farms and creating the families needed to run them. The concept of the yeomanry, after all, was built as much on promise as on performance. To an optimistic nation of individualists accustomed always to "looking ahead," the family farm was a natural attraction. It still is. Resting deep in many American minds even today is the appeal of family farming and its accoutrements of independence, liberty, and self-reliance.

Yet even before the Civil War, tiny and incipient cracks were developing in the facade of the North's peculiar institution. A clear and recurrent theme emerged from our research: a tension existed between the national pursuit of a new social order of individualism and independence based upon a family farm system, on one side, and the drive toward an interdependent capitalistic market economy in which farming functioned largely as a business enterprise, on the other. Eventually, of course, the latter prevailed, but the ambivalence and the tension have continued to the present.

Even by 1860 a basis for yeoman farming's decline already existed—in the financial ambitions of the surplus-producing farmers, who wanted commercial sales, as well as in the business organizations that were developing outside. Although these and other forces would sweep it ultimately into the realm of nostalgia, as of the 1860s a landowning yeomanry, the North's appealing but nevertheless peculiar institution, still seemed possible.

## NOTES

1. These data were collected by Fred Bateman and James Foust, with the support of the National Science Foundation under grant AS-27143. The sample is publicly available under the title, *Agricultural and Demographic Records of 21,118 Rural Households Selected from the 1860 Manuscript Censuses* from the ICPSR in Ann Arbor, Michigan. All substantive results in this paper that are not otherwise cited are derived from our book *To Their Own Soil* (1987).

2. Some tests of the sample are reported in Bateman and Foust (1974). More elaborate tests are reported in Atack and Bateman (1987). One important caveat with respect to the sampling procedure was that all the townships selected were drawn from nonurban counties, where an urban county was defined as one in which at least 90 percent of the population lived in the largest city. Moreover, a township was rejected if 90 percent or more of the gainfully employed population reported nonfarm occupations.

3. There were a number of farms unmatched with population as well as households in which someone reported their occupation as farmer but for whom no farm could be located. The problem of farms without farmers and farmers without farms is one frequently encountered by researchers using the manuscript censuses. See, for example, Curti (1959) and Bogue (1963).

4. Copies of the instructions to Assistant Marshals for the collection of data at the Eighth Census, long thought to be lost (see Wright 1900) are to be found in the Library of Congress, the Widener Library, Harvard University, the Library of the American Philosophical Society, and the University of Illinois Library. See U.S. Census Office. Eighth Census. 1860. *Eighth Census, United States—1860. Act of Congress of twenty-third May 1850. Instructions to U.S. Marshals. Instructions to Assistants,* Washington, D.C.: G. W. Bowman.

5. For histories of public land law and disposal of the public domain, see Hibbard (1965) and Gates (1968).

6. There is a long historiography on this proposition beginning with Thompson and Whelpton (1953) and Taeuber and Taeuber (1958), but perhaps most closely associated with the work of Yasuba (1961), Forster and Tucker (1972), and Easterlin (1976a, 1976b; Easterlin et al., 1978). See also the work by Leet (1975 and 1976) who develops the measure of land availability upon which our results are based.

7. The issue of tenancy is treated in greater detail in a forthcoming paper by Atack (1988).

8. Danhof (1941, 349) assumed that a 40-acre farm had 24–28 acres under cultivation. The *Country Gentlemen* 5(1855):213 set seed costs at $75 for 100 improved acres. Regan (1859, 353) set seed costs at $20 for a 40-acre farm and food for 4 for six months at $50, while Newhall (1846, 59) allowed $5 for 10 acres of corn, 5 of wheat, and 5 of potatoes, turnips, and garden produce.

9. See, for example, *Wisconsin Farmer* 8(1850):265 or Minnesota Commissioners of Statistics, *Minnesota: Its Place among States Being the First Annual Report of the Commissioner of Statistics for the Year Ending January 2, 1860.* (Hartford, 1860, 88).

10. The maximum wealth by age is attained when $\partial WEALTH/\partial AGE = 0$, that is at $[AGE/2(|AGESQ|)]$. See Table 2.2 for AGE and AGESQ.

11. Assuming specific generous livestock and human dietary allowances. The livestock diet was consistent with modern best practice farming techniques and almost certainly exceeded the contemporary feeding practices. The human diet was based upon a wide variety of contemporary accounts and provided about 5,000 calories a day per adult male. While this may seem more than generous, not to say excessive, these allowances

include the value of food wasted or spoiled. See Atack and Bateman (1984a, 1984b, 1987).

12. Our estimates of the surplus, however, were quite sensitive to the level of livestock feed assumed.

## REFERENCES AND FURTHER READINGS

Atack, Jeremy. 1982. "Farm and Farm-Making Costs Revisited." *Agricultural History* 56(Oct.).

______. 1988. "Tenants and Yeomen in the Nineteenth Century." *Agricultural History* 62(Summer).

Atack, Jeremy, and Fred Bateman. 1979. "The Measurement and Trend of Inequality: An Amendment to a Basic Revision." *Economic Letters* 4(Oct.).

______. 1981a. "The 'Egalitarian Ideal' and the Distribution of Wealth in the Northern Agricultural Community: A Backward Look." *Review of Economics and Statistics* 63(Feb.).

______. 1981b. "Egalitarianism, Inequality and Age: The Rural North in 1860." *Journal of Economic History* 41(Mar.).

______. 1984a. "Self-Sufficiency and the Marketable Surplus in the Rural North, 1860." *Agricultural History* 58(July).

______. 1984b. "Marketable Farm Surpluses: Northeastern and Midwestern United States, 1859 and 1860." *Social Science History* 8(Fall).

______. 1987. *To Their Own Soil: Agriculture in the Antebellum North.* Ames: Iowa State Univ. Press.

Bateman, Fred, and James D. Foust. 1974a. *Agricultural and Demographic Records of 21,118 Rural Households Selected from the 1860 Manuscript Censuses.* Data tape available from the ICPSR, Ann Arbor, Michigan.

______. 1974b. "A Sample of Rural Households Selected from the 1860 Manuscript Censuses." *Agricultural History* 48(Jan.).

Bogue, Allan G. 1963. *From Prairie to Corn Belt: Farming on the Illinois and Iowa Prairies in the Nineteenth Century.* Chicago: Univ. of Chicago Press.

Curti, Merle. 1959. *The Making of an American Community: A Case Study of Democracy in a Frontier County.* Stanford: Stanford Univ. Press.

Danhof, Clarence. 1941. "Farm-Making Costs and the 'Safety Valve': 1850–1860." *Journal of Political Economy* 49(June).

______. 1969. *Change in Agriculture.* Cambridge: Harvard Univ. Press.

Davis, Lance E., et al. 1972. *American Economic Growth.* New York: Harper and Row.

Dennen, R. Taylor. 1957. "Some Efficiency Effects of Nineteenth Century Federal Land Policy: A Dynamic Analysis." *Agricultural History* 51(Oct.).

Easterlin, Richard A. 1976a. "Population Change and Farm Settlement in the Northern United States." *Journal of Economic History* 37(Mar.).

______. 1976b. "Factors in the Decline of Farm Family Fertility in the United States: Some Preliminary Research Results." *Journal of American History* 63(Dec.).

Easterlin, Richard A., George Alter, and Gretchen Condran. 1978. "Farms and Farm Families in Old and New Areas: The Northern States in 1860." In *Family and Population in Nineteenth Century America,* ed. Tamara Hareven and Maris Vinovskis. Princeton: Princeton Univ. Press.

Faragher, John M. 1986. *Sugar Creek: Life on the Illinois Prairie.* New Haven: Yale Univ. Press.

Fogel, Robert W., and Jack Rutner. 1972. "The Efficiency Effects of Federal Land Policy, 1850–1910: A Report on Some Provisional Findings." In *The Dimension of Quantitative Research in History,* ed. William O. Aydelotte. Princeton: Princeton Univ. Press.

Forster, C., and G. S. C. Tucker. 1972. *Economic Opportunity and White Fertility Ratios, 1800–1960.* New Haven: Yale Univ. Press.

Gates, Paul W. 1960. *The Farmer's Age: Agriculture 1815–1860.* New York: Holt, Rinehart and Winston.

______. 1968. *History of Public Land Law Developments.* Washington, D.C.: Public Land Law Commission.

Hibbard, Benjamin H. 1965. *A History of the Public Land Policy.* Madison: Univ. of Wisconsin Press.

Leet, Donald. 1975. "Human Fertility and Agricultural Opportunities in Ohio Counties: From Frontier to Maturity, 1810–60." In *Essays in Nineteenth-Century Economic History,* ed. D. Klingaman and Richard Vedder. Athens: Ohio Univ. Press.

______. 1976. "The Determinants of the Fertility Transition in Antebellum Ohio." *Journal of Economic History* 36(June).

Macaulay, Frederick R. 1938. *Some Theoretical Problems Suggested by the Movement of Interest Rates, Bond Yields and Stock Prices in the United States Since 1856.* New York: National Bureau of Economic Research.

McInnis, Marvin. 1977. "Childbearing and Land Availability: Some Evidence from Individual Household Data." In *Population Patterns in the Past,* ed. Ronald Demos Lee. New York: Academic Press.

Newhall, J. B. 1846. *A Glimpse of Iowa in 1846.* Burlington, Iowa: W. D. Skillman.

Rasmussen, Wayne D., ed. 1960. *Readings in the History of American Agriculture.* Urbana: Univ. of Illinois Press.

Regan, John. 1859. *The Western Wilds of America.* Edinburgh: John Mengies.

Saloutos, Theodore. 1948. "The Agricultural Problem and Nineteenth-Century Industrialism." *Agricultural History* 22(Oct.).

Schlebecker, John T. 1975. *Whereby We Thrive.* Ames: Iowa State Univ. Press.

Secretary of the Treasury. 1845. *State of the Finances.* Washington.

Shannon, Fred A. 1936. "The Homestead Act and the Labor Surplus." *American Historical Review* 41(July).

Soltow, Lee. 1982. "Male Inheritance Expectations in the United States in 1870." *Review of Economics and Statistics* 64(May).

______. 1983. "Land Fragmentation as an Index of History in the Virginia Military District of Ohio." *Explorations in Economic History* 20(July).

Steckel, Richard H. 1980. "Antebellum Southern White Fertility: A Demographic and Economic Analysis." *Journal of Economic History* 40(June).

Taeuber, C., and I. Taeuber. 1988. *Changing Population Trends in the United States.* New York: John Wiley.

Thompson, W., and P. Whelpton. 1953. *Population Trends in the United States.* New York: McGraw-Hill.

U.S. Census Office. 1860. *Eighth Census, United States.* U.S. Patent Office. 1853. *Report 1852.* Washington.

Wright, Carroll D. 1900. *History and Growth of the United States Census.* Washington: Government Printing Office.

Yasuba, Y. 1961. *Birth Rates of the White Population in the United States, 1800–1860.* Baltimore: Johns Hopkins Univ. Press.

CHAPTER 3

# Northern Agriculture after the Civil War, 1865–1900

R. DOUGLAS HURT

AMERICAN AGRICULTURE underwent revolutionary change during the Civil War. That conflict not only ended slavery in the South but also caused fundamental economic and social readjustment in the North. High wartime prices and labor shortages, for example, encouraged farmers to purchase a host of laborsaving implements. Improved plows, reapers, and threshing machines helped northern farmers increase production and improve their profits. The agricultural prosperity the war created, however, did not last long. The availability of cheap western lands under the Homestead Act of 1862; the completion of a transcontinental railroad system; and the development of extensive, one-crop agriculture in the Midwest soon brought even more substantive economic, political, and social changes to northern agriculture (Rasmussen 1965, 195; Rasmussen 1962, 578).[1]

Although agricultural historians have given modest attention to New England, New York, New Jersey, and Pennsylvania during the antebellum period, the time from the end of the Civil War to the turn of the twentieth century has not attracted as much interest as it merits. Fortunately, what history has been written about late nineteenth-century northern agriculture includes the best of both the old as well as the new scholarship. By synthesizing and analyzing this work, one can not only gain a clear overview of the major developments in that region but also discover additional areas for research.

## THE HISTORICAL CONSENSUS

Older historians, whose work dates from the 1930s to the 1970s, reached a consensus concerning the major causes for agricultural change in the North after the Civil War. In general, they stressed economic determinism as the key. Specifically, they believed that cheap western lands, primarily offered by the Homestead Act, caused great agricultural change in the North. As those lands produced abundantly, western grain, beef, and pork flooded eastern markets by rail. Large shipments of hogs from Ohio, beef from Nebraska, corn from Illinois, and wheat from Kansas to eastern markets drove prices below the competitive level for northern farmers. As agricultural profitability declined precipitously, land values quickly dropped. Those farmers who remained on the land necessarily adopted new agricultural practices to survive. Northern farmers began to emphasize specialty production, such as dairying, truck gardening, fruit growing, and haying, which enabled them to capitalize on nearby urban markets. Western farmers could not compete with these perishable commodities until the perfection of refrigerated cars, and they never threatened the northern hay crop. Northern agriculturists who could not afford to make these adjustments often left their farms for better paying jobs in the cities or moved West in search of cheap, productive lands.

Despite these problems and adjustments, these historians argued that economic conditions never were severe enough to drive northern farmers to radical acts. Because northern agriculturists had better and cheaper access to needed goods, they had a higher standard of living than farmers in other regions. Northern farmers also were more diversified than southern and western farmers. As a result, northern agriculturists were not subject to the economic problems of one-crop agriculture. In addition, they received higher prices than farmers in other regions because they did not face steep shipping costs when marketing their produce. Although the Patrons of Husbandry, commonly known as the Grange, was strong in the North during the 1870s, no agrarian revolt occurred similar to that experienced in the Midwest. The needs of northern farmers were different. They wanted high not low railroad rates for interstate shippers in order to gain protection from western competition.

Among the older generation of historians, Harold Wilson (1936, 27) set the standard for the narratives that followed. Wilson believed that despite the economic advantages the Civil War provided for the northern agricultural community, New England experienced a "protracted autumn" by the war's end. Although a thriving sheep industry brought a temporary economic boom to some areas, northern agriculture entered a period of long decline as cheap western lands drained away young farmers and as the railroads linked those newly productive lands with northern and eastern markets. Northern farmers could not compete with the western agriculturists, and their sons and daughters fled the land for higher paying jobs in the rapidly developing industrial cities.

Indeed, Wilson argued that the Midwest had a twofold influence upon northern New England's agricultural economy. First, the cheap western lands, particularly those of the trans-Mississippi West, lured the "ambitious youth of the hill country as well as its discontented farmers. . . ." Secondly, the Midwest provided an "increasingly dangerous competition in agricultural production." Although the western lands and improved railroad transportation affected northern farmers before the Civil War, by 1870, Wilson (1936, 56–57, 61, 65, 71–72) contended, both began to undermine the northern agricultural economy.

Northern farmers were already commercial agriculturists by the end of the Civil War, when sheep raising was their most profitable endeavor. The rocky land of New England particularly was well-suited for sheep. By 1870, however, this aspect of northern agriculture had declined significantly. Since 1840 sheep production in Vermont (measured by the size of flocks) had dropped 64 percent, while it fell 39 percent in New Hampshire and 33 percent in Maine. A low protective tariff and more cost-efficient production in the Rocky Mountain West kept prices below profitability, and northern flockmasters sold many of their high-grade Merinos to the slaughter houses. Those who continued to raise sheep maintained relatively high wool production with fewer head because improved breeding practices and better care increased the weight of the fleeces. While the number of sheep declined rapidly between 1840 and 1870, for example, the wool clip fell only 8 percent in Vermont and 10 percent in New Hampshire while it increased 21 percent in Maine. Put differently, between 1840 and 1870 the weight of fleeces increased 143 percent in Vermont, 124 percent in New Hampshire and 77 percent in Maine. As a result, the best flockmasters were able to make some profit despite western competition, particularly if they shifted to dual purpose breeds for the production of both wool and mutton (Wilson 1935, 18–19; Wilson 1936, 86–88).

Northern farmers, however, were unable to gain adequate tariff protection against wool shipped from Australia and South America and, confronted with increasing land values, they abandoned sheep raising. The sheepmen who survived resided in the Far West, that is, in the Rocky Mountains and beyond, where they grazed their animals on cheap public lands and used the railroads to ship large quantities of wool to eastern markets at prices below the competitive level for northern farmers. By 1870, Wilson argued, agriculture in New England, particularly sheep raising, had become increasingly unprofitable and thousands of farmers fled the land for a better life elsewhere. As the rural population rapidly declined, abandoned farms grew in number. Wilson recognized, however, that not all population loss from the countryside was bad, because many farms had been founded on marginal or submarginal lands far removed from adequate transportation facilities. Few of these farms had any potential for commercial agriculture. Consequently, they were best abandoned (Wilson 1936, 74–75, 81–82, 94, 102, 107, 128–29).

Many of the successful farmers who remained on the land during the last quarter of the nineteenth century did so by developing other specialties, such as dairying. The cool climate, rocky terrain, and lush grasses of New England, together with the proximity of many urban markets, made dairying an ideal agricultural pursuit for northern farmers. Dairy cows provided more profit than sheep, and dairying was one of the few agricultural endeavors that enabled northern farmers to maintain profitable and competitive operations during the late nineteenth century. As early as 1868, Maine farmers could keep one cow for the cost of caring for eight sheep. Over a year's time, five dairy cows returned an estimated $357 from the sale of calves, cheese, and butter, while forty sheep earned the farmer an income of only $210 from the sale of lambs and wool. Cheese and butter manufacturers provided the primary market for dairy farmers in northern New England between 1870 and 1900. By the turn of the twentieth century, however, improved refrigeration technology and increased urban demands caused dairy farmers to emphasize the production of fluid milk that brought higher prices than did the sale of that commodity to secondary markets for the manufacture of butter and cheese (Wilson 1936, 191, 207; Wilson 1935, 36).

By the turn of the twentieth century northern farmers had shifted from general, diversified agriculture to intensive, specialized production in order to survive. Northern New England, for example, had changed from a meat-wool-grain–producing region to a dairy-fruit-truck crop area during an age of rapid scientific, technological, and economic change. By 1900 Wilson (1936, 209–11) wrote, the long "winter" of northern agriculture had passed. It had taken thousands of farmers with it, and it had forced those who remained to adjust their agricultural practices to meet the new demands of the urban market place.

A decade after Wilson reached his general conclusions, Fred A. Shannon (1945, 245, 247) supported his interpretation. Using Connecticut as a case study, Shannon concluded: "While the agricultural frontier was moving westward with more speed than ever before in the history of the country, farmers in the North Atlantic states often had to make adjustments quite as violent as those of the homesteaders on the Plains." High land values, small farms, and poor soils prevented the northern agriculturist from competing with the abundant production of meat and cereal grain on cheap western farms. Relatively low transportation rates from West to East made matters worse, and farm labor costs were prohibitive. While a midwestern farmer spent $9 for labor to raise an acre of corn at the turn of the twentieth century, New England farmers paid $17 for the same production.

Shannon was more explicit than Wilson regarding the benefits gained from the abandonment of unproductive lands. Farms on submarginal lands, he contended, were best suited for growing timber or for pasture rather than for crops. Consequently, the migration from New England farms removed the surplus of people from the lands that were incapable of supporting them. The most inefficient farmers left as well, so those who remained often were the best farmers working the lands most suitable for agriculture. The result was not the decay of the region's economy, but rather readjustment and maturation. Shannon also contended that while the sons and daughters of many farmers flocked to the cities where life supposedly was less toilsome and more profitable, the urbanization which they helped promote, in turn, created new markets for those farmers who remained on the land. Moreover, Shannon (1945, 248, 251) correctly contended, as did Wilson, that the migration from the northern farms to the cities was not just a phenomenon of the post–Civil War age but had begun much earlier in the nineteenth century.

In Connecticut, Shannon (1945, 252–54) argued, farmers shifted to truck crops to take advantage of the urban markets. Improved techniques, which many young farmers learned at the Storrs Agricultural College, also helped make farming more profitable than ever before. Potato, apple, peach, and tobacco production increased significantly in addition to dairying for the production of fluid milk. Profits from dairying encouraged farmers to return croplands to pasture, where cows grazed on the lush grass while farmers provided supplemental feed of cheap corn from Illinois and Iowa. Cultural limitations, however, prevented Connecticut and other New England farmers from responding to those new economic conditions by making maximum adjustments in their agricultural activities. Those farmers, for example, still regarded poultry production, that is the raising of chickens, as women's work. Instead of emphasizing the production of eggs for the nearby urban markets, they defaulted that activity to western farmers.

Nearly a quarter century after Shannon made his observations about Connecticut and New England agriculture, Paul Wallace Gates (1969, 117) offered a similar interpretation to analyze the agricultural history of New York during the late nineteenth century. Gates maintained that this state experienced its greatest prosperity between 1850 and 1870. Prior to the collapse of its agricultural economy in 1870, New York was the preeminent agricultural state in the nation. New York led the country in the number of farms, in the most improved acres, and in the value of farms as well as in the production of livestock, orchard products, dairy cows, fluid milk, butter, cheese, hay, potatoes, hops, and maple sugar. Farmers expanded their cultivated acreage, enlarged and improved their dairy herds, and invested heavily in new technology in the form of reapers, plows, mowers, and threshing machines, which the state produced from a host of implement factories. This latter economic endeavor also placed New York in the forefront of all other states (Hurt 1986, 38–51).

Despite this prosperity, however, all was not well with agriculture in New York. Thousands of farmers left the state in search of cheap western lands. With the average cost of New York farmland at $8 per acre, after passage of the Homestead Act farmers could purchase 160 acres for $1.25 per acre in the Great Plains or even acquire those lands at no charge if they met the residency requirements. At the same time many farmers in New York evidently speculated in lands, that is, they were more interested in farm making than in the long-term development of their lands. With land prices high, they were content to sell their farms and invest their profits in western lands where their earnings would buy even more acreage for speculation. After 1870 Gates (1969, 128–30, 132–33) maintained New York suffered an agricultural decline that approached economic depression in its severity.

Herbert G. Schmidt (1973, 166–67, 182–90) applied a similar interpretation to the agricultural history of New Jersey for the period of the late nineteenth century. Schmidt contended that the number of small farms devoted to specialty production, such as poultry and vegetables, increased. Most of the agricultural change in New Jersey, however, involved matters of emphasis, because the major field crops remained. Farmers merely intensified the cultivation of some crops, in part, with the aid of improved varieties. Forage crop production, for example, increased to feed the numerous horses and mules that farmers had acquired to pull their implements. Fruit production also expanded, particularly for apples, to meet the demands of the urban markets. In addition, Schmidt noted that the farmers who experienced economic stress in the Garden State, in contrast to New York and New England, responded not by selling out and moving west or by abandoning their lands, but by farming more intensively, efficiently, and scientifically than ever before.

Confronted with rising labor costs after the Civil War, farmers in New Jersey also turned to technology to reduce costs and to increase production. Schmidt (1973, 134–38, 148–49) observed: "The adoption of labor saving machinery constituted one of the most revolutionary breaks with the past, and certainly the most spectacular." By adopting reapers, mowers, threshing machines, binders, hay rakes, and fodder choppers, New Jersey farmers boosted production and capitalized on commercial markets. Technology not only helped the farmers in New Jersey solve the problems caused by inefficient labor but also enabled them to emphasize production for commercial sale.

Not all of the results, however, were positive. Increased capital costs for machinery slowed the rise from tenancy to full ownership. In addition, investments in technology and science, such as commercial fertilizers and improved seeds, made farmers more dependent on bankers and other moneylenders. Increased production also required farmers to use middlemen to arrange transportation and sale of their crops. By the late nineteenth century New Jersey farmers were so productive and the market system so changed that they could not take their commodities to market themselves or exercise close control over the sale of their products. Efficient production placed them at the mercy of others. They were no longer masters of their own destinies once they became commercial farmers. In this respect, however, the farmers in New Jersey were little different from others in New York, Pennsylvania, or the Midwest (Schmidt 1973, 148–49).

Schmidt also observed significant changes in the livestock industry in New Jersey during the late nineteenth century. Increased specialization for commercial crop production and the adoption of new technology forced livestock producers to adjust their operations. Horse and mule production increased to provide animals that would power the new machinery. Although beef cattle, swine, and sheep production declined, dairy cattle improved in quality and increased in quantity. The introduction of corn ensilage or silage with the underground pit silo, and later with expanded use of the upright silo, substantively improved feeding practices. The first practical incubators and brooders also revolutionized poultry production. As a result, flocks increased in size, and farmers were able to meet the market demands for eggs, pullets, and newly hatched chicks. Many farmers now became full-time poultrymen, specializing in the production of broilers, fryers, or eggs (Schmidt 1973, 150, 161–64).

New economic conditions and rapid technological change, then, enabled major readjustments in the agriculture of New Jersey. Western competition and the proximity of urban markets made New Jersey an ideal location for emphasizing the production of perishables, particularly dairy products and eggs. At the same time, improvements in the nation's transportation system caused a decline in the production of swine, sheep, and beef cattle because the farmers in New Jersey could not compete with western livestock producers (Schmidt 1973, 164–65).

Pennsylvania farmers made adjustments as well. By 1870 the local flour mills essentially sat idle because the milling industry had moved to Minneapolis. Pennsylvania farmers, however, purchased more machinery and improved their livestock through up-breeding to help meet increased competition from western farmers. Specialized breeds for dairying, such as Jerseys, Guernseys, Holsteins, Aryshires, and Brown Swiss, replaced dual purpose Shorthorns and Devons after 1880. A decade later silos also helped improve livestock feeding practices and enabled dairymen to extend the period of lactation into the winter months, thereby making dairying a year-round endeavor. Dairy farmers in Pennsylvania were little different from others in the North by the turn of the century, emphasizing the production of fluid milk for urban markets (Fletcher 1955, 2:59, 167–69, 170, 179).

Not all of those adjustments, however, saved the family farm. Between 1880 and 1890 a high rate of farm abandonment plagued the Keystone State. Stevenson Whitcomb Fletcher (1955, 366–67) concluded that most of the abandoned farms were naturally unprofitable, largely because of poor location and infertile soils. In any case, lost homes, broken families, and economic hard times were realities for many farmers in Pennsylvania. Little wonder that young men and women drifted away from the farms once the "flush times" ended with the Panic of 1873. Thereafter, new economic conditions in American agriculture kept the attrition rate high among Pennsylvania farmers.

These broad and jarring economic changes in northern agriculture caused some farmers to champion political, social, and economic readjustment outside the Democratic and Republican parties. Cutthroat competition among the railroads for the business of the western farmers meant that northern agriculturists lost the geographical advantage they had recently enjoyed. Western commodities drove prices down on the eastern markets and the collapse of land values soon followed. At the same time local freight rates remained high, often based on whatever the traffic would bear. Frequently farmers paid more for a short haul than for a long one, particularly on the trunk lines. The problems of interstate and intrastate railroad commerce produced an antirailroad backlash among many northern farmers which, as mentioned above, fostered the growth of the Patrons of Husbandry or Grange. The question is, how radical was the response of northern farmers compared to those in the Midwest?

Frederick Merk (1949, 7–8) and Fred A. Shannon (1945, 254) contended that the granger movement was weak in the North compared to the Midwest because severe price declines did not plague northern farmers, and because the northern legislatures and courts alleviated many of the transportation problems that so annoyed westerners. Consequently, unity with the western grangers threatened to make the northern agricultural situation worse rather than to improve it.

Lee Benson (1955, 80–83) agreed with Merk when he analyzed the agricultural and railroad problems in New York. Although New York farmers experienced a "golden age" between the end of the Civil War and the Panic of 1873, Benson contended, "directly, or indirectly, in terms of farming operations, the extension and improvement of transportation facilities after 1865 adversely affected the economic interests of almost every agricultural locality within the state." Indeed, rapid change in transportation, which the war fostered, provided the foundation for the agricultural problems which northern agriculturists confronted during the late nineteenth century. For Benson, "the loss of highly important location advantages told heavily on even the best farm lands in the state and deserves to be assigned first place in explaining its agrarian problems." New York farmers on marginal and worn-out lands particularly faced significant economic problems with the completion of a national railroad system. Western competition, however, hurt all farmers in New York economically. Had the farmers in New York been able to organize politically, they might have resolved many of their economic problems.

During the 1870s, however, New York farmers gave relatively little attention to the Grange, which was the one organization that could have enabled them to marshal their numerical strength to gain economic and political reform. Upon the formation of the New York Grange in 1873, George D. Hinckley, master of the state body, proclaimed that the purpose of the organization was to gain favorable agricultural legislation and to secure protection "from the aggressive evils of concentrated capital, and monied corporations, to which the agricultural interest, the most important interest of the country, is justly entitled." Beyond these high moral principles, however, Hinckley suspected that the organization's most immediate success would come from cooperative buying. In this respect, too, the New York Grange was little different from the organization in the Midwest. It also had the same problems and disappointments, notably the lack of business experience and capital (Benson 1955, 88–91).

Whatever the Grange's rhetoric, farmers in New York failed collectively to oppose abusive railroad practices because they wanted the railroad problems handled in a vastly different manner than the farmers demanded in the Midwest. New York farmers wanted pro rata freight rates, that is, legislation that would protect them from western competition. In contrast, western farmers wanted legislation that would reduce freight rates in order to make them even more competitive in the eastern market place. When the bottom fell out of the agricultural economy in 1876, dropping to a price index of 87 after peaking at 132 in 1873, New York farmers were more diversified than western agriculturist who relied primarily upon livestock and grain for their incomes. Diversification and relatively greater prosperity limited the ability of New York farmers to organize for the achievement of social, political, and economic reforms. Moreover, less financial need prevented the Grange from eliminating the isolation and localism farmers experienced in New York. In contrast, financial necessity forced unity and social interaction among western farmers. Benson concluded that most farmers in New York, who retained Grange membership, operated above-average farms and engaged in business apart from agriculture. Among those farmers, the most important reasons for maintaining membership were the social and educational opportunities the Grange provided (Benson 1955, 84–86, 91–93).

During the early 1870s farmers also organized local and state granges in Pennsylvania and New Jersey (Schmidt 1973, 111; Woodward 1927, 247–48). By 1880 the Grange in Pennsylvania had become one of the strongest in the nation. There it supported the social and educational goals of the organization, but the members were far more interested in cooperative buying and selling, banking services, the formation of a fire and life insurance company, the manufacture of farm and home implements, and the operation of retail stores. By 1880 most of these ventures with the exception of cooperative buying had failed. Many patrons left the organization, and a period of decline lasted until the eve of the Populist revolt, when the Grange began to rebuild its membership and to emphasize the social and educational features of the organization that founder Oliver Hudson Kelley had originally intended (Fletcher 1955, 411–12; Woodward 1927, 247–48). In Connecticut, the Grange apparently functioned primarily as a social and educational organization from its beginning in 1873. Nevertheless, the Connecticut Grange rapidly declined in membership during the 1880s prior to experiencing a mild rebirth during the next decade (Jenkins n.d., 384; Tontz 1964, 147).

In Vermont cooperative buying attracted the most interest in the Grange, which organized in the early 1870s. The grangers in the Green Mountain State, however, met problems identical to those of other patrons in the North—business failure followed by a protracted decline—until the organization began to attract new members about 1890. Although the Grange in Vermont gave some attention to railroad regulation, poor leadership and different concerns about railroad transportation from patrons in the Midwest prevented major reform. Overall, the grangers in Vermont were content, aside from developing cooperative buying practices, to attack the State Agricultural College for not providing adequate agricultural education for their children (Rozwenc 1957, 276–78).

In New Hampshire the Grange organized on the state level in 1873. Economic dissatisfaction stimulated many farmers to join the organization, but poor organization and inadequate leadership caused it to falter and membership to decline by the early 1880s. For the most part, the Grange in New Hampshire served as a social rather than an activist reform organization, although it did establish fire and life insurance companies. Like grangers in other states, patrons in New Hampshire never were able to marshal their resources sufficiently to engage effectively in cooperative buying. Merchants undersold the cooperative stores, and patrons failed to support the organization if they could obtain cheaper prices elsewhere. Moreover, beyond advocating increased taxation of the railroads, the Grange in New Hampshire did not become involved with the movement to gain greater governmental control of the railroads. Indeed, New Hampshire passed a railroad law in 1883, but not because of granger activity. Railroad attorneys wrote that legislation, which created a railroad commission designed to provide greater regulation of the lines in the state. With only one farmer in ten a member of the Grange during the late 1870s, patrons in New Hampshire lacked the numerical strength to pursue substantive economic, political, and social reforms (Metcalf 1921, 517; Sherman 1971, 9–16, 20–22).

In Maine a small group of men also organized the first Grange in 1873. The organization grew rapidly where patrons believed the capitalists and middlemen exploited them. To break from the middlemen, they too organized cooperative stores and fire and life insurance companies. When economic hard times befell the Grange after 1875, only the fire insurance company remained. Insufficient business experience, inadequate capital, mutual distrust, and the unwillingness to pool resources caused the cooperative stores to fail, and many patrons drifted away. Although grangers in Maine believed that most of their problems could be solved by eliminating the middlemen, like patrons in other states, they too advocated a broad program of social reform, which included such matters as the adoption of uniform textbooks for the public schools, abolition of the school-district system, liberal appropriations for the Maine State College, elevation of the secretary of the United States Department of Agriculture to cabinet status, increased regulation of the railroads and monopolies, and a graduated federal income tax. They did not fear the wealthy, however, as much as the western grangers. In 1881, for example, they elected Frederick Robie as master of the state Grange. Robie was a physician by profession and a large-scale farmer by avocation. He also held directorships with a railroad and bank. By this action alone, patrons in Maine revealed their indifference to the problems of the western farmers. Under Robie's leadership the organization grew as it gave major attention to the support of the Maine State College of Agriculture and to the establishment of an experiment station at that institution (Day 1963, 183–93).

Aside from these various state Granges, northern farmers did not form many organizations in the late nineteenth century. One exception, though, was the 1877 creation of the Farmers' Alliance in Rochester, New York, by members of the Grange and other agricultural organizations. This Farmers' Alliance was far different from that which spread through the Great Plains and South little more than a decade later. It was founded to combat the problems that low interstate railroad rates caused cattle and wheat producers in western New York. As freight rates fell through the late nineteenth century, an abundance of western meat and grain continued to depress eastern prices. At the same time, intrastate freight rates remained high. Commodities often could be shipped cheaper by rail from Chicago to New York City than from Rochester to that same destination. Moreover, land values fell with the drop in agricultural prices, and taxes remained high. New York farmers grew restive in the newly emerging capitalistic-corporate world, where landed property owners became increasingly subservient to the powerful business forces of the industrial age. Those who formed the Farmers' Alliance hoped the organization would serve as the mouthpiece for all farmers across the state as they worked for economic and social reform (Benson 1955, 94).

The farmers who controlled the Alliance, however, were large-scale landowners who kept the organization conservative in its goals and methods. They wanted nothing more than equity and equality of taxation, that is, the Alliance men did not want property taxpayers to be assessed at higher rates than those who held intangible property, particularly in the form of stocks and bonds. They also wanted major reductions in local railroad rates and a corresponding increase in interstate charges. They were not interested in the debate over greenbacks, the remonetization of silver, or other seminal issues that mobilized western farmers into the Alliance. The Alliance men and women in New York, Benson aptly noted, showed remarkable unity not only about what should be accomplished, but also about what should not be attempted. The Farmers' Alliance descended from the Grange, but it did not catapult New York farmers into the rough and tumble politics that characterized the agrarian revolt further west.[2]

Ultimately, insufficient political experience and lack of mutual cooperation caused the Alliance to fail in the North as well as the Midwest. Also, most New York farmers did not suffer enough economic hardship to drive them into the Alliance in the first place. The organization was unable to eliminate railroad control of the legislature or to achieve pro rata freight rates and equitable taxation. Still, it played a role in the creation of the New York Railroad Commission in 1882, and it helped farmers understand the power of political organization along agricultural lines (Benson 1955, 94–109).

## NEWER STUDIES

Beyond these broad, consensus conclusions about northern farming and farmers' attitudes, several recently published new studies provide additional scope and perspective for understanding agricultural developments during the late nineteenth century. Though these findings in many cases complement the older work, they also challenge the sweeping traditional interpretation that the late nineteenth century was a time of severe economic decline and mass migration to cheaper western lands and more lucrative urban jobs.

Paul Glenn Munyon was the first historian to take issue with the consensus interpretation of the past. Using New Hampshire as a case study, Munyon made an extensive analysis of the agricultural economy of the Granite State during the last three decades of the nineteenth century. He discovered that the number of farmers and agricultural workers fell approximately 20 percent from 1870 to 1900, but the decrease in the work force did not significantly influence the agriculture of New Hampshire because the decline centered in the sparsely populated areas of the state. Munyon noted that "while there may have been definite cause for personal concern in some areas of the state, the proclamation of widespread rural decay by the popular press was without empirical justification" (Munyon 1978, 51–52, 64–65, 84).

Munyon also contended that those who remained on the land expanded production with science, technology, and larger farms. While the average size of farms nationwide decreased slightly, dropping from 153.3 to 146.4 acres between 1870 and 1900, those in New Hampshire increased from 121.7 to 123.1 acres. In addition, no substantive change in the control or use of the land took place at that time because of economic conditions. Indeed, farmers raised crops on more acres at the turn of the twentieth century than they cultivated immediately following the Civil War. Although cereal grain and beef production dropped more than 40 percent during that time, farmers in New Hampshire increased production of horses and gave more attention to raising poultry and swine. The greatest gains, however, occurred in dairying where the sale of fluid milk increased tenfold during the last thirty years of the century. By the turn of the twentieth century, fluid milk production dominated the state's agriculture (Munyon 1978, 71–72, 78–81).

Munyon argued that the growth of the dairy industry indicated that agriculture in New Hampshire expanded rather than declined during the late nineteenth century. Rather than succumb to failure, Munyon contended, New England farmers in general and New Hampshire farmers in particular were flexible. They quickly adapted to the new realities of the marketplace and, by so doing, took advantage of their comparative geographical proximity to urban markets for the production and sale of a highly perishable commodity. As a result, gross farm income increased by more than $3,000 between 1869 and 1899, while net income rose approximately $1,500 (Munyon 1978, 92,97).

Farmers in New Hampshire also enjoyed better economic conditions than farmers nationwide because they had less indebtedness and the rates of interest they paid on their mortgages remained below the national average. Indeed, interest rates took less gross income in New Hampshire than in any other state. Moreover, the annual income of most agricultural workers generally was higher in New Hampshire than in other states. Even so, the earnings of agricultural laborers in New Hampshire were less than those of workers engaged in nonfarm employment. In 1880 farm laborers who drew wages averaged $331 annually compared to $367 for textile mill employees and $459 for bakers, all of whom worked six days a week for fifty weeks per year. By 1900 farm workers averaged $438 annually in earnings, while textile workers and bakers averaged $612 and teamsters $765 per year. Put differently, in 1880 off-the-farm workers needed only forty-one weeks of employment per year to equal the annual wages paid to agricultural workers based on a rate of $8 per week and only thirty-one weeks of work at $14 per week in 1900 (Munyon 1978, 126, 129–31, 168–70).

With higher wages in the cities and with little land remaining for expansion or the development of new farms, many young men and women left the land, but this did not mean that farming itself declined. In New Hampshire more land was in farms in 1900 than in 1869. And, while approximately 121,500 fewer people worked the land at the turn of the twentieth century than thirty years before, the per acre gross value added to New Hampshire's farmland had increased from $3.30 to $4.07 per acre. Similarly, the gross market revenue from fluid milk grew by more than $3.5 million between 1870 and 1900, while egg sales increased from $345,000 to $1.2 million. The late nineteenth century, then, was a time of increased prosperity for farmers in the Granite State (Munyon 1978, 184, 192, 201–3).

Overall, Munyon argued that agricultural productivity increased in New Hampshire during the last thirty years of the nineteenth century, and that Granite State farmers enjoyed a standard of living well above the national average for agriculturists. Consequently, New Hampshire farmers had little reason to look to the West for better agricultural opportunities, even though some of them left for higher paying jobs and a different kind of life in urban America. For Munyon, the late nineteenth century may have been a time of "continuous adjustment" in New Hampshire agriculture, but it was not as Wilson believed a "severe winter season" (Munyon 1978, 214–15).

Hal Seth Barron was the second historian to significantly challenge the traditional interpretation that the late nineteenth century was a time of severe economic decline in New England. By focusing on the community of Chelsea, Vermont, Barron provided a tightly focused study of those who stayed behind. He argued that the region's agricultural decline has been greatly exaggerated, and that the loss of farm population and the slow growth of agriculture were "normal" features of older communities. Barron wrote that between 1840 and 1900 "the local economy slowed down and stabilized as both farmers and villagers adjusted to outside economic changes as well as to local demographic realities." He found that few farms were abandoned during that time. Between 1860 and 1900, he noted, at least two-thirds of the male heads of households were farmers or agricultural workers. The number of farms remained constant at about two hundred, while farm size and value averaged approximately 125 acres and $2,000 respectively. Moreover, the township's main agricultural production of wool, butter, and maple sugar remained constant from 1850 to 1900 (Barron 1984, 29–30, 52, 55–56).

Still, Chelsea's population declined significantly over this period. Barron reported: "Between 1860 and 1880, the township replaced only 46 percent of those whose names disappeared from local enumerations; and between 1880 and 1900, it replaced only 34 percent." The major reason for the population loss, Barron contended, was that no additional land existed for the development of new farms; so young men, in particular, left in search of better opportunities. With the young leaving, the farmers who remained in Chelsea could not shift to more profitable, labor-intensive dairying. Instead, local farmers raised fewer but more highly productive sheep for wool to offset unfavorable market conditions. As a result, Barron concluded: "Chelsea's farm economy was stable, but the number of opportunities to farm available to the younger generation remained more or less fixed." In time, those who stayed behind on the farms surrounding Chelsea were bound to the township by strong social and economic ties that were not found among settlers in the trans-Mississippi West. Moreover, these Vermonters were more persistent and culturally homogeneous than those who resided in the urban areas or who settled on the advancing frontier (Barron 1985, 330, 333; Barron 1980, 331).

These conclusions led Barron to take issue with Wilson who contended that northern agriculture declined during the late nineteenth century *because* the young were attracted to cheap western lands and more lucrative jobs in the cities, and that low transportation rates enabled western farmers to overwhelm the eastern markets and to damage severely the northern agricultural economy. Instead, Barron contended that those who left the farms simply were surplus population. The rural population declined, but little negative change occurred in the agricultural economy. At worst, the late nineteenth century was a period of "economic stagnation" because those farmers who remained were left without adequate labor to adjust their operations to more lucrative endeavors, such as dairying. Rather than declining, the agriculture of Chelsea merely reached its limits of growth and leveled off thereafter. The late nineteenth century was a time of stability, not disintegration for northern agriculture. (Barron 1980, 321–22; Barron 1984, 55–56, 63–64.

In addition to these perspectives, David C. Smith provided another interpretation for the course of agricultural history in Maine. Smith conceded that cheap western lands contributed to the decline of agriculture in that state. He argued, however, that agriculturists in Maine might have been able to remain profitable by shifting to new crops had not the climate cooled after the Civil War. Smith contended that periodic, severe weather fluctuations created "climatic stress" that often proved ruinous. Smith held that between 1860 and 1880 the climate of Maine generally was cooler and wetter than usual. He succinctly declared: "The weather variability was at the bottom of Maine's agricultural history problem." Bad weather meant poor crops. Farmers became discouraged and left the land for better opportunities in the towns and cities. Efforts to improve agricultural education and farming practices could not stem that flight. Those who survived adopted the best new agricultural practices such as cattle raising that were not damaged so quickly or significantly when climatic changes occurred. If bad weather damaged the hay crop, for example, farmers merely marketed their cattle sooner. By pursuing livestock raising, farmers in Maine were not completely at the mercy of the weather or outside competition (Smith et al. 1982, 179–80, 190–91, 193; Smith and Bridges 1982, 15).

Other historians have emphasized the progressive nature of northern farmers who willingly adjusted to changing economic conditions. On Long Island, contended Richard Wines, farmers took advantage of new market opportunities and willingly changed their agricultural practices if they could benefit from doing so. Truck farming, for example, replaced wheat production. For Wines, "it was still the opportunities offered to the farmer and not the constraints placed on him that were most important in bringing about change." If the technology and capital were available, northern farmers quickly adjusted their mode of operation and emphasis (Wines 1975, 602; Wines 1985).

Richard A. Easterlin (1976, 67, 70–81) furnished still another perspective. He argued that "nonpecuniary advantages" kept northern farmers on their lands. Despite the potential for greater agricultural wealth in the West, emigration depended upon birth order and ties to parents, relatives, and friends. Easterlin believed that the eldest sons probably reached adulthood while their fathers were in the prime of life and able to operate the family farm. He suggested that the older sons moved west because they did not have immediate possibilities to begin farming. As a result, the younger sons eventually gained control of the family farm. Easterlin also argued that the agricultural population declined in New England because farm families intentionally limited the birth rate to have sufficient wealth to provide their children with an adequate start in life, no matter whether they chose to remain on the farm or move to the city. The obligations that a farmer felt to his children, rather than out-migration opportunities, determined the decline of the rural population during the late nineteenth century. Western competition and the resulting drop in prices diminished a farmer's prospects to accumulate wealth, and northern farmers responded by limiting the size of their families.

Certainly, the history of agriculture in the northern United States is complex, and it has been shaped by a multiplicity of causes. To place northern agricultural history in the broadest and most accurate perspective more scholarly research needs to be completed. We must know, for example, the effects of science and technology on the environment. Scholars must evaluate more intensively and critically the work of the agricultural experiment stations. Did the work of the agricultural scientists at those institutions aid the family farm, or did it primarily benefit the largest, most capital-intensive farms or the agricultural implement, food processing, and chemical industries? The brewing, food-processing, implement, and nursery industries also merit attention. More research also needs to be completed on the contribution of the agricultural colleges to the improvement of regional farming. The inheritance and intergenerational-assistance traditions as well as credit sources merit greater investigation in relation to agriculture. Topical articles on almost any agricultural subject will provide both deeper and broader understanding, while book-length studies of the agricultural history of Vermont, New Hampshire, and Massachusetts remain to be written and those of New York State can be greatly improved upon. Research in these areas, together with the agricultural history already published, will help provide the basis for a much needed synthesis of northern agricultural history since the Civil War.

## NOTES

1. For this study I have defined North as the area encompassing New England, New York, Pennsylvania, and New Jersey. The Midwest includes Ohio, Indiana, Illinois, Michigan, and Wisconsin.

2. See Benson (1955, 94–109). Between 1880 and 1885, many agriculturalists joined the Farmers' Alliance and Industrial Union in Pennsylvania. The history of this organization in the Keystone State, however, remains unstudied. See Fletcher (1955, 414).

## REFERENCES AND FURTHER READINGS

Adams, T. M. 1944. "Prices Paid by Vermont Farmers for Goods and Services Received by Them for Farm Products, 1790–1940; Wages of Vermont Farm Labor, 1790–1940." University of Vermont and State Agricultural College, *Bulletin 507,* Feb.

Barron, Hal. S. 1980. "The Impact of Rural Depopulation on the Local Economy: Chelsea, Vermont, 1840–1900." *Agricultural History* 54(Apr.):318–35.

______. 1984. *Those Who Stayed Behind: Rural Society in Nineteenth Century New England.* Cambridge: Cambridge Univ. Press.

______. 1985. "Staying Down on the Farm: Social Processes of Settled Rural Life in the Nineteenth-Century North." In *The Countryside in the Age of Capitalist Transformation,* ed. Steven Hahn and Jonathan Prude, 327–43. Chapel Hill: Univ. of North Carolina Press.

Benson, Lee. 1955. *Merchants, Farmers and Railroads: Railroad Regulation and New York Politics, 1850–1887.* Cambridge: Harvard Univ. Press.

Black, John Donald. 1950. *The Rural Economy of New England.* Cambridge: Harvard Univ. Press.

Brunger, Eric. 1955. "Dairying and Urban Development in New York State, 1850–1900." *Agricultural History* 29(Oct.):169–74.

Colman, Gould P. 1963. *Education and Agriculture: A History of the New York State College of Agriculture at Cornell University.* Ithaca, N.Y.: Cornell Univ. Press.

______. 1965. "Government and Agriculture in New York State." *Agricultural History* 39(Jan.):41–50.

Danhof, Clarence H. 1969. *Change in Agriculture: The Northern United States, 1820–1870.* Cambridge: Harvard Univ. Press.

Day, Clarence. 1963. *Farming in Maine, 1860–1940.* University of Maine Studies, 2d ser., no. 78, 183–93.

Donald, Adrain Francis. 1936. *The History of Tobacco Production in Connecticut.* n.p. Published for the Tercentenary Commission of the State of Connecticut by the Yale University Press.

Easterlin, Richard A. 1976. "Population Change and Farm Settlement in the Northern United States." *Journal of Economic History* 36(Mar.):45–75.

Fletcher, Stevenson Whitcomb. 1955. *Pennsylvania Agriculture and Country Life, 1840–1940.* Vol. 2. Harrisburg: Pennsylvania Historical and Museum Commission.

Gates, Paul W. 1969. "Agricultural Change in New York State, 1850–1890." *New York History* 50(Apr.):115–41.

Graves, Eugene M. 1967. "A History of Maine's Nineteenth Century Beef Industry." M.A. thesis, University of Maine.

Hedrick, Ulysses Prentiss. 1966. *A History of Agriculture in the State of New York.* New York: Hill and Wang.

Hendrickson, Clarence Irving. 1931. "A History of Tobacco Production in New England." Storrs Agricultural Experiment Station *Bulletin 174,* October.

Hurt, R. Douglas. 1982. *American Farm Tools: From Hand Power to Steam Power.* Manhattan, Kan.: Sunflower Univ. Press.

______. 1986. "Out of the Cradle: The Reaper Revolution." *Timeline* 3(Oct./Nov.):38–51.

James, Henry Francis. 1928. *The Agricultural Industry of Southeastern Pennsylvania: A Study in Economic Geography.* Philadelphia.

Jenkins, Edward H. Nd. *Connecticut Agriculture.* Reprinted from *History of Connecticut.* N.p.

Marcus, Alan I. 1985. *Agricultural Science and the Quest for Legitimacy: Farmers, Agricultural Colleges, and Experiment Stations, 1870–1890.* Ames, Iowa: Iowa State University Press.

Merk, Frederick. 1949. "Eastern Antecedents of the Grangers." *Agricultural History* 23(Jan.):1–8.

Metcalf, Henry H. 1921. "The New Hampshire State Grange." *Granite Monthly* 53(Dec.):517–26.

Munyon, Paul Glenn. 1978. *A Reassessment of New England Agriculture in the Last Thirty Years of the Nineteenth Century: New Hampshire, A Case Study.* New York: Arno Press.

Nordin, D. Sven. 1974. *Rich Harvest: A History of the Grange, 1867–1900.* Jackson, Miss.: Univ. of Mississippi Press.

Ramsey, Elizabeth. 1930. *The History of Tobacco Production in the Connecticut Valley.* Smith College Studies in History, vol. 15.

______. 1965. "The Civil War: A Catalyst of Agricultural Revolution." *Agricultural History* 39(Oct.):187–95.

Rasmussen, Wayne D. 1962. "The Impact of Technological Change on American Agriculture, 1862–1962." *Journal of Economic History* 22(Dec.):578–91.

Rozwenc, Edwin C. 1957. "Agriculture and Politics in the Vermont Tradition." *Vermont History* 25(July):268–87.

Russell, Howard W. 1982. *The Long Deep Furrow: Three Centuries of Farming in New England.* Abridged. Hanover, N.H.: Univ. Press of New England.

Schmidt, Hubert G. 1973. *Agriculture in New Jersey: A Three-Hundred-Year History.* New Brunswick, N.J.: Rutgers Univ. Press.

Shannon, Fred A. 1945. *The Farmers' Last Frontier: Agriculture, 1860–1897.* New York: Farrar and Rinehart.

Sharrer, G. Terry. 1980. "Farming in Saratoga County, 1860–1920." *Grist Mill* 14(Winter).

Sherman, Rexford. 1971. "The New Hampshire Grange, 1873–1883." *Historical New Hampshire* 26(Spring):2–25.

Smith, David C. 1980. *The Maine Agricultural Experiment Station: A Bountiful Alliance of Science and Husbandry.* Orono: Life Sciences and Agricultural Experiment Station.

______. 1981. "Climate Stress and Maine Agriculture, 1785–1885," 450–64. In *Climate and History: Studies in Past Climates and Their Impact on Man,* ed. T. M. L. Wigley, M. J. Ingram, and G. Farmer. Cambridge: Cambridge Univ. Press.

______. 1984. "Agriculture in New York State to 1880: The Impact of Public Funds." In *Agricultural Artifacts in Museums: Historical Context and Curatorial Practice,* 14–27. Cooperstown, N.Y.: The Farmers' Museum.

Smith, David C., and Anne E. Bridges. 1982. "The Brighton Market: Feeding Nineteenth-Century Boston." *Agricultural History* 56(Jan.):3–21.

Smith, David C., et al. 1982. "Climatic Fluctuations and Agricultural Change in Southern and Central New England, 1785–1800." *Maine Historical Quarterly* 21(Spring):179–200.

Tarr, Joel A. 1975. "From City to Farm: Urban Wastes and the American Farmer." *Agricultural History* 49(Oct.):598–612.

Tontz, Robert L. 1964. "Memberships of General Farmers' Organizations, United States, 1874–1960." *Agricultural History* 38(July):143–56.

Wilson, Harold Fisher. 1934. "Population Trends in Northwestern New England." *Geographical Review* 24(Apr.):272–77.

______. 1935. "The Rise of the Sheep Industry in Northern New England." *Agricultural History* 9(Jan.):12–40.

______. 1936. *The Hill Country of Northern New England: Its Social and Economic History, 1790–1930.* New York: Columbia Univ. Press.

Wines, Richard A. 1981. "The Nineteenth-Century Agricultural Tradition in an Eastern Long Island Community." *Agricultural History* 55(Jan.):50–63.

______. 1985. *Fertilizer in America: From Waste Recycling to Resource Exploitation.* Philadelphia: Temple Univ. Press.

Woodward, Carl Raymond. 1927. *The Development of Agriculture in New Jersey, 1640–1880.* New Brunswick, New Jersey Agricultural Experiment Station.

Zelinsky, Wilbur. 1962. "Changes in the Geographic Patterns of Rural Population in the United States, 1790–1960." *Geographical Review* 52(Oct.):492–524.

Zinar, Ruth. 1983. "Educational Problems in Rural Vermont, 1875–1900: A Not So Distant Mirror." *Vermont History* 51(Fall):197–220.

CHAPTER 4

# *The Economics of Midwestern Agriculture, 1865–1900*

DONALD L. WINTERS

AGRICULTURE was the principal sector of the United States economy until near the end of the nineteenth century. Throughout most of the postbellum period, farming employed a larger portion of the work force, accounted for more production, and generated greater income than any other economic sector. Although lagging manufacturing in rates of growth, agriculture expanded rapidly in number of farms, amount of capital investment, and value of output. As a result, American farmers were able to satisfy most of the increasing food and fiber needs of the United States and at the same time to contribute substantially to the export trade.

The Middle West was one of the country's major agricultural suppliers. This region developed steadily during the nineteenth century, receiving large numbers of settlers who moved west to take up farming on its fertile soils. By the outbreak of the Civil War, the Middle West had already become a source of much of the food consumed by Americans, a role that would grow in significance in the postbellum years. Because of the importance of midwestern agriculture in the national economy, scholars have devoted considerable attention to this aspect of the region's history. They have investigated patterns of settlement and landownership, the adoption and diffusion of farm technology, the emergence of regional specializations, the development and consequences of commercial agriculture, and the relationship between prairie farming and the national economy. If researchers can agree that these are crucial historical questions, they often disagree in their interpretations of them.

This essay will explore the major economic issues in the historiography of agriculture in the postbellum Midwest.

The following discussion assumes that the states of Ohio, Indiana, Illinois, Michigan, Wisconsin, Iowa, and Minnesota comprise the Middle West. Although the terrain varied across this region, at the time of settlement it was primarily natural prairie. Ohio and eastern Indiana, which were mainly forests broken by patches of grasslands, represented the major exceptions. Moreover, the prairies extended into the eastern third of the Dakotas, Nebraska, and Kansas, and into the northern half of Missouri. These exceptions notwithstanding, "prairie" will be used synonymously with "middle western" and "midwestern." The area between the western boundary of the Midwest and the Rocky Mountains will be considered the Plains.

Much of the Middle West had been settled by the time the country became embroiled in civil war (Atack and Bateman 1987, 71–85). Farmers had already taken up most of the arable land south of the Great Lakes and east of the Mississippi River, and settlement had reached across the river into eastern Iowa and southeastern Minnesota. The land policy of the federal government contributed greatly to the antebellum occupation of the Midwest, and it would encourage movement onto the region's remaining lands during the postbellum era. Congress not only continued to release large amounts of land for sale at enticingly low prices, it also found other ways to make the national domain available for private development. Parts of the Midwest were offered for settlement under the 1862 Homestead Act, which provided up to 160 acres of land free to qualifying households. The Land Grant College Act, also passed in 1862, transferred federal lands within the region to individual states, which in turn sold them to private holders. These modifications of an already liberal land policy hastened completion of the settlement of the Middle West in the years following the Civil War.

Scholars, like members of Congress in the nineteenth century, have debated the wisdom of such a generous land policy.[1] Some maintain that it encouraged too rapid an expansion of the agricultural sector and drew into farming individuals who lacked the necessary managerial skills and knowledge. For this reason, they claim, it was in large measure responsible for the overproduction, low agricultural prices, and widespread bankruptcy that plagued midwestern farmers throughout the late nineteenth century (Dennen 1977; Hibbard 1924, 550–56; LeDuc 1963; Saloutos 1962). Federal policies, according to Theodore Saloutos, "encouraged people to take to the land without regard for the fact that they accelerated agricultural production beyond all reasonable market demand. . . . The net result was the overexpansion of an already overexpanded agriculture" (Saloutos 1948, 161). The government, in the judgment of these scholars, should have released its land more slowly.

Robert W. Fogel and Jack L. Rutner challenge this contention, however. Their estimates of returns on investments in land from current production (as distinguished from capital gains) were relatively high, leading the authors to conclude that chronic overproduction and low commodity prices were not serious problems. Their findings cast "doubt [on] the proposition that there was a substantial and general overallocation of resources to agriculture. . . ." (Fogel and Rutner 1972, 391–98; Bowman 1965). Douglass C. North agrees with this assessment. In his opinion, the more restrictive approach advocated by the critics of federal land policies would not only have reduced agricultural income; it would also have slowed the country's overall economic growth (North 1966, 133).

Other scholars have attacked the government's land disposal programs on quite different grounds. Paul W. Gates, in railing against an "incongruous land system," charges that the practice of permitting unlimited purchase at low prices by a single buyer enabled speculators and large landholders to acquire massive amounts of the national domain and to drive up land prices by holding their property off the market.[2] The resulting land monopolization, he contends, had unfortunate consequences. Farmers had to go deeply into debt, frequently at exorbitant interest rates, to purchase land at inflated prices. Unable to meet mortgage obligations from the income of their farms, many were forced into bankruptcy by unscrupulous loan sharks and frontier bankers (Gates 1936, 1942, 1945). Furthermore, the practices of speculators were responsible for the early appearance and rapid expansion of midwestern tenancy. For new farmers who could not afford land at monopoly prices and bankrupt farmers whose land had been seized by lenders, tenancy—often on speculator land withheld from sale—was the only recourse if they were going to make a living from agriculture (Gates 1939, 1941, 1945).[3]

Gates's interpretation has attracted a wide group of challengers. Allan G. Bogue and Robert P. Swierenga argue that the enormous amount of land released by the federal government made it impossible for a single buyer or even a group of buyers acting in collusion to gain control of enough property to manipulate prices (Bogue 1963, 29–46; Swierenga 1968, 210–27). Competition, not monopoly, prevailed in the land business, and the market, not speculators, determined prices. Moreover, speculators in Swierenga's investigation of the land business in nineteenth-century Iowa rarely kept their land off the market for extended periods or rented it out to tenants. They learned soon that rates of return were inversely proportional to the time property was held; accordingly, the interval between purchase and sale averaged less than three years (Swierenga 1968, 192–96). Fogel and Rutner concur, insisting that theoretical considerations and empirical evidence support the conclusion that speculators "would generally maximize their profits if they resold the land in their possession quickly" (Fogel and Rutner 1972, 411–18). Seddie Cogswell's study and my own study of agricultural tenancy in nineteenth-century Iowa uncovered little evidence that farm renting was associated with land speculation (Cogswell 1975, 25–27; Winters 1978, 12–28). Areas of high tenancy rates did not systematically coincide with areas of high speculative activity, and few large-scale speculators could be found among the state's landlords. In his study of Trempealeau County, Wisconsin, Merle Curti discovered that only infrequently were landowners forced into tenancy as a result of foreclosure on their property (Curti 1959, 161). The incidence of foreclosure tenancy was equally rare in Iowa (Winters 1978, 27–28).

If the early appearance and steady increase of agricultural tenancy in the last half of the nineteenth century were not the result of land monopolization and farm bankruptcy, what were the causes?[4] Pioneer farmers sometimes chose to rent because they wanted to evaluate an area's potential before purchasing land, were hesitant to farm virgin soil in their first years after arriving, or were short of investment capital. Rising land values, reflecting farm improvements and the relationship of supply and demand rather than manipulation by speculators, and the heavy investments required for equipment and livestock created a barrier for those wishing to enter agriculture. Cogswell maintains that tenancy served as a "gateway through that barrier, and with each decade increasing numbers of young farmers passed through it" (Cogswell 1975, 153). Over time the anticipated future appreciation in land values came to represent an increasingly important determinant of price, which meant that meeting mortgage obligations from the current income of the land grew more and more burdensome for indebted farmers. Under such circumstances, some operators elected to rent rather than to purchase land (Winters 1978, 18–26). Tenancy was not, therefore, forced upon defenseless farmers; nor was it necessarily a sign of financial distress. Viewed from this perspective, it was a rational adjustment to the demands of a new frontier and to the changing conditions in prairie agriculture.

Disagreement also exists over the utility of the agricultural ladder thesis, which posits that tenancy served as a rung on an economic ladder that enabled farmers to move from laborer to tenant to encumbered landowner and, finally, to unencumbered landowner. Gates, as already observed, believes that the tenant rung was used more frequently by bankrupt farmers on the way down than by enterprising farmers on the way up. LaWanda F. Cox concurs, noting that "historical studies of the economic conditions of the western farmer suggest that backsliding may have been a major factor in many localities" (Cox 1944, 100–101). In our separate studies, however, Cogswell and I discovered that after the pioneer stage of settlement, tenancy became increasingly associated with younger and landownership with older farmers, the pattern one would expect if the ladder were functioning properly (Cogswell 1975, 30–34; Winters 1978, 79–88; Bogue 1963, 56). Moreover, my analysis of sample Iowa counties turned up a sizable portion of tenants (about 20 percent) who went on to become owners in the counties in which they had previously rented farms; how many became owners elsewhere is impossible to determine. This career pattern doubtless occurred throughout the Middle West with comparable frequency. Even though the precise proportion of successful renters can never be known, tenancy clearly served as a stepping stone to ownership for many farmers.

Young men just entering farming often needed credit not only to buy land, but also to purchase necessary equipment and supplies, and to maintain their families until the first crop came in (Danhof 1941; Atack and Bateman 1987, 45). Established farmers sometimes needed credit to expand their operations, to replace wornout machinery, and to sustain them through periods of low prices and poor crops. Some historians maintain that land speculators and bankers took advantage of the prairie farmers' dependence on creditors. Moneylenders charged unfair interest rates and attached unreasonable conditions to loan contracts. Some even encouraged farmers to take on unmanageable debts. Their intent, critics of the credit system charge, was to force debtors into default so that they could seize—"confiscate" is the term Gates uses—the land they had put up as collateral (Gates 1942, 326–27; Gates 1945, 205–6; Shannon 1945, 388–89; Saloutos 1948, 163).

Swierenga's study of land speculation in nineteenth-century Iowa and Bogue's study of moneylending in the postbellum Midwest provide a much different perspective on the rural credit system. They agree that pioneer farmers were highly dependent upon credit, but contend that neither speculators nor bankers deliberately contrived to force debtors into bankruptcy.[5] On the operations of one large mortgage firm, Bogue notes that "never was a loan made knowingly when the borrower's ability to pay was in doubt" (Bogue 1955, 273). Swierenga puts it more strongly: "That speculators eagerly foreclosed . . . contracts when debtors fell delinquent, as agrarians long maintained, is . . . a myth." The speculators' principal source of profits was land sales, not credit transactions, and they advanced loans mainly to facilitate those sales (Swierenga 1968, 215–16). Since the costs of foreclosure and finding a purchaser for the confiscated property usually reduced profits, speculators and mortgage companies quite naturally preferred debtors who paid off their obligations on schedule. Bogue and Swierenga concede that frontier interest rates were high, but they deny that this condition resulted from collusion and manipulation on the part of moneylenders. Financial markets in recently settled areas were inefficient and poorly developed, and the risks of lending were considerable. In such an environment, high interest rates were common. Rates declined, however, as the settlement of the Middle West proceeded and banking institutions matured. By the end of the nineteenth century they were at levels close to those of established areas in the Northeast (Bogue 1963, 169–81; North 1966, 141–42).

That moneylenders preferred payment over foreclosure and that competition rather than the machinations of creditors determined interest rates do not imply the absence of debtor distress in the postbellum Midwest. Some farmers had difficulty meeting their mortgage obligations, and many of them lost their property. But foreclosure was a serious problem primarily during downturns in the agricultural economy, suggesting that it was associated more with low farm prices than with the deceptive practices of moneylenders (Murray 1933, 371–96; Bogue 1963, 179–80; Winters 1978, 27–28). Even some historians who are critical of the rural credit system acknowledge a causal relationship between agricultural depression and foreclosure (Murray 1946, 13; Shannon 1945, 146).

Much less disagreement exists among scholars with regard to the adoption of new technology in midwestern agriculture in the late nineteenth century. The widespread use of horse-drawn equipment began in the years before the Civil War and accelerated in the postbellum era (Atack and Bateman 1987, 186–200). Without the steel plow, introduced in the 1830s, the opening of the prairies would have been a slower and more tedious process. Improvements in plows continued over the next half-century, and by the 1870s well-equipped farms in the Middle West had a sulky riding plow (Bogue 1963, 151–52). Just as important was the reaper, which farmers used in rising numbers from the 1850s on. As Paul A. David explains, the harvest of small grains, especially wheat, was a major bottleneck in prairie agriculture (David 1966; Atack and Bateman 1987, 194–200). Done by hand, harvesting was the one activity requiring more labor than the farm family could normally supply. Even though reapers were available as early as the 1830s, the costs of purchase and operation exceeded the costs of the hired labor they replaced; accordingly, few farmers invested in them. In the late 1840s and early 1850s, reaper prices declined because of greater efficiency in manufacturing, and labor costs increased because of competition from industry and railroad construction. For many farmers, therefore, the balance shifted in favor of mechanized harvest. Furthermore, as Alan Olmstead points out, a spate of improvements and technological refinements resulted in better and more durable machines, thereby making investments in reapers that much more attractive (Olmstead 1975). In addition to sulky plows and reapers, midwestern farmers in the late nineteenth century used a wide array of horse-drawn equipment, including wagons, seed drills, broadcast seeders, mowers, and cultivators. Their mounting capital investment reflected this increasing commitment to mechanization (Rasmussen 1962). Between 1860 and 1900 the average real value of equipment per farm in the Middle West more than doubled. It was the time when, as Bogue observes, farmers learned to "farm sitting down" (Bogue 1963, 148–68).

Adoption of new technology was not confined to mechanization; prairie farmers looked for other ways to improve their operations. They experimented with seed selection and new strains as they searched for the crop varieties best suited to their particular soil and climatic conditions. Although most farmers were not as attentive to soil conservation and restoration as they might have been, some employed contour cultivation, manured their fields, and even practiced limited crop rotation. They also attempted to upgrade their livestock by selective breeding and the introduction of blooded lines, such as Poland China swine, Jersey and Guernsey dairy cattle, Aberdeen Angus beef cattle, and Cotswold sheep (Bogue 1963, 86–147). More important, as Fred Bateman demonstrates, was the improvement in animal care and feeding. Farmers constructed fences to confine and barns to shelter their animals. They also increased the ratio of feed grains to grasses in livestock diets, which improved dairy yields, accelerated fattening, and produced better quality meat (Bateman 1968; Whitaker 1975, 115–26).

To innovate required information and farmers in the Middle West had several sources from which they might draw (Bogue 1963, 193–215; Bogue 1958; Throne 1951; Whitaker 1975, 85–121). In the late nineteenth century local newspapers and agricultural journals regularly offered advice on mechanization, crop varieties, livestock strains, and farming methods. In addition to providing recreational relief from the drudgery and boredom of rural life, county and state fairs held demonstrations of the latest agricultural techniques and products. Agricultural colleges, funded in part through the sale of grants under the Land Grant College Act (1862), and agricultural experiment stations, established under the Hatch Act (1887), developed new techniques and disseminated information. The federal government's annual report on agriculture also included counsel on farming practices. Representatives of implement companies and of seed and livestock distributors introduced new products and gave instruction in their uses. It is uncertain to what extent prairie farmers drew upon these sources of information and knowledge and to what extent they learned by trial and error or by observing more resourceful neighbors. Whatever the mode of education, most eventually participated in a process of technological innovation and diffusion that significantly altered midwestern agricultural practices in the postbellum era.

The adoption of technology was not the only way farmers endeavored to improve their operations; they also modified patterns of production. Household needs constrained production choices, as farmers normally placed high priority on satisfying subsistence requirements. But land and labor not used for this purpose could be employed in a variety of ways. Based on experimentation, observation, and outside information, each farmer settled on a combination of crops and livestock that would, in his judgment, generate the most income after meeting household needs (Atack and Bateman 1987, 201–24). Since farmers throughout the Midwest reacted to similar geographical and market determinants, their production decisions resulted in regional and subregional specializations.

During and immediately following the Civil War, wheat was the most widely grown cash crop. The soils, terrain, and climate (at least in the well-settled southern part of the Middle West) were suited to the cultivation of winter wheat, which was preferred over spring wheat because of its superior milling qualities. It required modest capital investment—a reaper was the main expense—and, outside of the relatively short periods of planting and harvesting, minimal labor. Although wheat prices fluctuated widely in the latter half of the nineteenth century, there was always a ready market. Much of the grain was sold to the Northeast or to foreign countries; some was sold to flour mills or distilleries in the Midwest. Since wheat could be consumed as well as marketed, its commercial production had the added advantage of providing a measure of insurance against failure to meet household subsistence needs.

The preference for wheat as a commercial crop eroded from the 1870s on, however. Prairie farmers discovered what northeastern farmers had known for some time, that soil placed in wheat over several years became infected with parasites and pests that seriously reduced yields. Winter-kill, always a threat throughout the Middle West, was particularly troublesome in the last-settled northern parts of the region. Improved milling processes, which increased the demand for spring wheat, seemed for a time to create a better alternative. But competition from the high-yielding grain belts of the recently opened Plains soon eliminated any commercial advantage the cultivation of spring wheat offered (Bogue 1963, 123–29).

Within a decade after the Civil War, then, farmers in the Midwest began exploring other commercial possibilities. The alternative they finally settled on was a feed-grain and livestock system. Livestock had, of course, been part of prairie agriculture since the pioneer period. Farmers normally kept a few swine and cattle for household consumption, which they grazed inexpensively on the open ranges, but raising animals for the market was not an important element in their business strategy. Developments in the postbellum years radically changed the market for meat products, which in turn altered the farmers' evaluation of commercial livestock production. Railroads linked the rural Middle West with meatpackers in cities such as Cincinnati, Chicago, and St. Louis, and linked the meatpackers with the principal urban markets in the Northeast. The invention of the refrigerated box car in the 1870s made possible the transport of fresh beef and pork from the interior for sale on the East Coast. Midwestern farmers responded to the resulting rise in demand for meat animals in two ways. They increased their commercial commitment to livestock, and they switched from wheat to feedgrain (mainly corn) cultivation. The open-range method for feeding livestock was far too land intensive for large-scale production. Besides, corn was superior to grass for fattening animals because it produced more rapid weight gain and better quality meat. Farmers bred their own swine and sometimes their own beef cattle. Increasingly though, they purchased western grass-fed calves, delivered to their areas by rail, which they fed out on grain in enclosed lots before marketing. By the 1890s the process was complete: the Middle West had been transformed from the country's major source of wheat flour into its major source of fresh meat (Ross 1951, 71–77; Whitaker 1975, 122–26).

Midwestern farmers pursued other commercial opportunities, though none was as important as the production of wheat and, later, meat animals. The sharp increase in wool prices during the Civil War attracted many prairie farmers to sheep production (Gates 1965, 157–76; Bogue 1963, 228–29). Some continued in this line after the postwar decline in wool prices, particularly in the hillier parts of Ohio and eastern Indiana, but most sold off their flocks. Operators located near the developing urban manufacturing centers along the Great Lakes found dairy production profitable (Bateman 1968; Bogue 1963, 229–30; Ross 1951, 78–83). Not only were they in close proximity to the market, but the area's natural endowments were better suited to dairy than to feedgrain and livestock farming. Farmers in a position to exploit special local situations sometimes produced oats, rye, flax, fruits, tobacco, or sorghum for sale to nearby processors.

The adoption of new technology and the specialization of production reflected the growing commercialization of prairie agriculture. Except on the newly opened frontier, as Rodney C. Loehr reminds us, agriculture in the United States always had some degree of commercial orientation (Loehr 1952). The Middle West was no exception. Even before the farm making was complete, farmers began searching for a food or fiber product to sell on the open market (Atack and Bateman 1987, 201–24). The same process had occurred in other regions and at other times reaching back to the colonial period. Yet, at least according to some historians, agricultural commercialization had a more profound effect on farmers in the postbellum period than it did on farmers in earlier times. Although commercialization still held out the promise of improving one's standard of living, it now entailed a relatively large investment in machinery and livestock, a heightened dependence on financial institutions, transportation facilities, and middlemen, and an increased vulnerability to market conditions. The farmer, as Anne Mayhew explains, became locked into a "system where, even if he was a 'good farmer' in pre-1860 terms, he might fail because he was a 'bad businessman' in late nineteenth-century terms" (Mayhew 1972, 475).

Many historians depict this expanding commercialization of prairie agriculture as essentially detrimental to the welfare of farmers (Hicks 1931; Saloutos 1948; Shannon 1945). They maintain that capital outlays for the necessary machinery, supplies, and livestock imposed an unmanageable financial burden on producers without significantly adding to their incomes. These required investments forced many operators to incur debts larger than would have been necessary under a simpler, more self-sufficient regimen, thereby increasing their dependence on an unfair and exploitative rural credit system. The broadened commercial network forced farmers into an unhealthy reliance on a battery of middlemen, each of whom took advantage of the hapless producer. Railroads inflated transportation rates, warehouses overcharged for storage, suppliers fixed prices for equipment and seeds, and commodity dealers and processors contrived to drive down farm prices (Hicks 1931, 54–95). The upshot, according to this interpretation, was that everyone except the farmer benefited from his commitment to the market. Ironically, those who produced the primary commodities upon which the whole system depended saw their welfare and status decline while others saw theirs improve.

Recently, scholars have suggested modifications in this traditional view of the farmer as victim of commercialization. They argue, first of all, that the suppliers of essential goods and services rarely possessed the control or market mechanisms required to exploit farmers. As pointed out above, expansion and competition in rural banking brought declining interest rates. Railroads sometimes held a monopoly in the early stages of settlement, but as population increased and the transportation network matured, manipulation of rates became increasingly difficult. As Douglass C. North notes, freight charges fell more rapidly than the general price level between 1865 and 1900 (North 1966, 139–41).[6] Likewise, competition among warehouses and equipment manufacturers placed severe limitations on their ability to overcharge farmers. Finally, given the enormous volume of agricultural products entering the market, it is inconceivable that commodity dealers and processors, no matter how sophisticated or complete their collusion, could have artificially depressed prices. Suppliers of goods and services were, to be sure, out to make as much money as possible, but they hardly exercised the kind of control over the market that has been attributed to them.

The truth of the matter is that farmers benefited greatly from those groups accused of exploiting them. For all of the criticism of their methods, rural bankers are a case in point. Farmers needed long-term credit for land, intermediate-term credit for equipment and livestock, and short-term credit for operating expenses. Bankers responded by developing a variety of ways to service those needs. If the rural credit system did not always function as well as debtors thought it should, it operated with improving efficiency and sophistication over the last half of the nineteenth century. Railroads, for their part, opened up previously inaccessible markets and serviced at lower costs existing markets for midwestern agricultural products, thereby creating incentives for western settlement and opportunities for expanded farm income (Fogel 1964, 208–19). The same can be said of warehouses and commodity exchanges, which negotiated increasingly complex transactions between sellers and distant buyers of farm goods (Lurie 1974, 116–17). American farmers, as Morton Rothstein (1960) observes, enjoyed a competitive advantage in foreign trade in large part because of the country's superior marketing system. In short, these commercial relationships were not a zero-sum game in which suppliers' profits had to be extracted from farmers' equity; the commercialization of agriculture enhanced income in both sectors.

The treatment of these issues is not meant to imply that farmers experienced no financial difficulties in the late nineteenth century. Critics of agricultural commercialization are quite correct when they assert that the producers' well-being became more and more dependent on uncertain market conditions. Prices in the postbellum years were highly volatile, largely because of the heavy dependence of American farmers on the foreign market.[7] Poor crops or armed conflict abroad drove prices up, while bountiful harvests abroad drove them down. Cyclical trends in the American economy contributed to the volatility. In general, high agricultural prices accompanied the periods of economic expansion and prosperity in the early 1870s and the 1880s, and low prices accompanied the periods of depression in the mid-1870s and the mid-1890s. That the agricultural sector in the nineteenth century was subject to easy entry and difficult withdrawal compounded the problem. When prices rose, new farmers bought up relatively cheap land and put it into production. When prices dropped, ownership of land, equipment, and livestock, in addition to recently acquired debt obligations, made it difficult for them to leave agriculture. Instead, they usually increased output in an effort to offset price declines, thereby aggravating the overproduction that was already causing problems (North 1966, 142–44). Furthermore, the late nineteenth century was a period of general deflation. The debt that many farmers carried was therefore especially burdensome, as they paid off their obligations in dollars of increasing value. Unquestionably, midwestern farmers periodically encountered hard times and distress, but their difficulties derived from basic conditions in the agricultural economy rather than from exploitation by special interests.

The record of the agricultural sector in the postbellum Middle West indicates that despite these financial problems, the welfare of farmers improved (Bogue 1963, 165–66, 283–86; Bowman 1965; Fogel and Rutner 1972, 395–404; Parker and Klein 1966). Even though food and fiber prices underwent a secular decline between 1865 and 1897, they fell at a slightly slower rate than did the general price level; real prices to the farmer thus increased modestly. More importantly, agricultural efficiency as measured by output per worker improved significantly, due primarily to specialization and the adoption of new technology (Atack and Bateman 1987, 186–200; Parker and Klein 1966). Landowners also reaped long-term capital gains, as their property steadily appreciated in value. Fogel and Rutner (1972, 395) report that "there was no decade of that period [1849–1899] during which the average real income of farmers declined." Prairie farmers did even better than their counterparts in other regions; they saw their per capita incomes rise at a rate approximating the national average.

Agriculture played an important role in the nation's overall economic growth. Improved efficiency contributed not only to higher standards of living for farmers in the Middle West, but to gains in per capita income for the country as a whole. Increases in output per worker also enabled the agricultural sector to satisfy the growing demand for its products at home and abroad while employing a declining portion of the labor force. In 1860 farming used about 60 percent of the labor force; by the end of the century it used less than 40 percent. Workers released from agriculture were available for employment in other sectors, with many of them entering manufacturing. The United States ran a favorable balance of trade in practically every year from 1865 to 1900, thanks in large measure to the export of surplus agricultural goods. Earnings from farm exports flowed into investments in other sectors, especially industry. Even though middlewestern meat was less important in foreign trade than either western wheat or southern cotton, prairie farmers still made a substantial contribution to the country's strong position in the international economy.

Despite rather considerable disagreement among scholars of agricultural history during the postbellum period, their views can, I believe, be divided into two schools of interpretation. One group of historians begins with the assumption that midwestern farmers failed to share in the fruits of the country's economic development and expansion. This group's principal objective is to explain why this was so. Implicitly, at least, they identify with the agrarian reformers of the late nineteenth and early twentieth centuries, who claimed that farmers were victims of land monopolists, loan sharks, greedy railroad owners, unscrupulous commodity speculators, and unethical suppliers and processors. Paul W. Gates, one of the most influential adherents of this position, revealed more than he realized when he stated, as he looked back over a long and fruitful career, that his main objective in studying federal land policy had been to uncover the "methods . . . used to establish the great holdings of timber, mining, grazing, and agricultural lands that have long troubled Populists, Progressives, trust busters, antimonopolists, conservationists, and advocates of single-family farms" (Gates 1973, 48). Not surprisingly, scholars of this persuasion adopt the same class interpretation, as well as much of the rhetoric, found in the critique of the earlier agrarian reformers. Their explanation indicts a variety of "moneyed interests" as the primary cause of rural distress in the post-Civil War period. They accuse those moneyed interests of engaging in collusion and conspiracy, usually without explaining the mechanisms, to expropriate property that rightfully belonged to defenseless farmers. It was a simple case of capitalist exploitation of an underclass.

The second group of historians has a much different agenda and perspective. They are primarily interested in understanding the development of the agricultural sector and its relationship to the general economy. In a very real sense, they are responding to a criticism issued by Thomas LeDuc almost thirty years ago. "Most historians," he charged, "have been so deafened by the noise of agrarian politics that they have never taken a hard and disinterested look at the economic realities in a swiftly changing enterprise" (LeDuc 1959, 51). Rather than accept the agrarian reformers' critique that farmers failed to participate in the advancing prosperity of the late nineteenth century, these historians elect to test that proposition. Their investigations indicate, for the most part, that although many midwestern farmers suffered periodic financial distress, as a group they enjoyed an improvement in their standard of living comparable to that of the average American. These historians also maintain that the hardships farmers did encounter were due more to the nature of the agricultural sector and to overall economic conditions than to the machinations of special interests. In fact, they assert, those special interests who have been criticized so severely by agrarian reformers and many later historians actually provided goods and services without which commercial farmers would have been unable to operate. In short, farmers were neither exploited nor left behind. Like many other groups in society, they were subject to the vicissitudes of the country's rapidly changing economy, but they also participated in its growing prosperity.

The scholarly dispute between these two groups of historians, however, runs deeper than their contrasting readings of agricultural conditions and the fate of postbellum farmers. They begin with conflicting conceptions of the nineteenth-century economy. The group sympathetic with the plight of farmers views the economy as essentially controlled and manipulated by powerful economic and political interests for the sole purpose of enhancing their own welfare. They accomplished their purpose with staggering success. Their interference in the market, though, so disrupted and distorted development that economic growth fell far short of its full potential. Those without economic or political leverage—for instance, farmers, workers, blacks, and immigrants—realized little, if any, improvement in their situations. Scholars advancing a more benign account of prairie agriculture, on the other hand, see a relatively free market system operating in accordance with economic forces. It was an environment in which no group had the power to exploit another. The actions of the historical participants were economically rational and constrained primarily by supply and demand. Some individuals suffered financial distress and economic dislocation, but that was in part the price of economic progress. For the most part, though, most groups shared in the benefits of that progress. Gains in welfare may not have been allocated evenly—they seldom are in a dynamic and expanding economy—but they were distributed widely.

Part of the reason for these conceptual differences doubtless derives from the economic environment in which the scholars worked. Many of those who see the nineteenth-century economy as controlled and manipulated wrote during the 1930s and the 1940s, decades when either the realities or the vivid memories of the Great Depression profoundly influenced attitudes. They were appalled at the rural devastation wrought by the economic crisis. Bogue, who was a Gates's student at Cornell University in the 1940s, has commented that his mentor was "profoundly moved by the foreclosures, the dispossession of farmers, low agricultural prices, and the erosion of farm lands, which were so integral a part of the great depression. . . ." (Aydelotte et al. 1972, 52). Many scholars who agree with Gates's historical interpretations must have suffered similar anguish. It was in this context that they began searching for the scapegoats for what they saw as the country's long history of rural distress.[8]

The work of those who argue that a free market system best describes the economy of the nineteenth century first appeared in the late 1950s and the 1960s. With the depression of the 1930s a fading memory and the United States once again enjoying prosperity, their attention was drawn to the country's economic growth and development. They were more interested in explaining economic change than in laying blame. For them, the nineteenth century was the time when the groundwork was set in place for what they saw as the long-term improvement in the welfare of all Americans.[9]

The agenda for future research in middle-western agriculture in the years following the Civil War seems clear. Studies at the microlevel will, perhaps, further clarify the role of federal policy in western settlement and landownership patterns, and illuminate the dynamics of technological innovation and diffusion. Closer scrutiny of the determinants of production decisions, particularly with regard to the choice between household consumption and sale, will doubtless reveal more than we now know about the process of agricultural commercialization. Detailed investigations of transactions between farmers and the suppliers of the goods and services upon which they depended will help to elucidate the nature of their relationship. But such studies will not take us very far unless they deal with broader theoretical questions. Future research must address the dispute over which conceptual framework—the one that assumes manipulation by special interests or the one that assumes predominance of market forces—offers the most appropriate model of the nineteenth-century economy. A better grasp of this issue will enable us to order more effectively the vast amount of evidence we already have and to narrow our interpretive differences. Only then will we significantly advance our understanding of middle-western agriculture in the postbellum era.

## NOTES

1. Benjamin H. Hibbard (1924), Roscoe L. Lokken (1942), and Daniel Feller (1984) provide excellent accounts, albeit from different perspectives, of the formulation of federal land policy.

2. It is interesting to note that some critics of federal land policy complain that land was released too quickly, thus causing overproduction; others complain that it enabled speculators to hold land off the market, thus keeping it out of production.

3. Several of Gates's most important articles have been reprinted in his *Landlords and Tenants on the Prairie Frontier: Studies in American Land Policy* (Gates 1973). Many historians have advanced some version of Gates's thesis (Billington 1949, 701; Dennen 1977; LeDuc 1950; Lokken 1942, 131–53; Murray 1946; Okada 1971; Robbins 1942, 268–71; Shannon 1945, 146–47; Saloutos 1948).

4. Tenancy rates in some areas of the Middle West were very low. Merle Curti notes that the rate in Trempealeau County, Wisconsin, in 1880 was only 4.3 percent, from which he concludes that tenancy must have been virtually nonexistent during the pioneer period (Curti 1959, 444–45). For an evaluation of the literature on land speculation and tenancy, see my article, "Agricultural Tenancy in the Nineteenth-Century Middle West: The Historiographical Debate" (Winters 1982).

5. Midwestern farmers sometimes secured credit in ingenious ways. Swierenga (1976) discovered that they often deliberately allowed their land taxes to become delinquent and used the funds for development or operating expenses. County officials in turn sold the land to tax buyers for the delinquent taxes. But farmers had a three-to-four-year grace period in which to pay back taxes and penalties before the title was transferred. Since the penalties were normally less than prevailing interest rates, farmers had use of their tax funds for up to four years at below credit market costs.

6. Theodore Saloutos, who sees farmers as victimized by special interests, acknowledges that railroad rates declined sharply in the postbellum period, but he complains that this only contributed to the intense competition among farmers that kept farm prices low (Saloutos 1948, 167). However, Robert Higgs claims that freight rates in relation to corn, wheat, and cotton prices remained roughly level between 1865 and 1893, and then increased significantly in the mid-1890s (Higgs 1970).

7. A larger portion of the country's wheat and cotton than of its meat was exported. Farmers in the South and the Plains were, therefore, more affected by conditions in the foreign market than were farmers in the Middle West.

8. It is interesting to note that many politicians of the time blamed "moneyed interests," who they accused of taking advantage of the economic favoritism of the federal administrations of the 1920s, for the economic problems of the 1930s.

9. There are also methodological differences between these two schools of historians. The more recent group makes systematic use of economic theories, models, and concepts, and of quantitative techniques. The earlier group, with a few exceptions, employs a more traditional approach.

## REFERENCES AND FURTHER READINGS

Ankli, Robert E. 1974. "Farm-Making Costs in the 1850s." *Agricultural History* 48(Jan.):51–70.

______. 1976. "The Coming of the Reaper." In *Business and Economic History: Papers Presented at the Twenty-Second Annual Meeting of the History of the Business History Conference,* ed. Paul A. Uselding, 1–24. Urbana: Univ. of Illinois Press.

Atack, Jeremy. 1982. "Farm and Farm-Making Costs Revisited." *Agricultural History* 56(Oct.):663–76.

Atack, Jeremy, and Fred Bateman. 1987. *To Their Own Soil: Agriculture in the Antebellum North.* Ames: Iowa State Univ. Press.

Aydelotte, William O., Allan G. Bogue, and Robert William Fogel. 1972. *The Dimensions of Quantitative Research in History.* Princeton: Princeton Univ. Press.

Bateman, Fred. 1968. "Improvement in American Dairy Farming, 1850–1910: A Quantitative Analysis." *Journal of Economic History* 28(June):255–73.

Bidwell, Percy W., and John I. Falconer. 1925. *History of Agriculture in the Northern United States, 1620–1860.* Washington: Carnegie Institution.

Billington, Ray H. 1967. *Westward Expansion: A History of the American Frontier.* New York: Macmillan.

Blodgett, James H. 1903. *Wages of Farm Labor in the United States: Results of Eleven Statistical Investigations, 1866–1902.* Washington: U.S. Department of Agriculture Miscellaneous Series Bulletin 26.

Bogue, Allan G. 1955. *Money at Interest: The Farm Mortgage on the Middle Border.* Ithaca: Cornell Univ. Press.

______. 1958. "Pioneer Farmers and Innovation." *Iowa Journal of History* 56(Jan.):1–36.

______. 1963. *From Prairie to Cornbelt: Farming on the Illinois and Iowa Prairies in the Nineteenth Century.* Chicago: Univ. of Chicago Press.

Bogue, Margaret Beattie. 1959. *Patterns from the Sod: Land Use and Tenure in the Grand Prairie, 1850–1900.* Springfield: Illinois State Historical Society.

Bowman, John. 1965. "An Economic Analysis of Midwestern Farm Land Values and Farm Land Income, 1869–1900." *Yale Economic Essays* 5(Fall):317–54.

Buck, Solon J. 1913. *The Granger Movement: A Study of Agricultural Organization and Its Political, Economic, and Social Manifestations, 1870–1880.* Cambridge: Harvard Univ. Press.

Cogswell, Seddie. Jr. 1975. *Tenure, Nativity and Age as Factors in Iowa Agriculture.* Ames: Iowa State Univ. Press.

Cox, LaWanda F. 1944. "Tenancy in the United States, 1865–1900: A Consideration of the Validity of the Agricultural Ladder Hypothesis." *Agricultural History* 18(July):97–105.

Curti, Merle. 1959. *The Making of an American Community: A Case Study of Democracy in a Frontier Community.* Stanford: Stanford Univ. Press.

Danhof, Clarence H. 1941. "Farm-Making Costs and the 'Safety Valve': 1850–1860." *Journal of Political Economy* 49(June):317–59.

______. 1969. *Change in Agriculture: The Northern States, 1820–1870.* Cambridge: Harvard Univ. Press.

David, Paul A. 1966. "The Mechanization of Reaping in the Ante-Bellum Midwest." In *Industrialization in Two Systems: Essays in Honor of Alexander Gerschenkron,* ed. Henry Rosovsky, 3–39. New York: John Wiley.

Dennen, A. Taylor. 1977. "Some Efficiency Effects of Nineteenth-Century Federal Land Policy: A Dynamic Analysis." *Agricultural History* 51(Oct.)718–36.

Easterlin, Richard A. 1976. "Population Change and Farm Settlement in the Northern United States." *Journal of Economic History* 37(Mar.):45–75.

Feller, Daniel. 1984. *The Public Lands in Jacksonian Politics.* Madison: Univ. of Wisconsin Press.

Fogel, Robert William. 1964. *Railroads and American Economic Growth: Essays in Econometric History.* Baltimore: Johns Hopkins Univ. Press.

Fogel, Robert William, and Jack L. Rutner. 1972. "The Efficiency Effects of Federal Land Policy, 1850–1900: A Report of Some Provisional Findings." In *The Dimensions of Quantitative Research in History,* ed. William O. Aydelotte, Allan G. Bogue, and Robert William Fogel, 390–418. Princeton: Princeton Univ. Press.

Gates, Paul W. 1934. *The Illinois Central Railroad and Its Colonizing Work.* Cambridge: Harvard Univ. Press.

______. 1936. "The Homestead Law in an Incongruous Land System." *American Historical Review* 41(July):652–81.

______. 1939. "Land Policy and Tenancy in the Prairie Counties of Indiana." *Indiana Magazine of History* 35(Mar.):1–26.

______. 1941. "Land Policy and Tenancy in the Prairie States." *Journal of Economic History* 1(May):60–82.

______. 1942. "The Role of the Land Speculator in Western Development." *Pennsylvania Magazine of History and Biography* 66(July):314–33.

______. 1945. "Frontier Landlords and Pioneer Tenants." *Journal of the Illinois State Historical Society,* 38(June):143–206.

______. 1965. *Agriculture and the Civil War.* New York: Alfred A. Knopf.

______. 1973. *Landlords and Tenants on the Prairie Frontier: Studies in American Land Policy.* Ithaca: Cornell Univ. Press.

Hibbard, Benjamin Horace. 1924. *History of the Public Land Policies.* New York: Macmillan.

Hicks, John D. 1931. *The Populist Revolt: A History of the Farmers' Alliance and the People's Party.* Minneapolis: Univ. of Minnesota Press.

Higgs, Robert. 1970. "Railroad Rates and the Populist Uprising." *Agricultural History* 44(July):291–97.

Lampard, Eric E. 1963. *The Rise of the Dairy Industry in Wisconsin: A Study in Agricultural Change.* Madison: Univ. of Wisconsin Press.

LeDuc, Thomas. 1950. "The Disposal of the Public Domain in the Trans-Mississippi Plains: Some Opportunities for Investigation." *Agricultural History* 24(Oct.):199–204.

______. 1959. "Recent Contributions to Economic History: The United States, 1861–1900." *Journal of Economic History* 19(Mar.):44–63.

______. 1963. "Public Policy, Private Investment, and Land Use in American Agriculture, 1825–1875." *Agricultural History* 37(Jan.):3–9.

Loehr, Rodney C. 1952. "Self Sufficiency on the Farm." *Agricultural History* 26(Apr.):37–41.

Lokken, Roscoe L. 1942. *Iowa Public Land Disposal.* Iowa City: State Historical Society of Iowa.

Lurie, Jonathan. 1974. "Commodity Exchanges, Agrarian 'Political Power,' and the Anti-option Battle, 1890–1894." *Agricultural History* 48(Jan.):115–25.

Mayhew, Anne. 1972. "A Reappraisal of the Causes of Farm Protest in the United States." *Journal of Economic History* 32(June):464–75.

Murray, William G. 1933. *An Economic Analysis of Farm Mortgages in Story County, Iowa.* Research Bulletin no. 156. Ames: Iowa State Agricultural Experiment Station.

———. 1946. "Struggle for Land Ownership." In *A Century of Farming in Iowa, 1846–1946,* ed. Earl Ross, 1–17. Ames: Iowa State College Press.

North, Douglass C. 1966. *Growth and Welfare in the American Past: A New Economic History.* Englewood Cliffs, N.J.: Prentice-Hall.

Okada, Yasuo. 1971. *Public Lands and Pioneer Farmers: Gage County, Nebraska, 1850–1900.* Tokyo: Univ. of Tokyo Press.

Olmstead, Alan L. 1975. "The Mechanization of Reaping and Mowing in American Agriculture, 1833–1870." *Journal of Economic History* 35(June):327–52.

Parker, William, and Judith L. V. Klein. 1966. "Productivity Growth in Grain Production in the United States, 1840–1860 and 1900–1910." In *Output, Employment, and Productivity After 1800,* ed. National Bureau of Economic Research. New York: Columbia Univ. Press, 1966.

Rasmussen, Wayne D. 1962. "The Impact of Technological Change on American Agriculture, 1862–1962." *Journal of Economic History* 22(Dec.):578–91.

Robbins, Roy M. 1942. *Our Landed Heritage: The Public Domain, 1776–1936.* Princeton: Princeton Univ. Press.

Rogin, Leo. 1931. *The Introduction of Farm Machinery in Its Relations to the Productivity of Labor in the Agriculture of the United States during the Nineteenth Century.* Berkeley: Univ. of California Press.

Ross, Earle D. 1951. *Iowa Agriculture.* Iowa City: State Historical Society of Iowa.

Rothstein, Morton. 1960. "America in the International Rivalry for the British Wheat Market, 1860–1914." *Mississippi Valley Historical Review* 47(Dec.):401–18.

Saloutos, Theodore. 1948. "The Agricultural Problem and Nineteenth-Century Industrialism." *Agricultural History* 22(July):156–74.

———. 1962. "Land Policy and Its Relation to Agricultural Production and Distribution, 1862 to 1933." *Journal of Economic History* 22(Dec.):445–60.

Schrob, David. 1975. *Hired Hands and Plowboys.* Urbana: Univ. of Illinois Press.

Shannon, Fred A. 1936. "The Homestead Act and the Labor Surplus." *American Historical Review* 41(July):631–51.

———. 1945. *The Farmer's Last Frontier: Agriculture, 1860–1897.* New York: Holt, Rinehart and Winston.

Soltow, Lee. 1971. *Patterns of Wealthholding in Wisconsin Since 1850.* Madison: Univ. of Wisconsin Press.

———. 1975. *Men and Wealth in the United States, 1850–1870.* New Haven: Yale Univ. Press.

Stewart, Charles L. 1916. *Land Tenure in the United States with Special Reference to Illinois.* Urbana: Univ. of Illinois Press.

Swierenga, Robert P. 1968. *Pioneers and Profits: Land Speculation on the Iowa Frontier.* Ames: Iowa State Univ. Press.

———. 1976. *Acres for Cents: Delinquent Tax Auctions in Frontier Iowa.* Westport, Conn.: Greenwood Press.

Taylor, Carl C. 1953. *The Farmers' Movement, 1620–1920.* New York: American Book Co.

Throne, Mildred. 1951. "'Book Farming' in Iowa, 1840–1870." Iowa Journal of History 49(Apr.):117–43.

Towne, Marvin W., and Wayne D. Rasmussen. 1960. "Gross Farm Product and Gross Investment in the Nineteenth Century." In *Trends in the American Economy of the Nineteenth Century,* ed. National Bureau of Economic Research, 255–312. Princeton: Princeton Univ. Press.

Whitaker, James W. 1975. *Feedlot Empire: Beef Cattle Feeding in Illinois and Iowa, 1840–1900.* Ames: Iowa State Univ. Press.

Winters, Donald L. 1978. *Farmers without Farms: Agricultural Tenancy in Nineteenth-Century Iowa.* Westport, Conn.: Greenwood Press.

______. 1982. "Agricultural Tenancy in the Nineteenth-Century Middle West: The Historiographical Debate." *Indiana Magazine of History* 29(June):128–53.

CHAPTER 5

# Agricultural Issues in the Middle West, 1865–1910

DOROTHY SCHWIEDER

BETWEEN 1865 AND 1910 the Middle West was transformed from a partially settled, predominantly frontier region into the breadbasket of the nation. During that period, however, as population expanded and as productivity continually increased, farmers also faced a long agricultural depression. In addition to low farm prices, they faced high transportation rates, discriminating railroad practices, high elevator charges, and excessive interest rates. The depression brought on the first strong agrarian protest, including the Grange, the Farmer's Alliance, and eventually the Populist Party, all particularly evident in the western portion of the region.

The Long Depression did not prevent many advances from taking place. The period witnessed great change in farm technology and considerable advancement of scientific knowledge in the area of agriculture. In many ways the period from 1865 to 1910 was a time of "feast or famine" for the nation's most productive agricultural region; by 1898, after a thirty-year depression, agricultural prosperity returned and the period from 1900 to 1910 would be regarded as one of the most prosperous ever for the American farmer. By 1910, one thing was clear: the Middle West had clearly established itself as the premier agricultural region in the nation.

Throughout good times and bad, however, farm residents also expressed concern over problems not so easily articulated or so directly confronted as economic issues. Farm families lamented the increasing exodus from the farm to the city, particularly for farm youths, and the farm populations' declining influence and status within the nation.

By the end of the Civil War, approximately three-fourths of the Middle West, a region composed of Ohio, Indiana, Michigan, Wisconsin, Minnesota, Illinois, Iowa, Missouri, the Dakotas, Nebraska, and Kansas, had been settled. The eastern portion, particularly Ohio, Indiana, Michigan, had experienced settlement in the early 1800s. The middle portion, including Iowa and Minnesota, was in the process of settlement in 1865, while the western portion of the Middle West, particularly the Dakotas, would experience continued settlement in the following three decades.

Like most regions within the United States, the Middle West was composed of major subregions. The eastern states, particularly Ohio, had originally been heavily forested while the middle portion including Minnesota, Illinois, Missouri, and Iowa, had all contained significant prairie areas. Prairie regions were identified by an absence of trees and abundance of tall grass. Typically prairie regions had fewer undulating surfaces than woodland areas. While Minnesota, Illinois, and Missouri all contained substantial prairie land, Iowa had the greatest amount with approximately four-fifths of the state defined as prairie. The last tier of states in the region—the Dakotas, Nebraska, and Kansas—while containing small amounts of prairie, were identified as Plains states indicating that the land was flatter, with fewer trees and less abundant rainfall than in states to the east (Madson 1982, 21).

Because of variations in settlement patterns and differences in geography and climate, agricultural practices also varied. In effect, as settlement moved westward, agriculture became more unpredictable and subject to vagaries of climate. Moreover, agricultural options decreased as settlers moved farther west. Rainfall in Ohio averaged 40 inches annually while that amount gradually decreased the farther west one moved. Settlers in the western half of South Dakota, where settlement did not get under way until after 1900, soon learned that they could expect an average rainfall between 21 inches (along the Missouri River) and 13 inches (along the western border) each year. Equally important, however, was the fact that 40 percent of the time precipitation was below that mean (Nelson 1986, 11; Jordon 1943, 61).

Initially, farmers in each state planted corn and small amounts of wheat, but subregional specialties soon appeared. By the 1870s western Ohio farmers continued to raise wheat while farmers in eastern Ohio had diversified. Farmers in both Illinois and Iowa quickly discovered that the prairie soils were especially well suited to the growing of corn and, in turn, that profits increased when corn was fed to hogs rather than sold. Farmers in each area soon began to move toward specialized production that best suited their area's climate, soil, and general conditions. Corn represented the greatest specialization in the Middle West and identified much of the region as the Corn Belt. Corn farmers shipped some of their product to the East Coast and some overseas, but they converted most of it into pork and beef. Prairie farming quickly diversified even more as Iowa and Illinois farmers not only raised corn but also barley, oats, rye, and hay. Two other major specialties were dairy products, particularly in the area north of the Corn Belt, and wheat in the western portions of the region. Development of all these specialties meant that farmers in the Middle West soon came to depend upon marketing a money crop for their economic well being (Saloutos and Hicks 1951, 8; Cochrane 1979, 91; Jordan 1943, 63–64).

Land policies and labor practices in the Middle West were a legacy from the period before the Civil War. Beginning with settlement in Ohio in the early 1800s, the nation's land laws had become increasingly liberal. In 1800 settlers in Ohio could purchase a minimum of 320 acres for as low as $2.00 per acre. In 1820 Congress passed a law that reduced the minimum amount of land to be purchased to 80 acres and lowered the price to $1.25. This law permitted the sale of 87,538,346 acres and as agricultural historian John Schlebecker has pointed out, "stimulated land sales, settlement, and astounding growth of farming and farm population." "Such a mass distribution of land," Schlebecker adds, "was unique in the history of the world, and none had been so effective and equalitarian" (Schlebecker 1975, 58–60).

The most well-known American land law, the Homestead Act, especially affected the western Middle West. Under this 1862 act persons could file for ownership of 160 acres of federal land if they met certain restrictions including a minimum age limit—twenty-one years—and resided on the claim for five years. This law had particular significance for the Plains states of the Dakotas, Kansas, and Nebraska. By contrast, in Iowa only 2.5 percent of the state's total acres were taken up under the Homestead Act; most Iowa land had already been claimed by 1862 (Sage 1974, 70).

Variations in land laws prevailed in different states. In addition to the more visible land laws such as the Homestead Act, settlers in South Dakota obtained land through government lotteries. In 1904 the federal government announced that some Indian land in the western portion of the state would be made available to the general public. Lotteries were established whereby individuals had to appear in public to enter their names. Government officials then drew names to determine the winners. The lands were not free, ranging from $4.00 an acre for the first recipients to $2.50 for land obtained six months later. Land on several Indian reservations was disposed of in this manner (Nelson 1986, 71–81).

Middle Westerners could also obtain land in other ways. Since the early days of the nation, the federal government had given land grants to soldiers as a reward for serving in the military. Men who fought in the War of 1812, the Black Hawk War, and the Civil War, among others, were able to secure land free. Although scholars have charged that the liberalizing of this process led to land speculation, one agricultural historian believes that most of the warrants ended up in the hands of farmers and "small-scale farmer-speculators" (Schlebecker 1975, 62).

Railroads provided yet another way many farmers acquired land in the central and western Middle West. In 1852 the federal government awarded the Illinois Central 2.5 million acres of land, which marked the first such grant to railroads. Five years later the government gave Iowa 4 million acres of land which that state, in turn, divided between four major lines. The Plains states would later share in the same land-grant process. Once railroads acquired land, they began to advertise and hold land sales. Railroads became known as the great colonizers, bringing tens of thousands of people into the Middle West (Schlebecker 1975, 66).

The increasing liberality of this legislation meant that land could be had quickly and cheaply. In the eastern part of the Middle West the concept of the agricultural ladder worked in the anticipated fashion. In other words, people coming into the region, though starting as hired hands or tenants, could through hard work and frugal living save enough money to purchase land. However, this agricultural ladder did not always work as predicted in the Plains states. Once settlers had arrived in South Dakota, Nebraska, or Kansas, they faced far greater obstacles than settlers farther east. Drought was more frequent as were plagues of grasshoppers. The very act of creating a farmstead on the Plains cost considerably more than in the eastern part of the region.

Land policies were directly related to labor practices. The free-labor practices first evident in Puritan New England in the seventeenth century would be carried along by later settlers as they moved into the Middle Atlantic states and then beyond into the Old Northwest. The Northwest Ordinance of 1787 formally prohibited slavery in Ohio, Indiana, Illinois,

Michigan, and Wisconsin. The tradition of small freehold farmsteads established early in New England and carried westward by successive generations of northerners was, in turn, reinforced by federal land policies. With a holding of 80 acres fairly typical throughout the central and eastern Middle West, a farm family usually provided most or all of its own labor. Farm women traditionally had their enterprises of gardening, raising poultry, and dairying. At an early age farm children assumed some responsibilities—weeding the garden, helping with the milking, and gathering eggs. As children grew older they continued to assume greater responsibility for both farm and home chores and, in general, became an important component of midwestern farm labor. Some families did hire outsiders, but generally on a seasonal basis. Farm women, for example, needed extra help during the summer and farm men needed extra help at planting and harvesting times. The farm family itself actually provided a flexible work force to cope with these situations. Farm children were often kept out of school to help with work. The oldest daughter frequently missed school on Monday morning to help with the family wash, while the oldest son was often absent to help with planting and harvesting. In the prairie states the corn harvest brought community events to a standstill as all family members, with the exception of the very young and the very old, trooped into the fields to pick the crop. The use of children to "take up the slack" meant that little or no money had to be spent on wages. Nor did anyone have to be fired at the end of the seasonal work (see Fink 1986, 23–24).

Kinship ties were strong in these rural communities, providing another component of middle western labor practices. Married children where possible settled close to their parents thus providing for continued family assistance with farm work. Fathers and sons exchanged labor as did brothers and brothers-in-law. This exchange of labor meant money savings not only on wages but also on farm machinery. Relatives sometimes purchased equipment together, enabling families to split the cost. Sometimes agreements reached beyond family ties to involve the immediate rural neighborhood. Threshing rings, which became popular in the late nineteenth century, perhaps best illustrate the concept of mutual aid among nonrelated farm people (Bogue 1968, 167–68).

While land policies were typically determined by the federal government and labor was regarded as a concern to be handled by each family, capital often represented an entirely different matter. For individuals who wished to borrow money to buy land or to tide them over during difficult times, borrowing rates presented a major problem. In general, the farther west one traveled in the Middle West, the higher the interest rates. Farming in the Plains states with a higher incidence of drought was obviously less predictable than farming in the central or eastern Midwest. After 1887, especially, mortgage companies charged from 18 to 24 percent and sometimes as high as 40 percent. South Dakota historian Herbert Schell notes "the most serious single problem confronting the early Dakota settler was his need for credit." Typically the farm family arrived with just enough to make a claim and make their first planting. Often the male went to work at a nearby sawmill or military fort to earn enough to tide the family over until the first crop was harvested. The farmer had to depend on local businessmen such as the banker, the farm implement dealer, and the local storekeeper for the advancement of credit. Schell concludes that in the early 1870s, Dakota Territory was "cursed" with high interest rates; under legislation enacted in 1872, the maximum interest was 24 percent a year. High interest rates continued to be a problem for farmers in the western Middle West throughout the latter nineteenth century (Shannon 1945, 373; and Schell 1968, 122–23).

Even with variations between the prairie and Plains subregions, the farm population in the Middle West from 1865 to 1910 would exhibit many common interests and would face similar economic difficulties. Agricultural historian Fred Shannon has pointed out that the main grievances of American farmers in the post-Civil War period centered around their lack of parity with other economic groups. This was perhaps more evident in the Middle West than in other regions. The unhappiness over this lack of parity focused on a number of areas, including railroads, interest rates, and marketing facilities. Before 1870, for example, some officials determined that half of what was paid for wheat shipped from west of the Mississippi to the Atlantic Coast went to middlemen. Complaints were heard that farming hardly paid for itself, let alone providing a profit to farmers. Throughout the period from 1867 to 1898, given the continual agricultural depression, farmers often received lower than expected prices for their products (Shannon 1957, 49–52).

A major economic complication of the period after 1865 was overproduction. As Willard Cochrane observes, total farm production increased 53 percent between 1870 and 1880. At the same time, however, this country's population increased only 25 percent. Farmers had to find new outlets—either through exports, industrial use, or increased food consumption. Even though both exports and home consumption increased, that rise was not sufficient to keep farm prices and farm incomes from dropping (Cochrane 1979, 94).

Although all businesses experienced economic fluctuations during the post–Civil War period, farmers found it particularly hard to adjust. When a farmer faced a problem of surplus and low prices, he could not lay off employees—actually most were members of his family—and he could not cut back production by refusing to harvest his crops. He could merely try to raise even more corn, or oats, or pork, hoping to earn enough simply to break even (Shannon 1957, 51).

Perhaps the most troubling single issue of the period was transportation, which meant, in effect, railroads. As agricultural historian John Hicks has commented, for midwestern farmers, their major problem in the post–Civil War period *was* the railroads. Hicks explained that "When the average Middle Western farmer living west of Chicago talked about monopolies and trusts, he was thinking primarily of the railroads." Track construction was often the first economic activity to appear in the region; the railroad companies imported the settlers and often determined what they paid for land. In transporting farm products to market, railroads settled to a large degree what profit margin the farmers made. Moreover, as the first major businesses in the area, the railroads were most visible and therefore the most apt to draw fire when things went wrong on the farm (Hicks 1949, 1–3).

By 1870 farmers in the eastern half of the Middle West had access to well-developed train lines, assuring a year-round system of transportation, while farmers in the extreme western Middle West were still awaiting railroad development. In Iowa, located in the middle of the region, four major east-west lines had been completed by 1870, bringing rail service to all parts of the state. In South Dakota's West River country, by contrast, the Milwaukee Road did not begin laying track until after 1900. Nonetheless, by the end of the 1870s complaints against the lines were heard in all major parts of the Middle West. These complaints centered on a few fundamental issues. Basically farmers viewed the railroads as huge monopolies that had gotten out of hand, depriving farmers of a sizeable portion of their fair profits. Farm patrons deeply resented the railroads' practice of charging more for short hauls than for long hauls. Freight rates depended largely on the degree of competition, and where a single line served an area, freight prices typically would be higher. In Minnesota the wheat rate on the Northern Pacific from Fargo to Duluth was almost twice as high as that from Minneapolis to Chicago (about twice the distance of the former route). That shippers on the latter route had more than one railroad from which to choose was of considerable importance (Shannon 1945, 300).

Farmers in eastern Iowa experienced the long haul-short haul controversy because of a somewhat different situation related to competition between the Mississippi River shipping interests and the midwestern railroads centered in Chicago. Before the Civil War much of the agricultural produce of the Upper Mississippi Valley was sent down the river to St. Louis. The Civil War disrupted that pattern; stepping in to fill the void were four railroads that stretched westward from Chicago to the Mississippi. After the war, when St. Louis tried to regain its former position by reinstituting shipment of farm products by water, the railroads worked hard to counter this action. One practice engaged in by the railroads was to charge Iowa farmers cheaper rates if they shipped from an interior point in the state all the way to Chicago rather than shipping only as far as the Mississippi. Frequently the haul to Chicago was more than double the length of the haul to the river (Belcher 1947, Chap. 10).

After the Civil War, as farmers in the prairie and Plains states increased their production, they also experienced frustration with marketing problems. Earle Ross points out in *Iowa Agriculture* that "the farmer, both as producer and consumer, felt he was the victim of an unsound and unfair marketing system." Activities such as cooperative marketing (as proposed by the Grange) seemed to be the solution. Farmers were particularly suspicious of elevator owners, although they obviously had no choice but to deal with them. Through the latter nineteenth century farmers became increasingly angry over what they viewed as unfair treatment by elevator owners and other marketing agencies. John Schlebecker believes that farmers suffered in the 1870s, 1880s, and 1890s because they were unwilling to adjust to changing conditions in the marketplace. He believes that they attempted to solve their problems, which he sees as "really troubles of the marketplace" by joining new organizations such as the Grange. Schlebecker points out that sometimes efforts to solve these problems worked, but sometimes they did not. Moreover, Schlebecker explains that farmers believed the solutions lay in expansion of the money supply, railroad legislation, and other agrarian proposals. In fact, Schlebecker concludes, the solutions were not that simple (Ross 1951, 98; Shannon 1945, 179; and Schlebecker 1975, 161–63).

Another scholar, however, has pointed out that although farmers did turn to organizations for cooperative marketing and increased political power, they continued to deal diligently with the new problems posed by commercialism. Adam Ward Rome in "American Farmers as Entrepreneurs, 1870–1900" asserts that farmers understood the need for forecasting of crops and market potential. Rome writes: "Acknowledging that agriculture had become a business [in the 1870s and 1880s] farmers discussed the risks and responsibilities of enterprise. They recognized the need to anticipate the market, to forecast the future. They exalted forecasting skill as the critical determinant of success of failure. They urged improved forecasting as the remedy for unprofitable agriculture." But by 1890, Rome concludes that farmers' "faith in self-help was shaken." Even though many had worked hard at forecasting the market, many had failed (Rome 1982, 466–67).

In their response to both economic and social grievances of the post–Civil War period, particularly against the railroads, midwestern farmers joined a number of protest groups. They would flock first to the Grange in record numbers but would later join in substantial numbers the Farmers' Alliance and the Populists attempting to improve their economic standing. Their action marked the first major agrarian uprising in the nation (Cochrane 1979, 95).

The Grange, founded in 1867, initially worked toward regulating railroads. In the 1870s a number of midwestern states—Iowa, Minnesota, Wisconsin, and Illinois—passed the so-called Granger laws, which brought about limited regulation. In Iowa the bill that emerged from the General Assembly in 1874 established a schedule of maximum freight and passenger rates for the railroads within the state. Although labeled the Granger Law, this legislation was not what the Grangers had in mind. Rather, the organization had recommended the creation of a board of commissioners to supervise the railroads operating in the state. Grange members were not particularly pleased with the bill that did emerge. Strongly opposed by the railroads, this legislation was repealed in 1878. The Granger laws in other states suffered the same fate (Throne 1953, 979–98).

The Grange also worked to serve farm families beyond the area of regulation. Founded by Oliver Hudson Kelley, Grange programs were designed to improve the quality of farm life through educational and social activities as well as through economic programs. At its convention in 1876, the National Grange proposed "working together, buying together, selling together, and generally acting together for . . . mutual protection and advancement. . . ." In effect, the Grange proposed to eliminate the middleman to decrease prices farmers had to pay. The Grange instigated cooperative buying through a system of state agents (Shannon 1945, 329–31).

The Grange faced great difficulty, however, when it moved into the area of manufacturing farm implements. In Iowa in 1873 the state Grange purchased the Werner Harvester Factory. During the next year, the Grange sold 250 machines at about half the original price. Iowa Grange members also attempted to manufacture plows, binders, and harrows. It soon became apparent that this new undertaking would fail. According to one historian, Grange personnel knew too little about the operation and did not have sufficient capital to carry it out. Moreover, the market was not sufficiently large to support the program (Beinhauer 1959, 225–26).

The Farmers' Alliance represented the next major group organized by farmers. In 1880 Milton George, publisher of the *Western Rural* at Chicago, helped found the National Farmers' Alliance, often known as the Northwestern Alliance. George had been strongly influenced by the Grange, particularly its educational and social features, but stressed an economic and political approach. The group held immediate appeal for farmers in the central and western Middle West. A drought in Iowa, Nebraska, and Kansas in 1881 and 1882 brought even greater Alliance membership in those states. By 1890 Kansas alone contained 130,000 members, and other midwestern states like the Dakotas and Minnesota also had large memberships. By 1900 the power of the Northwest Farmers' Alliance was evident in the North Central states. Members had demanded federal ownership and operation of railroads and free coinage of silver. In addition, farmers called for an increase in the money supply, particularly the establishment of a flexible greenback system rather than the current system of national bank notes, and a graduated income tax (Shannon 1957, 63; Shannon 1945, 313–18).

In the 1890 congressional elections, Alliance members were active in both the Plains states and the South. In that year Mary Elizabeth Lease issued her famous statement that Kansas farmers should "raise less corn and more Hell." Two years later the People's Party or Populist Party came into existence. They continued to demand public ownership and regulation of railroads, free coinage of silver, and a graduated income tax. To this list they added the creation of postal savings, government ownership of telegraph and telephone systems, and abolition of land monopolies. The party selected James B. Weaver of Iowa as their presidential candidate. Although Weaver was not viewed nationally as a major candidate, he did receive 9 percent of the total votes cast. The Populists would benefit politically from the Depression of 1893 and would run candidates again in the 1894 election in which they would increase their votes by half. In 1896, however, the Populists joined with the Democrats in supporting William Jennings Bryan as their presidential candidate on the free silver issue and were clearly defeated at the polls. Although the Populist Party attempted to maintain their organization for the next decade, they had little success; they later joined with the Democrats and by 1910 had ceased to exist (Shannon 1945, 321–22. 326).

While middle western farmers suffered through drought and depression, changes were going on in the areas of scientific and technological development that would have major, far-reaching consequences. In 1862 Congress passed the Morrill Act, which gave federal support to the creation of land-grant agricultural and mechanical arts colleges. During the previous forty years, many states and local agricultural societies had pushed for such a development. Their efforts had some effects as an agricultural college in Michigan had been established in 1855. One year earlier, the Pennsylvania legislature had created a Farmers High School, now the Pennsylvania State University. The Morrill Act determined that each state accepting the terms of the act should receive 30,000 acres for each senator and congressman. In the Middle West seven states eventually opted to create separate agricultural and mechanical arts colleges rather than turn the land grant over to the state universities. Even after the land grants had been made, colleges still faced problems with staffing, endowment, support, and maintenance. Professors of agriculture had to be located, or perhaps more precisely, had to be trained. In some colleges, administrators and boards of trustees favored teaching only those courses that pertained to either agriculture or mechanical arts. In contrast to what Willard Cochrane has called the "narrow gaugers," there were those who wanted to carry over the broad, classical educational format of the state universities. Given these difficulties, the development of the agricultural colleges was slow in the 1870s and 1880s. Yet by 1900 many colleges had departments such as animal husbandry, dairy husbandry, agricultural chemistry, soil science, horticulture, and entomology (Cochrane 1979, 242–43).

In 1887 Congress passed the Hatch Act that initially gave each state $15,000 annually to support the work of agricultural experiment stations. This step might be viewed as the second stage in the development of agricultural science and technology. In some cases the money was given to independent stations, but most often it went to land-grant schools. Minnesota, for example, established an experiment station two years before federal funding became available. In 1885 the Minnesota Board of Regents established a station at the University. The objectives of the program included providing a laboratory where future farmers could be trained in practical agricultural techniques, providing employment to aid students financially with their schooling, and conducting agricultural research. Over the next twenty years the program expanded considerably. The first divisions or departments were Agriculture, Agricultural Chemistry, Entomology and Botany, Horticulture and Veterinary. By 1910 the program would also include the areas of Dairy Husbandry, Agricultural Engineering, and Forestry. By that date the program also included six branch stations for agricultural research. According to one historian, two features helped account for the early success of the program: the staff "strongly identified with the farmers of the state," making considerable effort "to learn of farm problems and to show farmers ways of solving these problems"; and as the shift occurred to a science-based institution, a "more in-depth analysis of problems" led to greater farm productivity (Cochrane 1979, 242–43; Fitzharris 1974, 215).

At the same time, midwestern farmers experienced considerable change in the type of farm equipment used. Fred Shannon observes that "the phenomenal expansion of Prairie-state agriculture was dependent on the latest improvements in farm implements." Moreover, the expansion of agriculture into the Great Plains "would not have been possible without new machines." In his study of farming on the Illinois and Iowa prairies, Allan Bogue entitles one chapter, "How to Farm Sitting Down," which properly reflects the changes under way. In his discussion of corn production, Bogue comments that corn planters, two-horse cultivators, and power shellers "were the major exhibits of corn technology down to the 1800s." He notes that Illinois and Iowa prairie farmers were using machinery under "extremely favorable conditions" with the result that farm productivity probably increased more in these states than elsewhere in the nation. Threshing machines were available by the mid-1800s, but most prairie farmers did not own one. The practice that developed was either custom threshing or threshing rings. Bogue makes the important point that although new machines were continually coming on the market during the latter half of the nineteenth century, farmers seldom "outfitted themselves completely at any one time. . . ." Rather, some bought used equipment and some purchased equipment jointly with relatives or neighbors. Improved farm equipment and its greater availability did produce significant change. Bogue has noted that these changes, along with other factors such as expanded acreage under cultivation and better transportation, increased agricultural productivity by 32 percent between 1870 and 1910 (Shannon 1945, 125; Bogue 1968, 162–68).

After 1900 farmers in the Middle West would experience what many historians have called the "new agriculture." The United States Secretary of Agriculture James Wilson commented in his 1909 annual report: "The value of farm products is so incomprehensively large that it has become merely a row of figures." The Country Life Commission in the same year observed, "There has never been a time when the American farmer was as well off as he is today. . . ." Farm prices were up and consequently most farmers, tenants included, were doing well. Farmers experienced what Saloutos has called "a phenomenal rise in the price of land." As Saloutos points out in *Agricultural Discontent in the Middle West, 1900–1939,* farmers made their money more from the unearned increment that came from the ownership of farmland than from good farming methods (Saloutos and Hicks 1951, 2–23).

Even in this prosperous period, farm tenancy in the Middle West was a cause for major concern. Theodore Roosevelt spoke of this issue in 1907 when he said: "Nothing is more important to this country than the perpetuation of our system of medium-sized farms worked by their owners. We do not want to see our farmers sink to the condition of the peasants in the old world, barely able to live on their small holdings, nor do we want to see their places taken by wealthy men owning enormous estates which they work purely by tenants and hired servants." During the later 1800s and early 1900s, tenancy rates in the Middle West rose at a rate faster than anywhere except the South. In Iowa the tenancy rates rose from roughly 24 percent in 1880 to roughly 38 percent in 1910. By 1914 authorities estimated that in the Corn Belt about 40 percent of the farms were operated by tenants. In parts of Illinois the rate exceeded 50 percent (Saloutos and Hicks 1951, 20–23).

While farm families faced almost continual economic difficulties from 1867 until near the turn-of-the-century, those difficulties do not tell the whole story of midwestern agriculture. During that period the region continued to be settled, and by 1900 the Middle West exhibited many common cultural characteristics. All states attracted not only native-born but also foreign-born settlers. Foremost among the former were people born in states to the north and east of the region. Iowa, for example, found most of its settlers coming from Pennsylvania, Ohio, and Indiana. Most foreign born came from western and northern Europe and the British Isles. Wisconsin had one of the highest percentages of foreign born, reaching 39 percent. Iowa included nearly 20 percent foreign born. By 1870 Nebraska contained 25 percent foreign born; if one includes both foreign-born parents and their children in one groups, however, the figure reached 54 percent. North Dakota had the highest rate of first and second generation immigrants, 78 percent (Sage 1974, 93; and Luebke 1977, 405–7). Immigration resulted in small enclaves of ethnic groups scattered throughout the region, many still visible today. Among foreign-born settlers, Germans dominated with Scandinavians following close behind. The Dakotas, for example, attracted large numbers of Germans and Norwegians. Nebraska contained a large number of Czechs along with Germans and Scandinavians.

Each rural ethnic enclave typically had its own church. Perhaps because of their numerical dominance, the Germans showed the greatest religious diversity. In Kossuth County, Iowa, some Germans attended German Lutheran churches while others attended Catholic or Dutch Reform congregations. About seventy-five miles away, the German-Americans who dominated Carroll County were almost 100 percent Catholic. In other parts of Iowa, rural churches bore names of German Methodist, German Baptist, and German Presbyterian. Norwegians and Swedes tended to establish Lutheran churches while many Irish and Czech settlers belonged to Catholic churches. The church would be one of the most durable immigrant institutions. As Frederick Luebke has observed: "Of all important institutions, the church was the easiest to establish, the most effective in its mission, and hence the most long-lived. Unlike most ethnic institutions, the church survived the transition to an English-speaking society" (Luebke 1977, 411).

While foreign-born farmers, particularly the Germans, have long been perceived as good farmers, many other questions about them have been raised: How quickly did they assimilate into the larger society? Did they continue to use their Old World farming methods or did they quickly adopt local techniques? Did they join protest groups such as the Farmers' Alliance? In general how much cultural conflict existed in the Middle West in the post–Civil War period?

Recent studies have shed some light on these questions, conveying a general sense that there was less cultural conflict than historians had earlier assumed. Immigrant farmers tended to shed their Old World views on farming quickly and adopted local methods. Geographer Robert C. Ostergren observed in his study of Swedish immigrants in Minnesota that although they maintained their "traditional ties with the church, family, and homeland . . .," when it came to occupations or tilling the soil, "traditions and customs of the past were abandoned almost immediately" (Ostergren 1981, 400).

Recent studies of immigrant farmers' involvement in agricultural protest groups have cast doubt on earlier assumptions. In his study of the Populists in Marshall County, South Dakota, John Dibbern has shown that foreign-born farmers were more apt to join the Farmers Alliance than native-born farmers, by 51 to 39 percent. Dibbern writes that in contrast to Richard Hofstadter's view toward Populism in *Age of Reform,* the movement in South Dakota was not regarded as nativistic. Rather than being cast as the scapegoats of the Populist movement, foreign men were its supporters (Dibbern 1982, 677, 682–83).

Public schools were established soon after settlers moved into any new area. The Land Ordinance of 1787 insured that at least one school would be established in each township. The one-room country schoolhouse became a common feature of the landscape as the frontier moved west from Ohio into the central and western parts of the region. Midwesterners put a high premium on education—or at least on the basic rudiments of learning. Public schools were not only to educate young children but also to instill in them the prevailing morality (Atherton 1954, Chap. 3).

Even though all areas of the Middle West established churches and schools, farm society in the Great Plains was considerably distinct in one respect. Out on the Plains a major consideration of farm families was their isolation. Newcomers in the Dakotas, Nebraska, and Kansas quickly discovered that to succeed in farming, one needed to farm larger acreages; that fact meant in effect fewer families on the Plains. Some accounts relate that Plains dwellers might go for several months without seeing anyone but members of the immediate family. This condition led to extreme loneliness, especially for women, and has been sensitively described by Middle Border writers like Hamlin Garland. In contrast, women in the more rapidly settled prairie states like Minnesota and Iowa did not experience as much isolation. Emily Hawley Gillespie, a farm woman who came to northeastern Iowa in the 1860s, found that almost every day she had the opportunity to visit with relatives or neighbors. She still sorely missed her parents in Michigan, but this was not loneliness caused by isolation (Schwieder 1980, 165).

Having close neighbors did not alter the harshness of farm life throughout the region and the excessive physical labor it entailed in the post–Civil War period. All members of the family worked hard. Farm women found their lives were changed least of all by the new technology and by the ever-growing interest in scientific agriculture. Farmers seemed generally to have accepted the view that an investment in new machinery or the construction of a larger barn were wise expenditures that would bring some economic rewards. But building a new house, taking on a hired girl, or buying equipment for the home did not fall into the same category. The result was that farm women for the most part continued to do their work in the same way as their mothers and grandmothers before them (see USDA 1915a, 1915b, 1915c).

This fact was significant in its own right, but it took on even greater importance in light of another major concern of farm families—many young people leaving the farm. In 1913 Herbert Quick, drawing on years of experience with farm life, would write that the flight to the cities was really a "woman problem." He explained that many farm women were discontented with their lives and, in turn, urged their husbands to move to town. Moreover, they counseled their children to leave the farm and find employment elsewhere. The Country Life Commission in 1909 would respond to women's concerns by advocating lighter work loads and more social outlets (Schwieder 1983, 108–9).

In at least some areas farm women worked to quell the flight to the cities. In Indiana farm women began addressing this problem in the 1890s. By then some families had organized the Indiana Farmers' Reading Circle, which had grown out of earlier farm institutes. At their meetings farm parents discussed the "difficulty of keeping the younger generation on the farm in the face of the growing attraction of city life and work." By the early 1900s one way parents dealt with the problem was by forming boys' and girls' clubs. Seed corn was often distributed to boys' clubs and prizes awarded for growing the best corn. Young women formed sewing, canning, and poultry clubs. By 1908 thirty-five Indiana counties recorded a combined boys' club membership of 5,000 (Phillips 1968, 144–45).

While some farm women in the older, more settled farm states like Indiana did not feel inclined to leave the farm, many other women did. During the long agricultural depression of the latter nineteenth century, farm women appeared not only disillusioned with farming, but with farm life. Unhappiness with farming always left room for some optimism—next year's crop might be better and prices might rise. Disenchantment with farm life, however, implied a different set of grievances that could not be eliminated so easily by an upturn in farm prices or a bumper corn crop. Disillusionment involved a whole set of concerns including limited education for children, lack of conveniences in the farm home, lack of opportunities for socializing, and the unrelenting routine of hard physical work from sun-up to sun-down.

By 1910 midwestern farm families had come to share the prosperity of the nation's general farm economy. Some nagging concerns remained, particularly increasing tenancy rates and declining farm populations, but for the moment at least the midwestern farm economy was doing well. Between 1865 and 1910, in effect, the region had come full circle: in 1865, farmers were thriving, largely through increased wartime demands. Forty-five years later, the region once again was experiencing prosperity. The period in-between, however, witnessed the first major agrarian uprising, a time when major agricultural issues such as debt, high interest rates, high transportation rates, and marketing problems usurped the interests of the region's farmers. Farmers spent countless hours organizing, meeting, and protesting their treatment at the hands of government and big business. Although farmers learned a great deal about the mechanics of organization and protest, it is difficult to see how much of this organization produced long-term change. When prosperity returned, it seemed to come through influences outside the farmers' control.

By 1910 all regions of the Middle West were settled and subregional variations would be evident in every midwestern state. As the eastern part of the region became increasingly industrialized, the focus on agriculture moved westward. Coming to the foreground were the regions of the Corn Belt, the Wheat Belt, and dairy production. From a national perspective the Midwest was the region most strongly identified with agriculture. Crop and livestock production would change and different farm problems would surface, but the people who resided here would continue to raise the majority of the nation's food supply. Farming had moved into a new era dominated by scientific agriculture including experiment stations, fledgling extension services, and agricultural departments in land-grant colleges. No longer was there any room for the nostalgic view of the sturdy yeoman farmer that had dominated the nation's consciousness in the nineteenth century. While it may be true as Theodore Saloutos has observed that "Agricultural discontent [in the Middle West] was chronic and endemic . . .," by 1910 at least, that discontent had been submerged and farm families proceeded to enjoy the fruits of their labor (Saloutos and Hicks 1951, 30).

## REFERENCES AND FURTHER READINGS

Atherton, Lewis. 1954. *Main Street on the Middle Border.* Bloomington: Indiana Univ. Press.

Beinhauer, Myrtle. 1959. "Development of the Grange in Iowa, 1868–1930." *Annals of Iowa* 34(Apr.):597–618.

Belcher, Wyatt Winston. 1947. *The Economic Rivalry between St. Louis and Chicago, 1850–1880.* New York: Columbia Univ. Press.

Bogue, Allan G. 1968. *From Prairie to Cornbelt: Farming on the Illinois and Iowa Prairies in the Nineteenth Century.* Chicago: Quadrangle Paperbacks.

Buck, Solon J. 1913. *The Granger Movement.* Cambridge: Harvard Univ. Press.

Cochrane, Willard W. 1979. *The Development of American Agriculture: A Historical Analysis.* Minneapolis: Univ. of Minnesota Press.

Danbom, David B. 1979. *The Resisted Revolution: Urban America and the Industrialization of Agriculture, 1900–1930.* Ames: Iowa State Univ. Press.

Dibbern, John. 1982. " 'Who Were the Populists?' A Study of Grass-Roots Alliancemen in Dakota." *Agricultural History* 56(Oct.):677–91.

Faragher, John. 1981. "History from the Inside-Out: Writing the History of Rural Women." *American Quarterly* 33(Winter):537–57.

Fink, Deborah. 1986. *Open Country, Iowa: Rural Women, Tradition and Change.* Albany: State Univ. of New York Press. SUNY Series in the Anthropology of Work, ed. June Nash.

Fite, Gilbert. 1977. "Great Plains Farming: A Century of Change and Adjustment." *Agricultural History* 51(Jan.):244–56.

Fitzharris, Joseph C. 1974. "Science for the Farmer: The Development of the Minnesota Agricultural Experiment Station, 1868–1910." *Agricultural History* 48(Jan.):202–14.

Garland, Hamlin. 1899. *Main Traveled Roads.* New York: Harper.

Hicks, John. 1949. "The Legacy of Populism in the Western Middle West." *Agricultural History* 23(Oct.):225–39.

______. 1961. *The Populist Revolt: A History of the Farmers' Alliance and the People's Party.* Lincoln: Univ. of Nebraska Press.

Hudson, John C. 1985. *Plains Country Towns.* Minneapolis: Univ. of Minnesota Press.

Jensen, Richard. 1971. *The Winning of the Midwest: Social and Political Conflict.* Chicago: Univ. of Chicago Press.

Luebke, Frederick. 1977. "Ethnic Group Settlement on the Great Plains." *Western Historical Quarterly* 8(Oct.):405–30.

Madison, James, ed. 1988. *Heartland: Comparative Histories of the Midwestern States.* Bloomington: Indiana Univ. Press.

Madson, John. 1982. *Where the Sky Began: Land of the Tallgrass Prairie.* San Francisco: Sierra Club Books.

Marcus, Allen I. 1985. *Agricultural Science and the Quest for Legitimacy: Farmers, Agricultural Colleges and Experiment Stations, 1870–1890.* Ames: Iowa State Univ. Press.

Nelson, Paula. 1986. *After the West Was Won: Homesteaders and Townbuilders in Western South Dakota.* Iowa City: Univ. of Iowa Press.

North, Douglass C. 1959. "Agriculture in Regional Economic Growth." *Journal of Farm Economics* 41(Dec.):943–51.

Ostergren, Robert C. 1981. "Land and Family in Rural Immigrant Communities." *Annals of the Association of American Geographers* 71(Sept.):400–11.

Phillips, Clifton J. 1968. *Indiana in Transition: The Emergence of an Industrial Commonwealth.* Indianapolis: Indiana Historical Society, vol. 4, History of Indiana Series.

Rome, Adam Ward. 1982. "American Farmers as Entrepreneurs, 1870–1900." *Agricultural History* 56(Jan.):37–49.

Ross, Earle. 1951. *Iowa Agriculture.* Iowa City: State Historical Society of Iowa.

Sage, Leland. 1974. *A History of Iowa.* Ames: Iowa State Univ. Press.

Saloutos, Theodore, and John D. Hicks. 1951. *Agricultural Discontent in the Middle-West, 1900–1939.* Madison: Univ. of Wisconsin Press.

Schell, Herbert S. 1968. *History of South Dakota.* Lincoln: Univ. of Nebraska Press.

Schlebecker, John. 1975. *Whereby We Thrive: A History of American Farming, 1607–1972.* Ames: Iowa State Univ. Press.

Schwieder, Dorothy. 1980. "Labor and Economic Roles of Iowa Farm Women, 1840–1880." In *Farmers, Bureaucrats, and Middlemen: Historical Perspectives on American Agriculture,* ed. Trudy Huskamp Peterson. Washington, D.C.: Howard Univ. Press.

______. "Rural Iowa in the 1920s: Conflict and Continuity." *Annals of Iowa* 47(Fall):104–15.

Shannon, Fred. 1945. *The Farmers' Last Frontier: Agriculture, 1860–1897.* New York: Farrar and Rinehart, vol. 5, Economic History of the United States Series.

______. 1957. *American Farmers' Movements.* Princeton: D. Vann Nostrand.

Swierenga, Robert. 1980. "Ethnicity and American Agriculture." *Ohio History* 89(Summer):323–44.

Throne, Mildred. 1953. "Repeal of the Iowa Grange Laws, 1878." *Iowa Journal of History* 51(Apr.):97–130.

U.S. Department of Agriculture. 1915a. "Social and Labor Needs of Farm Women," Bull. no. 103. Washington: Government Printing Office.

______. 1915b. "Domestic Needs of Farm Women," Bull. 104. Washington: Government Printing Office.

______. 1915c. "Educational Needs of Farm Women," Bull. no. 105. Washington: Government Printing Office.

Webb, Walter Prescott. 1931. *The Great Plains.* New York: Grossett and Dunlap.

Woodward, Margaret L. 1963. "The Northwestern Farmers, 1868–1876; A Tale of Paradox." *Agricultural History* 37(July):134–42.

# PART II
# The South

CHAPTER 6

# Staple Crops and Slave Plantations: Alternative Perspectives on Regional Development in the Antebellum Cotton South

DAVID F. WEIMAN

IN 1800 THE POPULATION of the Southern states was largely concentrated along the eastern seaboard.[1] The process of settlement had proceeded farthest in the Upper South. The western expansion of the original Tidewater colonies and the southern migration along the Shenandoah valley filled in the area east of the Appalachians and spilled over into the Ohio and Tennessee valleys. In the Lower South, in contrast, settlements hugged the coast from Charleston to the Savannah River, although the western expansion of this population and the continued southern movement along the valleys formed scattered communities in the interior.

Over the next sixty years the waves of economic and geographic expansion shifted the settlement frontier west of the Mississippi River and formed the Cotton South, the core of the antebellum plantation system and the Confederate States of America. Located on the periphery of the expanding Atlantic economy, the antebellum Cotton South depended critically on its economic ties to metropolis regions. The growth of markets and the flows of capital and population from the latter fueled the process of territorial and economic expansion. Moreover, the integration of the Lower South into the expanding world market and location within that division of labor structured its subsequent economic development.

To explain the economic development of regions of recent settlement, economic historians and development economists have offered several hypotheses, but by far the most popular is a variant of the staple thesis, which blends geographical determinism and staple arguments.[2]

Regional development, it implies, depends on the natural resource endowment, which determines the choice of crops, and the characteristics of the staple crop. The staple thesis has received rather wide currency among historians of the American South. More importantly, it has framed the issues that have occupied historians of antebellum slavery and the plantation economy, even those critical of its premises.

Despite its simple logic the staple thesis has been sharply criticized, even by its initial proponents. In evaluating the contribution of this tradition, both Richard Caves (1971, 423, 432) and Albert Hirschman (1975, 94–95) deny any direct link between a region's staple crop and its dominant labor system and organization of production. Instead, they invert its logic by asserting that the methods and organization of staple crop production depend on the class structure and political culture.[3]

This critique suggests an alternative, institutional thesis explaining the pattern of economic development in regions of recent settlement. Although it has not been formulated as schematically as the staple thesis, this perspective is implicit in two strands of the literature on the slave South. The first emphasizes the impact of slavery on Southern economic development and, in turn, delineates the South by its spatially "segregated" labor market (Wright 1986, 6–7; Genovese 1961, Chap. 1; Odum 1936, Pt. I). The second treats the plantation as a formative settlement institution or mode of production, which shaped the region's agricultural, economic, and political development (Thompson 1975, 17–19; Mandle 1978, Chap. 1).

In this chapter I develop the institutional perspective on regional development in the antebellum Cotton South through a critique of the staple thesis. The first section outlines the staple thesis as applied to the region and defines the principal issues addressed in the literature. The remaining sections reexamine these issues and show that the South's peculiar institutions—the plantation and slavery—rather than natural resources and technology, constituted a distinct regional economy and political culture.

These alternative theories of regional development, this analysis shows, lie at the center of the debates over slavery and the plantation system. They imply, moreover, two distinct conceptions of the "South" as a region. According to the staple thesis, the South is simply a distinct bundle of natural resources, while in the institutional hypothesis, the plantation and slavery made the South different.

## CLIMATE, STAPLE CROPS, AND REGIONAL DEVELOPMENT IN THE ANTEBELLUM COTTON SOUTH

### *The Staple Thesis*

In *Life and Labor in the Old South,* Ulrich Phillips synthesized his pioneering research on the economy and society of the slave South. He formulated, moreover, an enduring paradigm in Southern historiography. "The weather," Phillips (1929, 3) asserted, "has been the chief agency in making the South distinctive." Endowed with a "mild and even warm" climate, southern farmers were insured a long growing season varying from six to seven months in the states of the Upper South to nine months along the coast of the Lower South (Emerson 1911, 21–3. See also Hilliard 1986, 13–18). These conditions "fostered the cultivation of the staple crops" that delineated the agricultural subregions of the Old South and gave rise to the plantation system and slavery.[4]

This staple thesis, as subsequently elaborated by Robert Baldwin, Douglass North, and others explains the origins of the slave plantation, the agrarian class structure, and the enclave development of the antebellum South by the physical properties of its staple crops. In Baldwin's model the plantation takes root in more tropical climates, where ecological conditions favor the cultivation of "plantation crops" (Baldwin 1956, 166–67). These crops are characterized by their technical conditions of production: simple, repetitive tasks, which are intrinsically difficult to mechanize; and increasing returns to scale in cultivation or processing. They are often "multiple-day" crops with long periods of peak labor demands (Earle 1978, 55–56). Plantation crops, therefore, favor production on a large scale, using some form of bound labor.

Plantation crops, also referred to as "bad" staples in the literature, spawn few backward and forward linkages to those industries that give rise to local development (Perloff and Wingo 1961, 200–201. See also North 1955; Watkins 1963, 144–46; Caves 1971, 433–37; Hirschman 1975, 72–77). These crops have low volume and weight relative to their value, or can be considerably reduced in these dimensions with simple processing. Additionally, many plantation staples are nonperishable. These properties restrict investments in transport improvements and specialized processing and storage facilities, while the technology of staple production diminishes demands for improved agricultural implements.

Together, the agrarian class structure in plantation regions and the lack of technical linkages sharply curtail the local multiplier effects of staple exports and the growth of intraregional trade (Perloff and Wingo 1961, 203–4).

### *The Culture of Cotton*

With its shorter growing season, tobacco was cultivated in the states of the Upper South, while along the coast of the Lower South the more tropical climate favored the production of rice and sugar. These zones bounded the Cotton South, a region specializing in the cultivation of short-staple or upland cotton. Below the line dividing the Upper and Lower South, approximately 37° North latitude, the growing season was at least 200 days long, sufficient to cultivate the fleecy staple (U.S. Department of Agriculture 1896, 143–52; U.S. Department of Agriculture, 1912, 36, 92–100; Brown 1938, 266–72). Moreover, the uniform daily temperature and levels and distribution of rainfall during the growing season made the region "ideally suited" to cotton culture (Bennett 1921, 17). Within the climatic limits of the Cotton South, the cultivation of the staple was concentrated in subregions that had fertile soils, level-to-rolling lands, and natural transport routes (Vance 1932, 14; U.S. Census Office 1884; U.S. Department of Agriculture 1896; Ransom and Sutch 1971).

The cycle of cotton culture, from preparing fields for planting to hauling bales to nearby depots, lasted a year (Turner 1865, Chap. 1; Phillips 1969a, 207–10; Moore 1958, 42–49; Anderson 1974, 53–60, 76–79). Its most distinguishing characteristics, at least those emphasized by staple theorists, were the peak labor demands during the postplanting and harvest periods and the simple steps in preparing the crop for shipment to coastal ports. From late April or early May to the end of July, the growing plants were gradually "thinned" to a single stand and repeatedly "cultivated" with a scraper plow and hoed to stem the spread of grasses and to loosen the soil. Because the cotton plant did not ripen uniformly, the fields were harvested continuously until yields tapered off, usually from September to the end of November. Picking cotton by hand required both dexterity and stamina, as it was a race against the elements, harsh weather, and the various diseases and insects that damaged the lint or blooming plants.

After the harvested crop was dried, it was ginned to remove the seeds and "trash" from the lint (Phillips 1969a, 207; Gray 1958, 704–6; Moore 1958, 47–50, 173–74). The lint cotton was then pressed into bales and hauled to nearby ports or stations. According to contemporary estimates, three hands operating one cotton gin could produce between two to six bales of lint cotton daily. Improvements in cotton presses and the application of steam power increased the average weight of a standard bale from 250 pounds to over 400 pounds by the end of the period (Gray 1958, 704–6; Watkins 1969, 299). Compacting bales on or near the farm facilitated local transport and reduced the demand for further processing at coastal ports.

The graph in Fig. 6.1 presents the seasonal distribution of labor requirements in cotton production for several subregions of the Cotton South. These data, based on contemporary and postbellum sources, conveniently summarize the annual cycle of cotton culture.[5] The periods of peak labor demands were of roughly equal length, approximately three months, and required similar inputs of labor per acre of cotton. Processing and transporting the staple occurred during the slack period in December and January and used substantially less labor.

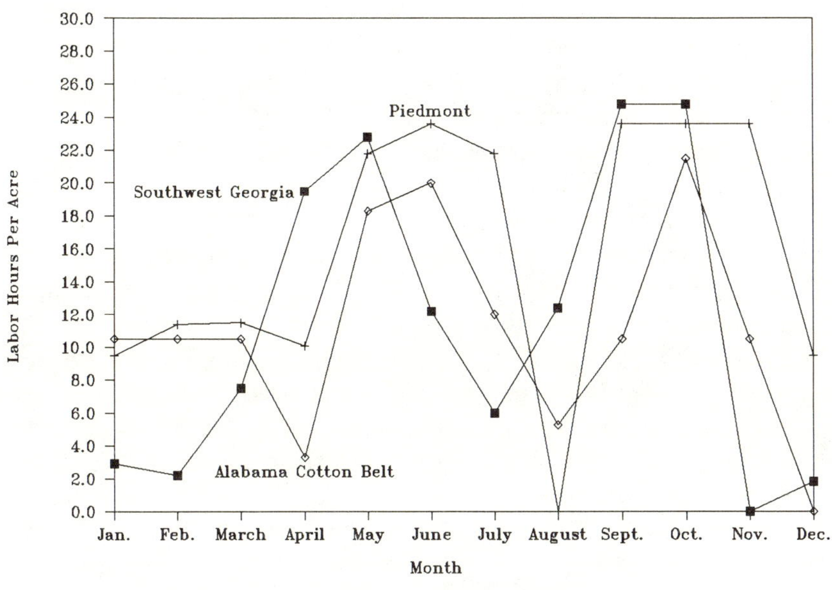


Fig. 6.1. Seasonal pattern of labor requirements in cotton production.

*The Natural Efficiency of the Slave Plantation*

The culture of cotton, according to the staple thesis, directly shaped the regional development of the Lower South through its effect on optimal farm size. Slave plantations, it asserts, reaped substantial productivity gains from the cultivation of cotton on a large scale. Their greater technical efficiency, relative to smaller-scale slave and free farms, explains the economic viability of the plantation and slavery as institutions.

This claim raises two distinct but related issues: whether the technical efficiency of the slave plantation derived from economies of scale in cotton production or from the system of labor control. The earlier works on the subject, most notably the writings of Phillips and Lewis Gray, clearly formulate the alternative positions, and so offer opposing explanations of the competitive advantages of the slave plantation. Through the greater division and specialization of labor, Phillips argues, slave plantations realized economies of "effort and expense of administration" in production.[6] Gray, in contrast, denies the existence of economies of scale in cotton culture and instead emphasizes the greater coercion of slave labor on plantations, especially the harsh discipline enforced by the gang system.[7] He squares this view with the staple thesis by suggesting that the economic viability of slavery depended on the characteristics of cotton, an "all year" crop whose simple production process was easily routinized.

The controversial reinterpretation of antebellum slavery by Robert Fogel and Stanley Engerman, in effect, synthesizes these views and further elaborates the staple thesis. Like Phillips, Fogel and Engerman explain the greater efficiency of slave plantations by economies of scale in the cultivation of the region's principal staples: sugar, rice, cotton, and tobacco.[8] But, like Gray, they root these scale economies in the organization of slave labor into "highly coordinated and precisely functioning" work gangs. While dependent on the entrepreneurship of the planter class as well as the work ethic of the slaves themselves, the efficiency of the gang system, Fogel and Engerman note, ultimately derived from the characteristics of the staple crop. The division of labor among work gangs, they argue, yielded greater returns precisely because it exploited the technical "complementarities and interdependencies" in the cultivation of staple crops (Fogel and Engerman 1977, 290–91).

Planters, they argue, applied the principles of "scientific" management to the production of field crops and replicated the technical efficiency of a well-oiled machine (Fogel and Engerman 1974, vol. 1, 205–6; 1977, 291; Aufhauser 1973, 811–24; Metzer 1975, 134–40, 142–43). In the first three stages of production, preparing, planting, and "cultivating" fields, field hands were divided into specialized work gangs and integrated into an interdependent assembly-line production process. In operations that could not be organized into a division of labor, such as picking cotton, planters used the task system. They assigned field hands quotas based on "time-motion studies" and devised incentive schemes to promote competition among individuals and work gangs.

By reducing the interdependent stages of production into separate tasks that required different levels of strength, stamina, and skill, planters could fully utilize their slave labor force in the cultivation of field crops. Field hands were rated according to their physical characteristics and assigned to work gangs and tasks for which they were naturally suited.[9] The larger, more heterogeneous plantation labor force also improved the allocative efficiency of field hands, as planters could gauge the abilities of their slaves with greater accuracy and so assign them to the appropriate gangs and tasks.

These organizational features account for the extremely high labor productivity on slave plantations and their greater productive efficiency, as measured by Fogel and Engerman.[10] The gang and task systems enforced a "speed-up" or "[g]reater intensity of labor per hour," and economized on the costs of supervision (Fogel and Engerman 1977, 293–94). Overseers and drivers set the pace of production, and the interdependence among gangs and competition between hands regulated their performance, while minimizing the amount of direct supervision and physical coercion.

These methods, in turn, determined the minimum scale of efficient production. Given the labor productivity in each operation, field hands worked at different rates. To coordinate the activities of work gangs and maintain the rhythm of production, planters had to adjust the size of each gang so that all tasks were completed at roughly the same time. The number of field hands necessary to effect this organization was approximately 10, or the equivalent of 16 slaves. This figure, in effect, defined a plantation, for only those slaveholders who exceeded the threshold could fully utilize an overseer and so be liberated from having to perform manual labor (Fogel and Engerman 1974, vol. 1, 200–201; Metzer 1975, 143–44; Gray 1958, 498–99; Scarborough 1966, 8–12).

In identifying the source of the plantation's efficiency, Fogel and Engerman also explain why slave labor was utilized exclusively on plantations. Generalizing Domar's thesis about the origins of coerced labor, they argue that the gang system imposed "large nonpecuniary disadvantages" on workers that exceeded the productivity differential between plantations and free farms.[11] Consequently, without some form of coerced labor, field hands would demand higher wages as compensation for these harsh conditions, and plantation agriculture would not be viable. By judiciously combining force and incentives, planters could operate the gang system "without having to pay a premium . . . [greater than] the gains from economies of scale" (Fogel and Engerman 1974, vol. 1, 236–38; vol. 2, 155–62). At the same time, they emphasize that force alone was not sufficient to realize the greater efficiency of large-scale production "if that potential was not inherent in the production process."[12]

Several independent studies of nineteenth-century southern agriculture have corroborated the general findings of Fogel and Engerman. Donald Schaefer and Mark Schmitz reproduce their analysis on slave and free farms in the regions of the Old and New South and show that only slave farms reaped greater productivity from production on a large scale.[13] In the context of Reconstruction, Ralph Shlomowitz (1978, 85–87, 94–96) and Gerald Jaynes (1986, 164–66, 243, App. B) offer evidence that large-scale, centrally organized free plantations were technically more efficient than small, tenant farms.

Following a different line of argument, in fact one originally formulated by Gray, Carville Earle develops a further rationale for the use of slave labor in the cotton culture. Slave labor, Earle (1978, 55–56) argues, was more profitable in the cultivation of "many-day" staples, like cotton, that required intensive labor inputs over most of the growing season. Free labor employed on a seasonal basis, in contrast, was more economical in the cultivation of those staples that required concentrated labor inputs for a short period of time (see also Fields 1985, 5–6).

Finally, planters gained additional advantages by internalizing necessary costs of production, and these benefits are often attributed to the properties of cotton. The "natural" complementarity of cotton and corn in production enabled planters to specialize in cotton up to the levels dictated by labor constraints and still produce enough food to feed their slaves and livestock (Gallman 1970, 19–23; Wright 1978, 57–61). Also, because of the staple's simple processing requirements and nonperishability and the slack period after the harvest, planters saved both time and money by processing and storing their own crops and shipping them to interior ports (Phillips 1908, 9–10; Earle and Hoffman 1987, 165–66).

### *The Natural Basis of the Agrarian Social Structure*

According to both Phillips and Gray, the greater efficiency of slave plantations was manifest in the "tendency . . . of geographical specialization" (Gray 1958, 444; Phillips 1968, 97–98, 109–15; Rothstein 1967, 373–82). Because of their competitive edge in the cotton market, planters displaced yeoman farmers from the lands suited to staple production, and the increasing concentration of large slaveholders in these regions formed what Phillips called the "Black Belts." Yeoman households, in turn, transplanted their "rude self-sufficing economy" to marginal lands on the periphery of these regions.

This version of the "planter dominance" thesis was initially challenged by Frank Owsley and his students (Owsley 1982), who documented the numerical superiority of yeoman households even in Cotton Belt regions. Subsequent scholarship, critical of the Owsley school, has emphasized the concentration of wealth among planters (Linden 1946; Wright 1970; Campbell 1974; Foust 1975, Chap. 7). Yet it finds no strict spatial segregation between a staple-producing planter class and self-sufficing yeomanry and no trend of increasing wealth concentration in the Cotton South during the 1850s (Wright 1978, 69–71; Schaefer 1987, 12–17).

An alternative staple interpretation of the agrarian class structure, however, has evolved alongside the Phillips-Gray thesis. Geography and the staple crop, it suggests, decomposed white society in the Lower South into distinct populations, not classes, that were segregated economically and spatially. In the Cotton Belt the distribution of wealth among free households was actually made up of two distinct distributions, one for slave and free farms (Yang 1984, 97–100; 1985, 230–31; Schaefer and Schmitz 1985, 223–25). These different distributions explain the greater concentration of wealth in the Cotton South compared to the rural North and are consistent with the findings of Fogel and Engerman and Schaefer and Schmitz, both of which imply that slave, but not free, farms benefited from economies of scale in the cultivation of staple crops. Slave and free farms, according to this interpretation, were separate populations, governed by different "economic, demographic and cultural forces" (Yang 1984, 100).

The regional division between the Cotton Belt and upcountry, in contrast, is explained by the parallel migration of distinct populations to the western frontier (Steckel 1983, 19–28; Owsley 1982, 50–58; McDonald and McWhiney, 1975; Jordan 1967; Otto 1985). The source populations, characterized by their geographic or ethnic origins, inherited different agricultural practices, and migrated along east-west lines in search of areas with familiar climate, terrain, soil texture, and vegetation. In these distinct ecological regions, they could transplant seeds, crops, and methods of cultivation without extensive experimentation and potential losses of income and wealth and so preserve their accumulated human capital and traditional way of life.

### *Staple Crops and the Natural Limits to Economic Development*

The slave South grew rapidly during the last two decades of the antebellum period. According to Richard Easterlin's estimates, income per capita in the South increased at an average annual rate of 1.7 percent over the period, as compared to the national average of only 1.4 percent (Easterlin 1961, 527–36; Easterlin 1975, 95–96; Fogel and Engerman 1974, vol. 1, 247–52; Fischbaum and Rubin 1968, 117–22). The geographic regions comprising the South, however, grew at substantially lower rates, between 1.0 and 1.3 percent, and in striking contrast to the North, regional incomes declined from east to west. This spatial pattern implies that economic growth in the South depended on the territorial expansion of staple production to the fresh lands of the Southwest, rather than on technological and structural changes in the older staple-producing regions.[14]

The staple thesis explains this extensive mode of economic growth by the very features that defined the Lower South as a distinct region: its natural resource endowment and principal staple crop. These conditions constrained agricultural innovation, improvements in cotton and food production, and soil conservation. They also fettered the growth of the region's urban manufacturing sector.

The southern soil and the properties of cotton prevented farmers from taking advantage of the mechanical technologies of the first industrial revolution. Wide variations in soil quality and terrain prevented the development of specialized implements, plows, seeders, cultivators, and harrowers, that could be standardized and produced on a large scale (Bonner 1964, 94). Harvesting cotton, in contrast, was simply difficult to mechanize. As William Parker (1972, 384–85) points out, the mechanical technologies of the period replicated the simple motions of the human arm and so were not well suited to the picking, plucking, or snapping motions of the hand required in cotton, corn, or fruits and vegetables (see also McMath 1987, 9–10).

Climate frustrated the efforts of farmers and planters to develop new crop rotations and to improve their livestock and soils. In particular, Rubin (1975, 364–65) argues, it created an inhospitable environment for the methods of "high farming" (see also Vance 1932, 155). Grains matured quickly in the warm, humid weather but produced lower yields and contained less protein. Root crops and legumes, on the other hand, suffered in the hot, dry weather of the late summer. Without the protection of cover crops, the soils were more vulnerable to erosion and leaching from heavy rains. This problem was particularly acute during the winter, because the ground did not freeze in the milder temperatures.

Livestock also fared poorly in the Southern climate. The inferior quality of forage crops and hot, humid weather decreased the yields of animal products and the reproductive efficiency of livestock. Southern livestock also endured repeated infections by animal parasites that survived the mild winters (Rubin 1975, 336–67, 371–72; Vance 1932, 159–61; Haywood 1986, 557–63). Native livestock developed immunities to these diseases, but at the expense of lower weights and greater vulnerability to other diseases and famine. The disease environment, moreover, was fatal to imported breeds and doomed experiments in livestock breeding. Like mechanization of the cotton harvest, Rubin concludes, the solution to these problems awaited scientific advances of the twentieth century.

Despite the greater value of Southern staples in total exports, the region lagged behind the North in the growth of cities. The slow pace of urbanization, the staple thesis maintains, followed mainly from the properties of cotton. The cotton culture generated few critical linkages to the wholesale trade, transportation, and manufacturing sectors, and at the same time drew scarce factors of production into agriculture.

The simple processing requirements and nonperishability of the cotton crop, Carville Earle and Ronald Hoffman (1987, 151–54, 163–66) argue, stunted local town development in the Lower South (see also

Earle and Hoffman 1976, 29–30, 40–43; North 1959, 948–49; Rothstein 1966, 94–95). Ginning cotton to remove the seeds and pressing the lint into compact bales substantially reduced the weight and volume of the crop. Planters or storeowners in crossroad towns could provide these simple services, and the high value of a bale of cotton relative to its bulk justified the cost of transporting the crop by wagon to interior depots or port cities. It was also profitable for steamboats to stop at wharves and collect the crops of nearby plantations, especially as the crop did not require specialized storage and transport facilities.

The staple crop and the region's resource endowment also weakened the demand for transport improvements and specialized intermediaries in port cities. The extensive system of navigable rivers provided cotton growers with an inexpensive and flexible means of transportation (Earle and Hoffman 1987, 151–54, 160–63; Phillips 1908, 6–8; Fogel and Engerman 1974, vol. 1, 255–56). Competition from water routes and the seasonality of cotton shipments made investments in railroads less profitable than in the North and so delayed their construction.

In port cities the cotton crop demanded only limited services of intermediaries and so generated little urban employment. It was compressed into smaller bales and briefly stored, but the bulk of marketing costs paid the commissions of factors, who arranged the sale and shipment of the crop (Earle and Hoffman 1987, 168–70). The inherent difficulty in establishing standard grades of cotton, Rothstein (1966, 97–98) further argues, limited the function and size of the merchant class in port cities by preventing the diffusion of marketing innovations to the cotton trade and hence the greater specialization of mercantile functions among intermediaries.

As the above discussion suggests, the backward, forward, and consumption linkages of the cotton culture created only limited markets for regional manufactures. The cultivation, processing, and transportation of cotton did not require large investments in machinery and, in particular, specialized farm equipment. The skewed distribution of rural income and small urban population created only scattered markets for consumption goods (North 1959, 948–49; Rothstein 1967, 373–75).

The cotton culture, it is proposed, further impeded manufacturing growth by increasing the wages of rural workers and hence the cost of a potential labor force. According to Claudia Goldin and Kenneth Sokoloff (1984, 471–74), women and children were more productive relative to men in cotton culture than in wheat or dairy production, and this difference in agricultural technology accounts for the higher wages of women and children relative to men in the South than in the North. Thus southern manufacturers were deprived of an important source of cheap labor that fueled early industrialization in the Northeast. Earle and Hoffman (1980, 1066–68, 1071–75), in contrast, emphasize the longer period of peak labor demands in cotton than in wheat. The rural proletariat in the South, they explain, could work more days a year and so earn higher annual incomes than their northern counterpart.

Goldin (1976, 89–102) and Fogel and Engerman (1974, vol. 1, 94–102, 255; vol. 2, 152–55) apply a similar explanation to the supply of slave labor. Economies of scale, they assert, could only be realized in the production of staple crops with slave labor. Thus slave and free labor were close substitutes in urban but not rural employments, and the demand for slaves was more elastic in the cities than in the countryside. In response to the growing demand for cotton in the 1850s, planters and farmers bid up the price of slaves relative to free labor, and the change in relative labor costs induced urban capitalists to replace the latter with the former.

In identifying the natural determinants of extensive growth in the Lower South—the resource endowment and the characteristics of cotton—the staple thesis ultimately implies a tendency of geographical expansionism. The boom in the cotton market may have made it more profitable for farmers and planters in the older seaboard states to sell or abandon rather than improve their worn lands and move westward (Engerman 1967, 88; Engerman 1975, 346–48; Lebergott 1985, 189–92, 201–4). Yet these natural conditions also locked Southern farmers into this pattern of growth by limiting their technological, market, and employment opportunities. Nature, in effect, impelled the territorial expansion of slavery and the political conflicts that eventually erupted into the Civil War.

## PECULIAR INSTITUTIONS AND AGRICULTURAL DEVELOPMENT IN THE ANTEBELLUM COTTON SOUTH

Although rural society in the antebellum Lower South was literally built on a natural substructure, the physical environment, as subsequent history illustrates, could support rather diverse patterns of agricultural and regional development (Fite 1984; Daniel 1985; Wright 1986). In grounding this development in geography and cotton, then, the staple thesis in effect confuses what are only necessary with sufficient conditions. It is certainly a truism that at the time cotton could not be cultivated above the Mason-Dixon line and that river transport offered a cheap alternative to overland routes. Yet, these "natural" facts do not explain the origins and viability of the plantation and slavery, nor did they condemn the region to extensive growth without the benefits of a dense network of cities and diverse manufacturing sector.

By drawing together the literature critical of the staple thesis, I expose the gaps in its arguments and at the same time, elaborate an alternative. The plantation and slavery, rather than nature, I propose, created the economic and social environment that structured the decisions and behavior of Southern planters, farmers, and capitalists, and ultimately the pattern of regional development. This alternative thesis does not attempt to account for the origins of these institutions, but instead identifies the conditions necessary for their reproduction over time.[15]

*Slavery, the Plantation, and the Rationale of Coerced Labor*

According to the institutional perspective, the cultivation of cotton, the South's principal staple, did not yield significant productivity gains to large-scale producers either because of conventional scale economies or because of technological change. In other words, plantations were not technically more efficient than smaller-scale producers. At one level this claim has been formulated as a critique of Fogel and Engerman's analysis of the relative efficiency of slave (i.e., plantation) agriculture and addresses problems of method and the interpretation of evidence. Paul David and Peter Temin (1976, 209–12, 214–18; 1979, 216–17) have questioned their estimates of land and labor inputs because of qualitative differences in factors used on plantations and free farms. Their index of relative efficiency, as Gavin Wright (1978, 74–85; 1979a, 222–24) has shown, varies directly with the degree of specialization in cotton and so provides only indirect evidence of economies of scale in staple cultivation on which the theoretical arguments hang.[16]

More generally, these critical evaluations imply that planters exploited the potential for extensive growth in the expanding market for cotton and in the favorable resource environment by instituting a labor system—slavery—that gave them greater control over their workforce

(David and Temin 1976, 209–12, 221–22; Wright 1978, 82–83, 90–97; Wright 1979a, 219–22; Ransom and Sutch 1977, 73–77; Anderson and Gallman 1977, 30, 41; Gallman 1979, 1018; Schaefer 1983, 859–65). Following the product cycle in the English textile industry, cotton demand grew briskly during the antebellum period at an average annual rate of 5 percent. Geography endowed southern farmers with a climate uniquely suited to the cultivation of cotton and with a longer working day and growing season. The mild, dry harvest period during census year of 1859 produced a bumper cotton crop, especially in the New South states, which biases productivity estimates in favor of large slaveholders who tended to specialize more in cotton. Finally Southern slaveholders, not constrained by the norms that governed free households, forced slave women and children to work in the fields and worked their slaves longer and harder than free laborers.

This perspective also questions the relationship between the seasonal labor requirements of cotton and the labor system on southern farms. The problem of seasonality, as Anderson and Gallman note, depends not only on the natural properties of the crop, but on the opportunities for alternative employment during the nonpeak seasons and for diversifying production into other cash crops.[17] The sparse network of interior towns and cities in the Cotton South limited both, and so cotton producers required a bound labor force to satisfy the peak demands of cotton culture during the postplanting and harvest periods. To satisfy this demand, however, planters required only an annual labor force. Thus the cotton culture was consistent with various forms of bound labor, such as the share arrangements that evolved during Reconstruction (Whatley 1987, 49–50). Slavery was only one of these alternatives and the most extreme.

The critique of Fogel and Engerman's analysis of slave agriculture implies an alternative logic of slavery and the plantation. Slavery as a labor system transformed labor into a commodity and entitled the master class to the immediate control over their labor force. This economic and political power yielded two important advantages.

First, by tying laborers to the individual producer rather than to the land slavery facilitated the redistribution of the labor force across farms and over space in response to market conditions. Although fixed in the short run, the aggregate supply of slaves could be reallocated through the slave trade. As the numerous studies of the profitability of slavery have shown, the market for slaves was competitively organized, and so individual farmers faced an elastic supply of slave labor (Fleisig 1976, 573–76; Wright 1978, 47–48; Aitken 1971). Consequently, those producers with adequate capital or the right connections could purchase additional laborers and expand the scale of production beyond the level dictated by the supply of family labor.

Slaves could also be readily transported to the frontier. Unlike land, slaves were a physically movable form of wealth, and their value was enhanced by the expansion of staple production to the more fertile, level lands in the states of the New South (Thompson 1975, 17–19, 89–90; Engerman 1973, 49; Wright 1986, 17–18; Wright 1987, 852–55; Passell and Wright 1972, 1188–1202). The exodus of slaveowning households to the frontier brought not only the necessary labor supplies to improve and cultivate these fresh lands, but the collateral to purchase supplies, land, and even additional slaves on credit. The interregional slave trade augmented the slave population on the frontier, especially in younger age groups (Fogel and Engerman 1974, vol. 1, 44–47, 50–51; Sutch 1976, 160–61, 178–80; McClelland and Zeckhauser 1982, 52–53, 159; Miller 1985, 16–17). It also enabled stationary planters to rationalize their slave labor force and to finance their agricultural reforms (Genovese 1961, 136–41; Fields 1985, 18).

Second, slavery afforded masters more direct control over the labor supply and standard of living of their work force. By forcing all able-bodied slaves to work in the fields and by working their slaves intensively over a longer working day and growing season, masters exacted greater effective supplies of field labor per capita of the farm population.[18] According to the estimates made by Ransom and Sutch (1977, 232–36), the field work of women and children alone increased the supply of slave labor, relative to that of the free population, by 19 to 25 percent (see also Jaynes 1986, 228–32; Engerman 1986, 333–36).

Masters adjusted the care and treatment of slaves to accommodate staple production and to insure the overall profitability of their farms. Able-bodied slaves were fed rations that contained enough calories to compensate for their toil in the fields (Sutch 1976, 266–81). The bulk of their food, however, came from corn and pork, a diet deficient in essential nutrients. Although neither product was a natural complement to cotton in production, they could be integrated into the annual cycle of cotton and returned ample, but less than optimal yields.[19] Masters also did not adequately provision growing slave households until the children were old enough to work in the fields and pay their own way.[20] Consequently, slave infants were often malnourished and suffered high rates of mortality.

The social and economic organization of the plantation compounded these benefits and made the plantation singularly efficient in settling sparsely populated "frontier regions" of the cotton economy and producing the staple for export. Plantations could rapidly settle areas where climatic and ecological conditions were best adapted to staple cultivation and quickly forge direct ties to metropolis regions through staple production. The regimentation of the slave population in staple cultivation, moreover, yielded greater effort per hand and supported the extensive methods of cultivation that fully exploited the abundant natural resources.

Slaves functioned as a surrogate community and made the plantation a relatively self-contained frontier settlement (Gray 1958, 438–39; Davis 1977, 81–83; Allman 1979, 108–9; Miller 1985, 16–17). Accompanied by their prime-age male slaves and with the purchase or rental of additional hands, migrating planters assembled the collective labor required to clear and improve land and build residential and farm structures (a barn, a grist mill, and the like). Through its extensive division of labor, plantations produced foodstuffs and simple manufactures consumed by slaves and provided the artisan and processing services necessary for the cultivation of cotton and its preparation and shipment to coastal ports.

An extensive network of intermediaries located in port cities integrated pioneering planters into the world market and reinforced their relative independence from the local economy. Planters established accounts with factors in the region where they intended to settle and relied on these commission merchants for credit and valuable information about the land market and the social milieu.[21] This relationship was secured by the personal reputation of the planter and slave wealth, as the value of each transcended the bounds of the local community.

Planters and factors could conduct direct transactions over long distances because of the specialized, wholesale nature of the trade. Factors realized economies of scale in handling the large, periodic shipments of the staple, and these savings more than offset the negligible increments to unit costs from trading over greater distances.[22] Moreover, variations in the quality of cotton shipments induced the centralization of trade over a vast area into port cities where the crop was graded and compacted for export. Factors also mediated the bulk purchases of plantation supplies, including slaves' consumption goods that were typically coarse or second hand. They could even cater to the more refined needs of planters and their families with whom they often had close, personal ties (Atherton 1968, 34–35; Woodman 1968, 30–32; Buck 1969, 68–75, 79; Genovese 1961, 163–72; Parker 1970, 117).

Finally, the centralized organization of slave households and field crop production, both of which required at least 15 or 20 slaves, enabled planters to exact even greater supplies of labor from their work force than could small slaveholders. Planters, rather than adult slave males, asserted patriarchal control over slave households, subsuming them into a plantation household, and determined the allocation of family labor between domestic and market production (Wright 1978, 82–83, 162–63; Gallman 1979, 1012–1013; Fox-Genovese 1983, 238–48). Planters mobilized all able-bodied slaves into the fields and internalized many of the reproduction costs of the slave population (Hilliard 1972, 62; Metzer 1975, 140; Garrett 1978, 63–66, 77–78). Older or infirm slaves cared for children, prepared meals, and mended and made clothes. This allocation of domestic tasks freed up the labor time of slave women and their older children for field work. Rated as partial hands, they were employed in the less strenuous tasks of seeding, hoeing, and picking, while prime-age males, who were full hands, performed the more physically taxing work. By putting-out household manufacturing to small shops, planters could purchase coarse and secondhand clothes, which would be finished on the plantation.

In the fields, as Fogel and Engerman emphasize, the gang system replicated the discipline of the nonmechanized factory and enhanced the productivity of field hands in planting and "cultivating" field crops. Yet they incorrectly explain its advantages by the natural properties of cotton and the unique qualities of Southern planters and slaves.[23] The gang system clearly had technical prerequisites. It was, after all, limited to the initial stages of production, which were composed of distinct, simple operations that could be sequentially ordered.

The economic rationale of the gang system depended on a system of rotation that paired cotton and corn and cultivated them with similar methods. The complementarity between the two crops, however, was not natural. Both corn and cotton were "summer" crops, planted in clean rows, and had similar seasonal patterns of labor inputs (Gray 1958, 707; Hammond 1966, 81; Hilliard 1972, 153–54; Anderson 1974, 62–68, 75–80; Anderson and Gallman 1977, 37–38; Gallman 1979, 1018). To minimize their conflicting labor requirements and to utilize fully their work gangs, planters staggered the planting and harvesting of corn around that of cotton and adopted more labor-intensive methods, requiring only simple, general-purpose implements in the corn culture. They were unable to avoid a labor bottleneck during the months of April to June, when the growing stalks were thinned and weeded. The greater productivity of the gang system eased this labor constraint and enabled planters to "cultivate" more acres of cotton and corn per hand than smaller-scale producers (Stampp 1956, 56; Gray 1958, 707–8; Moore 1958, 42–43). Consequently, they could plant more acres of the two crops without suffering lower yields, and by working their slaves until late December or January, they could pick the larger crops as well.

The greater productivity of gang labor, moreover, cannot simply be reduced to the work ethic of slaves or superior entrepreneurship of planters. The organization of field crop production into specialized, interdependent work gangs increased the "proficiency" of field hands and overseers by simplifying and routinizing their tasks (Gray 1958, 479; Genovese 1961, 26, 43–44; David and Temin 1976, 206–7; Fenoaltea 1984, 643–47). Equally important, it increased the planter's control over the pace of production and intensity of work effort. When combined with the lash, the gang system forced slaves to work harder than free workers would willingly submit and to maintain their work effort without excessive use of direct force and monitoring.[24]

These properties underwrote the unique advantage of the gang system. Coerced labor yielded greater productivity only in more "natural" production processes, those that required large doses of physical effort and natural resources rather than skill and sophisticated capital equipment (Fenoaltea 1984, 637; Fox-Genovese and Genovese 1984, 42–43, 56–57). Thus the gang system supported extensive methods of cultivation that exploited the region's natural resources. Additionally, as both Phillips (1968, 107–8) and Gray (1958, 459) argue, it restricted the range of activities in which slave labor could be effectively utilized and thus increased the rigidity of the plantation system.

### *Slave Plantations and the Agrarian Class Structure*

In the end, then, the competitive advantages of the plantation in staple production depended on coercion and hence the political power of planters over their bound slave laborers. Southerners justified this social order by a natural principle that segregated the population along racial lines. This ideology regarded blacks as genetically inferior people, condemned by nature to a life of servitude, which meant dependence and labor (Thompson 1975, 36–37, 236–37; Fields 1982; Fox-Genovese and Genovese 1984, 403; Davis 1984, 10–22; Fenoaltea 1984, 651–52).

By contrast the Southern creed implied an organic unity among whites, who, it was assumed, were naturally endowed with their freedom and independence. Yet despite this ideology, slavery and the plantation system forged a complex agrarian class structure that differentiated the free population both spatially and economically. The territorial expansion of the plantation system segregated the frontier population by geographical area of settlement and formed distinct regional economies in the Cotton Belt and upcountry. Within the Cotton Belt, wealth and in particular slave wealth determined the location of households in the staple economy and braced the economic and political power of the planter class.

The formation of the Cotton Belt and upcountry regions occurred through the preemptive displacement of yeoman communities by plantations. Geographically mobile planters quickly followed yeoman farmers to the frontier, and because of their greater wealth and autonomy they travelled longer distances over a wider range in search for prime cotton lands.[25] Thus, in contrast to yeoman households, they were not constrained to follow the east-west pattern of migration and quickly settled prime cotton lands located near natural transport routes, especially those in the alluvial delta and prairie regions.

The presence of these wealthier settlers on the frontier intensified the competition for land. Planters bid higher prices for prime cotton lands and often purchased the land outright, unlike yeoman farmers who relied on informal sources of credit, such as squatting (Allman 1979, 145; Reidy 1982, 59–61). They also procured large tracts of unimproved acreage that provided reserves to replace worn fields or to expand production. Land speculation limited yeomen households to smaller tracts of high-quality land, with little room for future expansion, or to less fertile, hillier lands that were more geographically isolated.

Plantations accelerated the integration of frontier regions into the world market. By settling along natural transport routes, planters could minimize their investments in commercial infrastructure, such as transport improvements and interior towns. Instead, they concentrated their capital and labor on clearing and improving their lands and began to specialize in staple cultivation within four to five years after settlement.[26]

The decentralized and highly commercialized pattern of settlement in Cotton Belt regions further alienated yeoman communities. Although farm families assisted each other throughout the year, they often had to rely on their wealthier neighbors for help in times of need as well as for essential town services (Genovese 1975, 337–38; Oakes 1982, 93; Russel 1941, 115; Ford 1983, 83–85). Many small farmers, especially those with personal or familial ties to planters, took advantage of their position to advance themselves economically and socially. For others, however, this form of dependence violated deeply held values of self-reliance and mutuality (Genovese 1975, 334–35; Hahn 1983, Pt. 1; Weiman 1988b, 8–13. See also Phillips 1969a, 179–80; Russel 1941, 121).

In the absence of a diversified local economy and communal social relations, yeomen farmers were under greater economic pressure to cultivate cotton for export. Specialization in cotton, however, involved greater risk for small producers who had to conduct their transactions through intermediaries and had limited access to credit (Wright and Kunreuther 1975, 562–52; Weiman 1988b, 13–16). Consequently, many yeoman either avoided Cotton Belt regions, or sold their land rights and moved to less congested areas.

Within the Cotton Belt, wealth and in particular slave wealth differentiated farm households by the scale and composition of crop production and spatial location. The first relationship reflected the role of slave labor in the cultivation of field crops. Within each subregion of the Cotton South, the ownership of slaves explains the distribution of farm size (Wright 1978, 17, 22–23; Russel 1941, 118; Yang 1984, 97–99). Except in the black prairie and alluvial regions, a majority of farm operators in the Cotton South owned no slaves. Median farm size in these regions, that is the size of a family farm, varied between 50 and 80 acres and was comparable to that in the Midwest. Slave labor enabled southern farmers to expand production beyond this level, and the largest farms in the region were cultivated exclusively with slave labor.

The distribution of slaveholdings, moreover, mirrored the distribution of staple production (Wright 1978, 27–29). Slaves supplied greater effective inputs of labor so that slaveholders and especially planters could cultivate more acres of cotton and corn per hand than free farmers. At the same time, the reproduction of the slave population required lower levels of food production per field hand.[27] Large slaveholders not only cultivated larger farms, but they devoted greater shares of their improved acreage to cotton and realized greater net incomes from specialization in the staple.

The variation in the crop mix by slaveholding class in each region of the Cotton Belt and in the Cotton South as a whole evinces this pattern of economic differentiation (Wright and Kunreuther 1975, 528–36; Wright 1978, 56–58). Farms in all classes, except the largest plantations, produced comparable levels of food crops per capita of the population. The degree of specialization in cotton, however, increased systematically with the size of slaveholdings. The cotton crop accounted for only a quarter of the value of field crops on slaveless farms but over half of the field crop output on plantations.

Within each region the wealthiest farm households settled on the most valuable lands. The distribution of farm value, consequently, was even more skewed than farm size, and was comparable to total agricultural wealth (Wright 1978, 27). The spatial distribution of large slaveholding farms and plantations in each region reinforced their tendency to specialize in staple production. More level, fertile lands were easier to cultivate and less vulnerable to erosion and produced higher yields. Also, direct access to navigable rivers reduced transport costs and the risk of delays and damage in shipping the crop to coastal ports (Smith 1958, 156–59; Abernethy 1965, 86–87; Atherton 1968, 153–56. See also Ford 1986; Weiman 1987).

Farm households were also differentiated spatially among the soil regions of the Cotton Belt. Slaveholders and planters quickly settled the extremely fertile, level lands of the brown loam, black prairie, and alluvial regions, and by 1860 they composed a majority of farm households in these regions. The spatial concentration of slaveholders was even greater in the latter two regions, because their soil and ecological conditions were poorly suited to the traditional practices of family farms and increased the capital requirements of farm construction (Gray 1958, 531–37, 894–97; Foust 1975, Chap. 6; Wayne 1983, 7–12). Consequently, the prairie and delta regions accounted for 40 percent of the staple output of the Cotton South, although they contained only 16 percent of its farms (Wright 1970, 71; Wright 1978, 18, 84).

## SLAVE PLANTATIONS AND ENCLAVE DEVELOPMENT IN THE ANTEBELLUM COTTON SOUTH

The institutional hypothesis rejects arguments that explain the extensive growth of the Lower South merely by natural and ecological conditions. Instead, it points to the impact of slavery and the plantation in impeding the development of a home market in the Lower South, local systems of commodity and information flows and of transport and communications. The absence of a home market slowed technical progress in cotton culture, the diffusion of agriculture improvements into the Lower South, and industrial and urban growth. Consequently, it tied the region's economy to external markets and sources of finance.

The slow pace of technical progress in southern agriculture cannot simply be explained by the availability of fresh lands on the frontier. By migrating to the fertile, level lands in the West, northern farmers also realized productivity gains in the cultivation of their principal staples, corn and wheat. Yet the diffusion of improved implements occurred most rapidly in the areas of recent settlement in the North and compounded productivity growth from the spatial redistribution of the farm population (Hammond 1966, 79–80; Parker 1971, 176–81; David 1975; Atack and Bateman 1987, 189–92, 250–55). In the face of western competition, moreover, farmers in the Northeast developed new systems of crop rotation and methods of livestock husbandry that restored the fertility of their worn soils and the profitability of their farms.

The negligible productivity growth in cotton culture after 1840 was instead the product of a labor system and pattern of settlement that limited the demand for mechanical innovations and the formation of a regional technological community capable of responding to the particular needs of southern farmers.[28] Slavery weakened the labor constraint on southern farms and hence the incentives to adopt these methods. It insured southern farmers adequate supplies of relatively cheap labor during the periods of peak labor demands and enabled them to expand the scale of production in response to the growing demand for cotton (Fleisig 1976, 576–81; Wright 1978, 51–54; Mandle 1978, 67–70; Ferleger 1985, 416–17). Thus slave labor functioned as a substitute for capital equipment in production.[29]

Additionally, slave labor was incompatible with more mechanized methods of production. These methods, as Genovese notes, required the greater supervision of smaller work groups to insure the proper use and treatment of more complex, expensive farm implements (Genovese 1961, 43–44, 90–96, 133–35; Moore 1958, 41; Fenoaltea 1984, 637–38). In contrast, heavier, less specialized implements, which could withstand abuse and could be used in the cultivation of other field crops as well, were better suited to the hard–driving gang system of the plantation. This argument does not imply that slaves "could not handle tools."[30] Rather, it recognizes the unique economic advantage of coerced labor, which achieved greater productivity only by using the gang system in simple, more natural production processes.

Slave plantations also short-circuited commodity and information flows between local artisans and agricultural producers that were essential to the development and diffusion of new methods of production. In the Lower South rural artisans, both slave and free, were confined to highly localized markets and tended to produce lower quality, less standardized implements (Gray 1958, 792; Bateman and Weiss 1981, 107, 156–58; Weiman 1984, 237–38; Ferleger 1985, 413–15; Meyer 1988, 15–16, 19–22). Slave artisans on plantations repaired and refined implements for their masters and neighboring farmers, but without geographical mobility they could not develop their skills and extend their markets.

Competition from slave artisans and the dispersed settlement of plantations and farms, in turn, limited the markets of free artisans, and few specialized in the production of agricultural implements. With better sources of information and greater wealth, planters could afford to purchase improved implements directly from northern manufacturers and adapt them to their specific needs (Gray 1958, 794; Moore 1958, 166–68, 184–85; Genovese 1961, 56–58). Farmers, on the other hand, depended on local transactions for technical information, repair services, and spare parts and were more reluctant to experiment with new methods in their absence.

Similarly, slavery and the plantation system, rather than the climate and soil, restricted the choice of crops and the diffusion of improved methods of cultivation and livestock husbandry. Ecological conditions certainly influenced agricultural practices in the Lower South, but as Sam Hilliard (1972, 34–36) persuasively argues, they supported the cultivation of a wide range of food crops and many breeds of livestock. Southern farmers, however, did not greatly diversify their production because it conflicted with the simple but rigid regimen of cotton culture. Winter wheat, for example, could be grown in the cooler regions of the Lower South, but planters often extended the cotton harvest into late

November and December, the planting season of the wheat crop (Hilliard 1972, 163–64). Breeding and fattening livestock and dairy production demanded a more steady stream of labor inputs over the year, as well as careful supervision of the labor force (Bonner 1964, 95–96; Genovese 1961, 90–95, 133; Gray 1958, 459; Hilliard 1972, 133–40).

Diversification of commodity production into food crops and livestock also required substantial investments in marketing infrastructure and proximity to market centers (Smith 1958, 74; Hilliard 1972, 198–200; Weiman 1987, 629–31). In the absence of these conditions, southern farmers produced food products almost exclusively for domestic consumption. Relegated to a secondary or residual activity, food production was organized around the cultivation of cotton (Gray 1958, 836–44, 940–42; Smith 1958, 71–74; Genovese 1961, 138–39; Parker 1971, 186; Anderson and Gallman 1977, 31–32; Weiman 1987, 630–32). Methods of corn cultivation, for example, produced lower yields, but planters had no incentive to improve them as long as they provided adequate supplies of food and livestock feed and their labor requirements did not conflict with those of cotton.

The rotation of cotton and corn also compounded the problem of soil exhaustion. Neither cotton nor corn greatly depleted soil nutrients as long as the stalks were plowed under (U.S. Department of Agriculture 1896, 91–95; Mathew 1985, 209–10; see also Grey 1958, 447–48; Earle n.d., 12–15). Yet because both crops were cultivated in clean rows, they exposed the soil to the damaging effects of heavy rains—erosion and leaching. Although planting cow peas between the corn stalks helped to restore the fertility of the soil, an effective program of soil restoration, such as the one advocated by Edmund Ruffin, required contour plowing, the intensive application of fertilizers, and planting grasses.

Many progressive planters experimented with these reforms, but most producers did not perceive their benefits and so they did not diffuse widely. The careful and intensive preparation of the soil and more complex crop rotations required greater supervision of field hands and so reduced the efficiency of the gang system (Gray 1958, 470; Moore 1958, 38–39; Genovese 1961, 90–96, 111–13; Fenoaltea 1984, 646–47). Also, planters avoided grasses, which could quickly spread to their cotton and corn fields and damage growing crops (Bonner 1964, 186; Vance 1932, 156–57). Finally, commercial fertilizers were expensive, and penning in livestock for their manure was only profitable if the entire system could be implemented and animal products sold in urban markets.

Smaller-scale producers, in contrast, had less exposure to these innovations. The paucity of agricultural periodicals and societies in the Lower South limited the formal circulation of information (Smith 1958, 91, 99; Mathew 1985, 211–17; Moore 1958, Chap. 4; Genovese 1961, 128–31; Bonner 1964, Chap. 8). Equally important, they were spatially and socially isolated from planters who were more likely to experiment with these methods and could demonstrate their effectiveness.

The slave plantation, moreover, created a self-reinforcing dynamic that constrained manufacturing and urban development and thus the growth of a home market in the Lower South. Through its impact on the structure of the regional economy, slave plantations limited the growth and diversification of the manufacturing sector in the Lower South (Genovese 1961, 163–72; Parker 1970, 116–17; Meyer 1988, 7–11, 15–22. See also Bateman and Weiss 1981). On the one side, slavery skewed regional markets for manufactured commodities. Slaveholders created sparse markets for local agricultural implement producers, as they equipped their farms with fewer and lower quality implements per worker than free farmers. In addition, because of their greater access to coastal ports, planters often purchased general purpose farm implements from northern manufacturers, and even licensed more experienced northern firms to produce specialized equipment (Moore 1958, 166, 184–85). The concentration of staple production on plantations and dispersed settlement of the farm population created a "dual" pattern of demand for consumption goods, notably lacking in the medium-quality products that fostered the growth of a denser, more diversified manufacturing sector in the North.[31]

On the other side of the equation, these institutions diminished local supplies of labor and finance for prospective manufacturers. In contrast to the northern experience, western competition did not displace agricultural laborers—slave or free—in the older cotton-producing regions who would have formed pools of cheap wage labor. In fact, because of the geographical mobility of slaves, the economic and territorial expansion of the cotton economy sharply increased the demand for slaves in agriculture and slave prices throughout the South (Parker 1970, 115; Wright 1978, 118–20; Wright 1979b, 667–75. See also Goldin 1976, Chap. 4; Goldfield 1986, 82). Thus manufacturers faced rising labor costs and substituted cheaper free labor for unskilled slaves.

Supplies of free labor were no more forthcoming. Yeoman households in the upcountry, only marginally integrated into the cotton economy, were less vulnerable to the economic pressures of falling cotton prices.[32] In the Cotton Belt marginal farmers resisted proletarianization because of the political and ideological stigma attached to unskilled labor.[33] Instead, they preferred to work in agriculture as tenants or croppers or sold their farms to encroaching planters and migrated to upcountry and frontier regions in search of better opportunities.

While immigration partially filled the gap, immigrants tended to avoid the South. In part, they were responding to the fewer employment opportunities for unskilled workers in urban centers (Dunlevy 1983, 225–32). They were also alienated by the institution of slavery and the political attitudes of slaveholders, who viewed them as a potentially subversive element within slave society (Wright 1979b, 667–78; Berlin and Gutman 1983, 1195, 1198; Miller 1985; Fields 1985, 50–52).

Slavery, it has been argued, biased southern capitalists against devoting their savings to real asset accumulation and allocating their capital to manufacturing production.[34] By presuming that domestic supplies of savings and capital were fixed *ex ante* and constrained levels of investment, these explanations abstract not only from international and interregional flows of capital, but from the endogenous factors determining the savings rate, including the extent of industrial development.[35] If the Lower South indeed suffered from low levels of savings, then one of the reasons is the limited investment opportunities—both past and current—in the manufacturing sector.

Potential capitalists may have lacked adequate supplies of finance provided through banks and credit markets. The South contained many of the largest banks in the antebellum United States, and the number of banks and their assets grew rapidly during the antebellum period. Yet southern banks primarily issued notes to finance the cotton trade, and the greater share of their liabilities in these note issues, rather than locally circulating deposits, made them vulnerable to fluctuations in world markets (Ransom and Sutch 1972, 643; Smith 1958, 103–4; Govan 1938, 170–76; Green 1972, 175–77; Schweikart, 1987a, 26–27; Schweikart 1987b, 227). The growth of the banking system, moreover, was concentrated in the pre-1840 period. Partially subsidized by foreign capital flows, it essentially financed the massive migration of planters to the frontier (Schweikart 1987a, 30–31; Schweikart 1987b, 257–62; Govan 1938, 166–67; Woodman 1968, 99–101; Green 1972, 109–11). The banking system contracted sharply in the ensuing panics of 1837 and 1839, and with the recovery of the cotton economy in 1845 banks grew slowly, mainly by discounting short-term notes of planters and factors.

The region's financial underdevelopment can perhaps best be understood by the failure of southern cities to evolve into higher-order metropolises that organize the circulation of commodities and information in their hinterland and gradually develop clearinghouse functions to coordinate local and regional financial flows.[36] Plantation slavery preempted this process in two ways. First, it weakened the incentives to promote local development, such as investments in town construction and improved means of transportation and communications, in areas of recent settlement. In the North local development was intimately tied to land speculation, and the returns were largely embodied in the capital gains on real wealth. Large wealth holders on the southern frontier had a greater share of their wealth tied up in personal property, slaves whose value did not depend on local development.[37] They, therefore, stood to make lower returns on their wealth holdings from such investments.

Second, during this period the growth of interior cities depended on their spatial location at the intersection of inter- and intraregional transport routes (Wade 1959; Rubin 1967, 11–12; Muller 1976, 183–85; Meyer 1890, 131–32). There, they formed stopping-off points where settlers purchased provisions, utensils, and implements before continuing their journey. The steady flow of migrants through these cities constituted an important source of external capital flows that enhanced local demands for the products and services of urban residents and financed their early growth. By establishing direct commercial ties to factors in port cities and providing town services for themselves and neighbors, pioneering planters diverted this trade from interior towns and directly competed with town residents.[38]

The self-sufficiency and dispersed settlement of plantations further diminished the density of local commodity circulation. Negligible intraregional trade limited the scale and degree of specialization of mercantile and manufacturing activities within and among cities and ultimately the size of cities and extent of regional urbanization. The staple trade dominated interurban commerce, which flowed almost linearly between interior depots (towns and even wharves), port cities, and metropolitan centers.[39] Southern cities in this hierarchical system functioned as transshipment points, and intermediaries at each level, including coastal ports, essentially served as agents, dependent on those at the next higher level for financial and commercial services.

The region's "colonial-style" rail network evinced this spatial pattern of passenger and freight traffic (Fox-Genovese and Genovese 1984, 50. See also Taaffe et al. 1963, 503–5; Vance 1970, 96–98; Conzen 1975, 370–73; Muller 1976, 183). Merchants and planters financed the construction of trunk lines, linking coastal ports to interior staple-producing regions, but the small scale of local traffic did not justify the formation of lateral ties between metropolises and smaller hinterland cities, "filling-in" the rail network. Even the major trunk lines required heavy state subsidies because the staple trade, characterized by large seasonal fluctuations in freight shipments and the greater volume of exports (down freight) over imports (up freight), resulted in the underutilization of capacity (Phillips 1908, 158–59; Heath 1954, 262–66, 286–87; Smith 1958, 178–89; Russel 1960, Chap. 4; Goldfield 1986, 76–77). Similarly, the dominance of the staple trade, relative to local and regional traffic, prevented intermediaries in coastal ports from extending their entrepôt functions, such as the establishment of shipping and related services and the development of correspondent relations with banks in their hinterland.

## EXTENSIVE GROWTH, CLASS CONFLICT, AND SECESSIONISM

As stated above, the extensive growth of the antebellum cotton economy depended on both the rapid expansion of cotton demand and the availability of fresh lands on the frontier (Fischbaum and Rubin 1968, 123–25; Wright 1978, 90–97; Ransom and Sutch 1979, 210–12). The greater value productivity of labor on these lands increased regional incomes per capita, despite the stagnant agricultural technology and small scale of the nonagricultural sector. At the same time territorial expansion of the staple economy tended to increase the concentration of wealth among slaveholders and planters and swelled the class of nonslaveholders, free farmers, and wage laborers within the Lower South.

The growth of cotton demand in the 1850s, which fueled the migration of slaveholders to fresh lands in the West, magnified the demand for slaves. Slave values rose sharply during the decade, and slaveowners realized substantial capital gains on their slave wealth (Wright 1978, 34–

35). Rising slave prices, however, increased the economic barriers to slaveownership. By 1860 a majority of farm households owned no slaves, and the gap between the wealth of slave and free-farm households increased in most regions.

In the older staple-producing regions, the increasing density of settlement and competition from western producers created shortages of land. Suffering from both an absolute and relative decline in yields, planters purchased additional lands to replace their worn soils and to graze their livestock. The engrossment of land by planters displaced many small slaveholders and free farmers and diminished the opportunities for younger generations to acquire land and build farms and plantations (Bonner 1944, 666–75; Foust 1975, 150–66; Wright 1978, 33; Barney, 1981, 34–35; Reidy 1982, 68–69; Ford 1983, 340–41).

The concentration of real and slave wealth sharpened the class divisions among whites and threatened to shatter the organic unity that preserved the social order. These political conflicts began to surface in the 1850s (Johnson 1977, Chap. 5; Barney 1982, 163–64; Hahn 1983, 96–116; Berlin and Gutman 1983, 1198; Oakes 1985, 560–68; Watson 1985, 275–98; Miller 1985, 46–53; Fields 1985, 57–62; Wallenstein 1987). The growth of nonslaveholders, especially upcountry households and a free urban proletariat made up largely of immigrants, challenged the hegemony of slaveholders in state legislatures and the legitimacy of slavery.

The fear of an internal revolution made slaveholders in the Cotton Belt more receptive to the political demands of upcountry farmers. At the same time they rejected programs to promote urban manufacturing development that would provide employment for the redundant free farm population. Instead, they sought to expand the frontier of slave society and create fresh opportunities for displaced farmers and new generations to accumulate wealth and become slaveholders. Thus, the viability of the plantation system as a social order depended on the extension of the legal boundaries of slavery. Yet at the same time, agitation for the expansion of the regime only served to heighten the sectional conflict between North and South that ultimately became the death knell of slavery itself.

## NOTES

1. (Paullin 1932, Plates 76, 77; Gray 1958, 652–55; Davis 1977, 29–33.) Settled areas are defined somewhat arbitrarily as those areas where population density exceeded 18 persons per square mile.

2. (Baldwin 1956; North 1955; Watkins 1963.) I am grateful to Paul David for suggesting this formulation. In the rest of the paper, I will simply refer to this argument as the staple thesis.

3. In reviewing the theoretical approaches to the colonial economy in America, John McCusker and Russell Menard (1985, 27), note the "difficulty of separating the impact of a given staple from that of the organization of labor." Richard Graham (1981, 620–55) reaches similar conclusions in his comparison of the impact of plantation slavery on Brazil and the United States.

4. (Phillips 1929, 3.) "Staple crops," he continues, "promoted the plantation system, which brought the importation of negroes, which not only gave rise to chattel slavery but created a lasting race problem."

5. In the case of the Piedmont region, William N. Parker (1979, 234, 238) presents contemporary estimates of labor requirements for each operation. The monthly distribution assumes that the labor requirements for each operation were evenly distributed across the time period as follows: (1) clearing land and cutting stalks, December through January; (2) breaking, February; (3) bedding, March; (4) planting, April; (5) hoeing and cultivating, May through July; (6) picking, September through November. The monthly labor requirements for the Alabama Cotton Belt and Southern Georgia are based on direct estimates by the Alabama Polytechnic Institute Extension Service (1919, 44) and Georgia State College Extension Division (1923). Agricultural practices did not change greatly after the Civil War, and the seasonal distribution of labor requirements for the two periods are remarkably similar; see U.S. Department of Agriculture (1912, 93–94; U. S. Department of Agriculture 1917, 4–14) and the seasonal pattern of labor requirements presented in Anderson (1974, 77), Metzer (1975, 129–30), and Whatley (1987, 48).

6. (Phillips 1968, 106.) Kenneth Stampp (1956, 38–42) claims that plantations with at least thirty slaves "achieved maximum efficiency." In later writings, Phillips (1969a, 162; Phillips 1929, 129) reverses his position and asserts that "[t]here was no vital need for large units of production" in cotton.

7. (Gray 1958, 478–80.) Slavery, he also argues, provided planters with a stable labor force, especially during the periods of peak labor demands, and enabled them to utilize fully all able-bodied hands in the field while maintaining them at or near the subsistence level. The apparent economies of scale in production, he notes, reflect the fact that planters required at least fifteen slaves to establish a division of labor among work gangs. Russel (1941, 114–18) offers a similar interpretation.

8. (Fogel and Engerman 1974, vol. 1, 192–94, 203–4; vol. 2, 142–49; 1977, 279, 290–92). The extent of these scale economies, they argue, decreased from sugar to tobacco.

9. (Metzer 1975, 134–35, 145; Fogel and Engerman 1977, 291–92.) During periods of peak labor demands, Metzer argues, planters allocated hands to different tasks according to the principle of comparative advantage.

10. (Fogel and Engerman 1974, vol. 1, 192–93; vol. 2, 126–42; 1977, 276–79.) Fogel and Engerman measure the "technical" efficiency of southern farms by total factor productivity, which is defined as total output relative to a weighted average of inputs. Because southern farms produced a range of outputs, total output is measured in value, not physical terms, where each product is valued at national prices in 1860. Inputs are weighted by

their share in national farm output. According to this index of total factor productivity, slave farms were 29 percent more efficient than free farms, and only plantations with at least sixteen slaves realized substantially greater productivity. Criticisms of their approach will be discussed below.

Donald F. Schaefer and Mark D. Schmitz (1979, 208–9) show that the greater efficiency of slave plantations was expressed in, if not caused by, the greater productivity of labor.

11. (Fogel and Engerman 1974, vol. 1, 236; Engerman 1973, 43–65; Domar 1970, 21–3.) The Domar thesis states that in the absence of economies of scale or large capital requirements in staple production, a rentier planter class can derive its income only through the exploitation of a coerced or bound labor force.

12. (Fogel and Engerman 1977, 294.) As evidence for the latter point, they argue that because small slave farms could not employ the gang system, they were only marginally more efficient than free farms.

13. (Schaefer and Schmitz 1979, 209–12; 1982, 91–96.) In contrast to the interpretation of Fogel and Engerman, they explain the productivity gains from large-scale production by economies of scale in the states of the New South and neutral technical change in states of the Old South. They also caution against explaining their results by the efficiency of slavery per se.

14. To illustrate the point, I calculated what southern income in 1860 would have been if there had been no change in the distribution of population among regions between 1840 and 1860 (see Easterlin 1975, 97–98). Assuming all other things equal except the regional distribution of the population, southern income per capita in 1860 would have been $93.9 or one-third less, and the rate of economic growth would have averaged only 1.2 percent per year. In contrast, under the same assumptions northern income per capita would have been almost 30 percent greater.

15. To address the former issue would extend the essay well beyond the historical scope of the volume. For an analysis of the origins of the plantation and slavery in colonial America that emphasizes the exigencies of settlement institutions and class structure, rather than geography and staple crops, see Morgan (1975), Thompson (1975, 232–42), and Nash (1979, 107–11).

16. (See also David and Temin 1976, 218–23.) To Fogel and Engerman (1977), the capacity of plantations to specialize in cotton and produce abundant supplies of food crops and other outputs demonstrates their greater technical efficiency. This evidence, however, is consistent with the alternative interpretation developed below: that the organization of the plantation forced slaves to work harder and reallocated their labor time from nonstaple production to the cultivation of cotton.

17. (Anderson and Gallman 1977, 42–45.) The "affinity" between slavery and wheat production in the Virginia Piedmont, documented by James Irwin (1988, 296–301), directly contradicts Earle's thesis as well.

18. Fogel and Engerman (1977, 285–88) claim that slaves on upland cotton plantations actually worked fewer days and hours a year than northern farmers. Their evidence on the length of the work year in the North, however, comes from surveys of farms in the early twentieth century and is of questionable relevance. Gallman (1975, 55–57) maintains that the average agricultural work year in the nineteenth century was "not . . . above two thousand hours per year," and that it increased between 1800 and 1900. Specifically, dairy production, which accounts for the long work year, was mainly a secondary activity on antebellum northern farms except in the Northeast (Atack and Bateman 1987, 151–53).

19. The system of crop rotation is discussed further below.

20. (Steckel 1986a, 732–35; Steckel 1986b, 427–65.) In particular, Steckel notes, mas-

ters forced women back into the fields within three months after delivery, and so infants were weaned early off breast milk.

21. (McMichael 1987, 251–53; Phillips 1969b, vol. 2, 197–218, 239–40, 252–54, 257–58; Abernethy 1965, 73–76; Woodman 1968, Chap. 4; Oakes 1982, 73–76; Miller 1985, 11–15.) McMichael (258) observes that the decentralized credit system, supported by the London discount market, "encompassed plantations in a global 'putting-out' system."

22. (Meyer 1980, 122–23; Atherton 1968, 28–30; Woodman 1968, 16–17, 158–59.) The extensive system of interior rivers in the Lower South and the simple processing requirements of cotton reinforced these advantages. They enhanced the value of land located near navigable rivers and of processing facilities to planters but did not determine the pattern of trade.

23. (David and Temin 1976, 206–8.) Fogel and Engerman (1977, 290–92) emphasize only the entrepreneurial abilities of planters and disregard their earlier arguments relating the efficiency of the plantation of the "high quality" of slave labor (1974, vol. 1, 209–23).

24. Thus Fenoaltea (1984, 645) concludes, "The superior productivity of gang slaves appears attributable specifically to their subjection to the lash, and not to conventional economies of scale."

25. (Phillips 1969a, 179; Allman 1979, 109–10; Campbell 1974, 376–77; Schaefer 1985, 569–71, 573–74.) The mobility of planters is measured by their low rates of persistence during the late antebellum period, between 40 and 60 percent (Wiener 1978, 10; Oakes 1982, 77–78; Barney 1982, 148–49).

26. Gavin Wright (1971, 116–17) estimates a two to four year lag between the sale of public lands and the growth in cotton output in the New South.

27. (Weiman 1984, 36–42.) Food production per field hand depended on the average rations allocated to the farm population and the dependency ratio, the ratio of the dependent population (those not employed directly in field work) to field hands. Formally, $F/H = (F/P) * (1 + D)$, where F equals food output; H, the number of field hands; P, the total farm population, and D, the dependency ratio. By forcing women and children to work in the fields, planters lowered the dependency ratio of the farm population and so reduced the required levels of food output per hand necessary to feed the work force.

28. Estimates of productivity growth in cotton culture by Foust and Swan (1970, 41–46) and in southern agriculture by Easterlin (1975, 96–97) are subject to the same criticism as Fogel and Engerman's estimates of total factor productivity (See pp. 132–33 above). Based on the trend in cotton prices, Wright (1978, 101–2) concludes that there was no productivity growth in cotton production after 1820, while Ransom and Sutch (1979, 214–15) estimate that per capita crop output increased by only 0.64 percent annually from 1839 to 1857.

29. This argument explains the empirical findings of Schmitz and Schaefer (1978, 330–32) that labor and capital are relative substitutes on free farms, but not slave farms.

30. Wright (1978, 54–55) quotes Genovese (1961, 54) out of context and so misinterprets his argument.

31. (Genovese 1961, 163–72; Parker 1970, 116–17.) Despite their criticism of Parker's hypothesis, Fred Bateman and Thomas Weiss (1975, 192–94) show that the Midwest had a "wider variety of industries where middle-sized firms were comparatively more important"; see also Meyer (1988, 15–18).

32. (Terril 1976, 92–98.) The position of the upcountry yeomanry did not change until after the Civil War (see Hahn, 1983, Chaps. 4–5; Ford 1984; Weiman 1985, 78–90).

33. The displacement of farmers from the Cotton Belt is discussed in the next section. According to Ira Berlin and Herbert G. Gutman (1983, 1186), "Most urban workers were alien to the dominant character of Southern culture" (see also Fields 1985, 50).

34. Fleisig (1975, 581–83), for example, argues that slavery enabled southern entrepreneurs to amass large fortunes in agriculture and in this way biased their investments against manufacturing, whereas Bateman and Weiss (1981, 117–27, 133–42) emphasize the more conservative attitudes of the planter class. Finally, Ransom and Sutch (1988, 10–14) suggest that capital gains on slave labor lowered the rate of savings in the South and in this way "crowded out" physical capital accumulation.

35. From a Keynesian perspective, for example, the rate of savings depends on the share of investment and net exports in regional income; that is, $S/Y = s = I/Y + (X - M)/Y$, where S, Y, I, and $(X - M)$ equal aggregate savings, income, investment, and net exports respectively, and s, the aggregate savings rate. According to this view, the growth of the capital goods sector and successful import-substitution in the consumption good sector—that is, industrial development—will increase the savings rate, without any change in the savings behavior of individuals (Robinson 1974, 40–41, 109–10; Terzi 1986–87; Heim and Mirowski 1987, 137–38; Knodell 1988).

36. According to Michael P. Conzen (1977, 99–100), cities in the Deep South, with the exception of New Orleans, did not develop strong financial hinterlands until after 1920. The dependence of southern cities on New York for capital and information and commodity flows is emphasized by Woodman (1968, Chap. 12; Pred 1980, 109–16, 119–31). Lance Davis (1965, 368, 388–92) and John A. James (1981, 444–47) document the regional segmentation of the South in national capital markets during the late nineteenth century.

37. (Wright 1986, 17–31; Wright 1987, 858–59.) Foust (1975, 94–95) shows that nonslaveholders in the Old South and New South states owned two to four times more real than personal wealth than the farm population as a whole (see also Hammond 1966, 79–80; Gray 1958, 539–40; Oakes 1982, 77–90).

38. Thus Conzen (1981, 331) explains the limited urban development in the South by the "town-usurping" character of plantations.

39. (Rothstein 1966, 98; Woodman 1968, 12–17, 78; Green 1972, 9–11; Pred 1980, 116; Goldfield 1986, 30–31.) An exception to this pattern is the growth of Atlanta, a wholesale distribution center located in the Georgia upcountry, in the 1850s (Weiman 1988a).

## REFERENCES AND FURTHER READINGS

This essay benefited from the comments of Michael Bernstein, Charles Calomiris, Paul David, Michael Edelstein, Stanley Engerman, Duncan Foley, Gerald Jaynes, Jane Knodell, Howard Lamar, David Levine, Philip Mirowski, William Parker, Gavin Wright, and the participants of the Committee on Historical Studies Seminar at the New School for Social Research, the Economics Department Seminar at Swarthmore College, the Social Science History Workshop at Stanford University, and the University of California Graduate Seminar in Southern History. I would especially like to thank Lou Ferleger for his encouragement, patience, and insights, as well as his capable editing of the manuscript. Financial assistance from the ENO Foundation is gratefully acknowledged.

Abernethy, Thomas. 1965. *The Formative Period in Alabama, 1815–1828.* University: Univ. of Alabama Press.

Aitken, Hugh G. J. 1971. *Did Slavery Pay?* Boston: Houghton Mifflin.

Alabama Polytechnic Institute Extension Service. 1919. "Man Labor Requirements of Some Principal Southern Crops." Circular no. 33(June).

Allman, John M. III. 1979. "Yeoman Regions in the Antebellum Deep South: Settlement and Economy in Northern Alabama, 1815–1860." Ph.D. diss., University of Maryland.

Anderson, Ralph V. 1974. "Labor Utilization and Productivity, Diversification and Self Sufficiency, Southern Plantations, 1800–1840," Ph.D. diss., University of North Carolina at Chapel Hill.

Anderson, Ralph V., and Robert E. Gallman. 1977. "Slaves as Fixed Capital: Slave Labor and Southern Economic Development." *Journal of American History* 64(June):24–46.

Atack, Jeremy, and Fred Bateman. 1987. *To Their Own Soil: Agriculture in the Antebellum North.* Ames: Iowa State Univ. Press.

Atherton, Lewis E. 1968. *The Southern Country Store, 1800–1860.* Westport, Conn.: Greenwood Press.

Aufhauser, Keith. 1973. "Slavery and Scientific Management." *Journal of Economic History* 33(Dec.):811–24.

Baldwin, Robert E. 1956. "Patterns of Development in Newly Settled Regions." *Manchester School* 24(May):161–79.

Barney, William L. 1981. "The Ambivalence of Change: From Old South to New in the Alabama Black Belt, 1850–1870." In *From the Old South to the New: Essays on the Transitional South,* ed. Walter J. Fraser, Jr., and Winfred B. Moore, Jr., 33–41. Westport, Conn.: Greenwood Press.

______. 1982. "Towards the Civil War: The Dynamics of Change in a Black Belt County." In *Class, Conflict, and Consensus: Antebellum Southern Community Studies,* ed. Orville Vernon Burton and Robert C. McMath, Jr., 146–72. Westport, Conn.: Greenwood Press.

Bateman, Fred, and Thomas Weiss. 1975. "Comparative Regional Development in Antebellum Manufacturing." *Journal of Economic History* 35(Mar.):182–208.

______. 1981. *A Deplorable Scarcity: The Failure of Industrialization in the Slave Economy.* Chapel Hill: Univ. of North Carolina Press.

Bennett, Hugh Hammond. 1921. *The Soils and Agriculture of the South.* New York: Macmillan.

Berlin, Ira, and Herbert G. Gutman. 1983. "Natives and Immigrants, Free Men and Slaves: Urban Working Men in the Antebellum American South." *American Historical Review* 87(Dec.):1175–1200.

Bonner, James C. 1944. "Profile of a Late Ante-Bellum Community." *American Historical Review* 49(July):663–80.

______. 1964. *A History of Georgia Agriculture, 1732–1860.* Athens: Univ. of Georgia Press.

Brown, Harry B. 1938. *Cotton.* 2d ed. New York: McGraw-Hill.

Buck, Norman Sydney. 1969. *The Development of the Organization of Anglo-American Trade 1800–1850.* Hamden, Conn.: Archon Books.

Campbell, Randolph B. 1974. "Planters and Plain Folk: Harrison County, Texas as a Test Case, 1850–1860." *Journal of Southern History* 40(Aug.):369–98.

Caves, Richard E. 1971. "Export-Led Growth and the New Economic History." In *Trade, Balance of Payments and Growth: Papers in International Economics in Honor of Charles P. Kindleberger,* ed. Jadish N. Bhagwati, Ronald W. Jones, Robert A. Mundell, 403–42. Amsterdam: North-Holland.

Conzen, Michael P. 1975. "A Transport Interpretation of the Growth of Urban Regions: An American Example." *Journal of Historical Geography* 1(Oct.):361–82.

______. 1977. "The Maturing Urban System in the United States, 1840–1910." *Annals of the American Association of Geographers* 67(Mar.):88–108.

Daniel, Pete. 1985. *Breaking the Land: The Transformation of Cotton, Tobacco, and Rice Cultures since 1880.* Urbana: Univ. of Illinois Press.

David, Paul A. 1975. "The Mechanization of Reaping in the Ante-Bellum Midwest." In *Technical Choice Innovation and Economic Growth: Essays on American and British Experience in the Nineteenth Century,* ed. Paul A. David, 195–219. New York: Cambridge Univ. Press.

David, Paul A., and Peter Temin. 1976. "Slavery: The Progressive Institution?" In *Reckoning with Slavery: A Critical Study in the Quantitative History of American Negro Slavery,* ed. Paul A. David, Herbert G. Gutman, Richard Sutch, Peter Temin, and Gavin Wright, 165–230. New York: Oxford Univ. Press.

______. 1979. "Explaining the Relative Efficiency of Slave Agriculture: A Comment." *American Economic Review* 69(Mar.):213–18.

Davis, David Brion. 1984. *Slavery and Human Progress.* New York: Oxford Univ. Press.

Davis, James E. 1977. *Frontier America 1800–1840: A Comparative Demographic Analysis of the Settlement Process.* Glendale: Arthur H. Clark.

Davis, Lance. 1965. "The Investment Market, 1870–1914: The Evolution of a National Market." *Journal of Economic History* 25(Sept.):355–99.

Domar, Evsey. 1970. "The Causes of Slavery or Serfdom: A Hypothesis." *Journal of Economic History* 30(Mar.):18–32.

Dunlevy, James A. 1983. "Regional Preferences and Migrant Settlement: On the Avoidance of the South by Nineteenth-Century Immigrants." *Research in Economic History* 8:217–51.

Earle, Carville. 1978. "A Staple Interpretation of Slavery and Free Labor." *Geographical Review* 68(Jan.):51–65.

______. N.d. "Tillage Capacity and Soil Maintenance in the Nineteenth-Century Cotton South: A New Theory of Crop Choice," Ms., Miami University.

Earle, Carville, and Ronald Hoffman. 1976. "The Urban South: The First Two Centuries." In *The City in Southern History: The Growth of Urban Civilization in the South,* ed. Blaine A. Brownell and David R. Goldfield, 23–51. Port Washington: Kennikat Press.

______. 1980. "The Foundation of the Modern Economy: Agriculture and the Costs of Labor in the United States and England, 1800–1860." *American Historical Review* 85(Dec.)1055–94.

______. 1987. "Regional Staples and Urban Systems in Antebellum America." In *Géographie du Capital Marchand aux Amériques 1760–1860,* ed. Jeanne Chase, 150–81. Paris: Editions de l'Ecole des Hautes Etudes.

Easterlin, Richard A. 1961. "Regional Income Trends, 1840–1950." In *American Economic History,* ed. Seymour E. Harris, 525–47. New York: McGraw-Hill.

______. 1975. "Farm Production and Income in Old and New Areas at Mid-Century." In *Essays in Nineteenth Century Economic History: The Old Northwest,* ed. David C. Klingaman and Richard K. Vedder, 77–117. Athens: Ohio Univ. Press.

Emerson, F. V. 1911. "Geographic Influences in American Slavery." *Bulletin of the American Geographic Society* 43(Jan.–Mar.):13–26, 106–88, 170–81.

Engerman, Stanley L. 1967. "The Effects of Slavery on the Southern Economy: A Review of the Recent Debate." *Explorations in Economic History* 4(Winter):71-97.

______. 1973. "Some Considerations Relating to Property Rights in Man." *Journal of Economic History* 33(Mar.):43–65.

______. 1975. "A Reconsideration of Southern Economic Growth, 1770–1860." *Agricultural History* 49(Apr.):343–61.

______. 1986. "Slavery and Emancipation in Comparative Perspective." *Journal of Economic History* 46(June):317–40.

Fenoaltea, Stefano. 1984. "Slavery and Supervision in Comparative Perspective: A Model." *Journal of Economic History* 44(Sept.):635–68.

Ferleger, Louis. 1985. "Capital Goods and Southern Economic Development." *Journal of Economic History* 45(June):411–18.

Fields, Barbara J. 1982. "Ideology and Race in American History." In *Region, Race and Reconstruction: Essays in Honor of C. Vann Woodward,* ed. J. Morgan Kousser and James M. McPherson, 143–77. New York: Oxford Univ. Press.

______. 1985. *Slavery and Freedom on the Middle Ground: Maryland during the Nineteenth Century.* New Haven: Yale Univ. Press.

Fischbaum, Marvin, and Julius Rubin. 1968. "Slavery and the Economic Development of the South." *Explorations in Economic History* 6(Fall):116–27.

Fite, Gilbert C. 1984. *Cotton Fields No More: Southern Agriculture, 1865–1980.* Lexington: Univ. of Kentucky Press.

Fleisig, Haywood. 1976. "Slavery, the Supply of Agricultural Labor, and the Industrialization of the South." *Journal of Economic History* 36(Sept.):572–98.

Fogel, Robert W., and Stanley L. Engerman. 1974. *Time on the Cross: The Economics of American Negro Slavery.* 2 vols. Boston: Little, Brown.

______. 1977. "Explaining the Relative Efficiency of Slave Agriculture in the Antebellum South." *American Economic Review* 67(June):275–96.

Ford, Lacy K., Jr. 1983. "Social Origins of a New South Carolina: The upcountry in the Nineteenth Century." Ph.D. diss., University of South Carolina.

______. 1984. "Rednecks and Merchants: Economic Development in the South Carolina upcountry, 1865–1900." *Journal of American History* 71(Sept.):294–318.

______. 1986. "Yeoman Farmers in the South Carolina Upcountry: Changing Production Patterns in the Late Antebellum Period." *Agricultural History* 60(Fall):17–37.

Foust, James D. 1975. *The Yeoman Farmer and Western Expansion of U.S. Cotton Production.* New York: Arno Press.

Foust, James D., and Dale E. Swan. 1970. "Productivity and Profitability of Antebellum Slave Labor: A Micro-Approach." *Agricultural History* 44(Jan.)39–63.

Fox-Genovese, Elizabeth. 1983. "Antebellum Southern Households: A New Perspective on a Familiar Question." *Review* 7(Fall):215–53.

Fox-Genovese, Elizabeth, and Eugene D. Genovese. 1984. *Fruits of Merchant Capital: Slavery and Bourgeois Property in the Rise and Expansion of Capitalism.* New York: Oxford Univ. Press.

Gallman, Robert E. 1970. "Self-Sufficiency in the Cotton Economy of the Antebellum South." *Agricultural History* 44(Jan.):5–23.

______. 1975. "The Agricultural Sector and the Pace of Economic Growth: U.S. Experience in the Nineteenth Century." In *Essays in Nineteenth Century Economic History: The Old Northwest,* ed. David C. Klingaman and Richard K. Vedder, 35–76. Athens: Ohio Univ. Press.

______. 1979. "Slavery and Southern Economic Growth." *Southern Economic Journal* 45(Apr.):1007–22.

Garrett, Richard D. 1978. "Primitive Accumulation in the Antebellum Cotton South." Ph.D. diss., New School for Social Research.

Genovese, Eugene D. 1961. *The Political Economy of Slavery: Studies in the Economy and Society of the Slave South.* New York: Vintage Books.

______. 1975. "Yeomen Farmers in a Slaveholders' Democracy." *Agricultural History* 49(Apr.):331–42.

Georgia State College Extension Division. 1923. "Systems of Farming for South Georgia." Bulletin no. 273(Jan.).

Goldfield, David R. 1986. *Cotton Fields and Skyscrapers: Southern City and Region, 1607–1980.* Baton Rouge: Louisiana State Univ. Press.

Goldin, Claudia D. 1976. *Urban Slavery in the American South, 1820–1860: A Quantitative History.* Chicago: Univ. of Chicago Press.

Goldin, Claudia, and Kenneth Sokoloff. 1984. "The Relative Productivity Hypothesis of Industrialization: The American Case, 1820 to 1850." *Quarterly Journal of Economics* 99(Aug.):461–87.

Govan, Thomas P. 1938. "Banking and the Credit System in Georgia." *Journal of Southern History* 4(May):164–84.

Graham, Richard. 1981. "Slavery and Economic Development: Brazil and the United States in the Nineteenth Century." *Comparative Studies of Society and History* 23(Oct.):620–55.

Gray, Lewis C. 1958. *History of Agriculture in the Southern United States to 1860.* 2 vols. Gloucester, Mass.: Peter Smith.

Green, George D. 1972. *Finance and Economic Development in the Old South: Louisiana Banking, 1806–1861.* Stanford: Stanford Univ. Press.

Hahn, Steven. 1983. *The Roots of Southern Populism: Yeoman Farmers and the Transformation of the Georgia upcountry, 1850–1890.* New York: Oxford Univ. Press.

Hammond, M. B. 1966. *The Cotton Industry.* New York: Johnson Reprint.

Haywood, Tamara M. 1986. "Cows, Ticks, and Disease: A Medical Interpretation of the Southern Cattle Industry." *Journal of Southern History* 52(Nov.):551–64.

Heath, Milton S. 1954. *Constructive Liberalism: The Role of the State in the Economic Development of Georgia to 1860.* Cambridge: Harvard Univ. Press.

Heim, Carol E., and Philip Mirowski. 1987. "Interest Rates and Crowding-Out in the Industrial Revolution." *Journal of Economic History* 47(Mar.):117–40.

Hilliard, Samuel B. 1972. *Hog Meat and Hoecake Food Supply in the Old South, 1840–1860.* Carbondale: Southern Illinois Univ. Press.

______. 1986. *Atlas of Southern Agriculture.* Baton Rouge: Louisiana State Univ. Press.

Hirschman, Albert O. 1975. "A Generalized Linkage Approach to Development with Special References to Staples." *Economic Development and Cultural Change* 25(supple.):67–98.

Irwin, James. 1987. "Exploring the Affinity of Wheat and Slavery in the Virginia Piedmont." *Explorations in Economic History* 25(July):295–322.

James, John A. 1981. "Financial Underdevelopment in the Postbellum South." *Journal of Interdisciplinary History* 11(Winter):443–54.

Jaynes, Gerald D. 1986. *Branches Without Roots: Genesis of the Black Working Class in the American South, 1862–1882.* New York: Oxford Univ. Press.

Johnson, Michael P. 1977. *Towards a Patriarchal Republic: The Secession of Georgia.* Baton Rouge: Louisiana State Univ. Press.

Jordan, Terry G. 1967. "The Imprint of the Upper and Lower South on Mid-Nineteenth Century Texas." *Annals of the Association of American Geographers* 57(Dec.):667–90.

Knodell, Jane E. 1988. "Mainstream Macroeconomics and the 'Neutrality' of Finance: A Critical Perspective." *Money and Production* 12(forthcoming).

Lebergott, Stanley. 1985. "The Demand for Land: The United States, 1820–1860." *Journal of Economic History* 45(June):181–212.

Linden, Fabian. 1946. "Economic Democracy in the Slave South: An Appraisal of Some

Recent Views." *Journal of Negro History* 31(Apr.):140–89.

McClelland, Peter D., and Richard J. Zeckhauser. 1982. *Demographic Dimensions of the New Republic.* New York: Cambridge Univ. Press.

McCusker, John L., and Russell R. Menard. 1985. *The Economy of British North America, 1607–1789.* Chapel Hill: Univ. of North Carolina Press.

McDonald, Forrest, and Grady McWhiney. 1975. "The Antebellum Southern Herdsmen: A Reinterpretation." *Journal of Southern History* 41(May):147–66.

McMath, Robert C., Jr. 1987. "Where's the 'Culture' in Agricultural Technology?" Ms., Georgia Institute of Technology.

McMichael, Philip. 1987. "Bringing Circulation Back into Agricultural Political Economy: Analyzing the Antebellum Plantation in its World Market Context." *Rural Sociology* 52(Summer):242–63.

Mandle, Jay R. 1978. *The Roots of Black Poverty: The Southern Plantation Economy after the Civil War.* Durham: Duke Univ. Press.

Mathew, W. M. 1985. "Planter Entrepreneurship and the Ruffin Reforms in the Old South, 1820–1860." *Business History* 27(July):207–21.

Metzer, Jacob. 1975. "Rational Management, Modern Business Practices, and Economies of Scale in the Ante-Bellum Southern Plantations." *Explorations in Economic History* 12(Apr.):123–50.

Meyer, David R. 1980. "A Dynamic Model of the Integration of Frontier Urban Places into the United States System of Cities." *Economic Geography* 56(Jan.):120–40.

______. 1988. "The Industrial Retardation of Southern Cities, 1860–1880." *Explorations in Economic History* (forthcoming).

Miller, Randall. 1985. "The Enemy Within: Some Effects of Foreign Immigrants on Antebellum Southern Cities." *Southern Studies* 24(Spring):30–53.

Miller, Steven F. 1985. "Transplanting Plantation Society: Migrant Planters and the Shaping of the Cotton South, 1815–1840," Ms., University of Maryland.

Moore, John H. 1958. *Agriculture in Ante-Bellum Mississippi.* New York: Bookman.

Morgan, Edmund S. 1975. *American Slavery, American Freedom: The Ordeal of Colonial Virginia.* New York: W. W. Norton.

Muller, Edmund K. 1976. "Selective Urban Growth in the Middle Ohio Valley, 1800–1860." *Geographical Review* 66(Apr.):178–99.

Nash, Gary B. 1979. *The Urban Crucible: Social Change, Political Consciousness, and the Origins of the American Revolution.* Cambridge: Harvard Univ. Press.

North, Douglass C. 1955. "Location Theory and Regional Economic Growth." *Journal of Political Economy* 63(June):243–58.

______. 1959. "Agriculture in Regional Economic Growth." *Journal of Farm Economics,* Proceedings, 41(Dec.):943–51.

Oakes, James. 1982. *The Ruling Race: A History of American Slaveholders.* New York: Alfred A. Knopf.

______. 1985. "From Republicanism to Liberalism: Ideological Change and the Crisis of the Old South." *American Quarterly* 37(Fall):551–71.

Odum, Howard W. 1936. *Southern Regions of the United States.* Chapel Hill: Univ. of North Carolina Press.

Otto, John Solomon. 1985. "The Migration of the Southern Plain Folk." *Journal of Southern History* 51(May):183–200.

Owsley, Frank L. 1982. *Plain Folk of the Old South.* Baton Rouge: Louisiana State Univ. Press.

Parker, William N. 1970. "Slavery and Southern Economic Development: An Hypothesis and Some Evidence." *Agricultural History* 44(Jan.):115–25.

______. 1971. "Productivity Growth in American Grain Farming: An Analysis of 19th-

Century Sources." In *The Reinterpretation of American Economic History,* ed. Robert W. Fogel and Stanley L. Engerman, 175–86. New York: Harper and Row.

______. 1972. "Agriculture" In *American Economic Growth: An Economist's History of the United States,* ed. Lance E. Davis, Richard A. Easterlin, and William N. Parker et al. New York: Harper and Row.

______. 1979. "Labor Productivity in Cotton Farming: The History of a Research." *Agricultural History* 53(Jan.):228–44.

Passell, Peter, and Gavin Wright. 1972. "The Effects of Pre–Civil War Territorial Expansion on the Price of Slaves." *Journal of Political Economy* 80(Nov.–Dec.):1188–202.

Paullin, Charles O. 1932. *Atlas of the Historical Geography of the United States.* Ed. John K. Wright. New York: American Geographical Society.

Perloff, Harvey S., and London Wingo, Jr. 1961. "Natural Resource Endowment and Regional Economic Growth." In *Natural Resources and Economic Growth,* ed. Joseph J. Spengler. Washington, D.C.: Resources for the Future.

Phillips, Ulrich B. 1908. *A History of Transportation in the Eastern Cotton Belt to 1860.* New York: Columbia Univ. Press.

______. 1929. *Life and Labor in the Old South.* Boston: Little, Brown.

______. 1968. "The Origin and Growth of the Southern Black Belts." In *The Slave Economy of the Old South: Selected Essays in Economic and Social History,* ed. Ulrich B. Phillips. Baton Rouge: Louisiana State Univ. Press.

______. 1969a. *American Negro Slavery.* Baton Rouge: Louisiana State Univ. Press.

______. ed. 1969b. *Plantation and Frontier, 1649–1863.* 2 vols. New York: Burt Franklin.

Pred, Allan. 1973. *Urban Growth and the Circulation of Information: The United States System of Cities, 1790–1840.* Cambridge: Harvard Univ. Press.

______. 1980. *Urban Growth and City-Systems in the United States, 1840–1860.* Cambridge: Harvard Univ. Press.

Ransom, Roger L., and Richard Sutch. 1971. "Economic Regions of the South in 1880." *Southern Economic History Project Working Paper Series,* no. 3. Berkeley: Univ. of California.

______. 1972. "Debt Peonage in the Cotton South After the Civil War." *Journal of Economic History* 32(Sept.):641–69.

______. 1977. *One Kind of Freedom: The Economic Consequences of Emancipation.* New York: Cambridge Univ. Press.

______. 1979. "Growth and Welfare in the American South of the Nineteenth Century." *Explorations in Economic History* 16(Apr.):207–36.

______. 1988. "Capitalists Without Capital: The Burden of Slavery and the Impact of Emancipation." *Agricultural History* (forthcoming).

Reidy, Joseph P. 1982. "Masters and Slaves, Planters and Freemen: The Transition from Slavery to Freedom in Central Georgia, 1820–1880," Ph.D. diss., Northern Illinois University.

Robinson, Joan. 1974. *Essays in the Theory of Economic Growth.* London: Macmillan.

Rothstein, Morton. 1966. "Antebellum Wheat and Cotton Exports: A Contrast in Marketing Organization and Economic Development." *Agricultural History* 40(Apr.):91–100.

______. 1967. "The Antebellum South as a Dual Economy: A Tentative Hypothesis." *Agricultural History* 41(Oct.):372–82.

Rubin, Julius. 1967. "Urban Growth and Regional Development." In *The Growth of the Seaport Cities 1790–1825,* ed. David T. Gilchrist. Charlottesville: Univ. of Virginia Press.

______. 1975. "The Limits of Agricultural Progress in the Nineteenth Century." *Agricultural History* 49(Apr.):362–73.

Russel, Robert R. 1941. "The Effects of Slavery Upon Nonslaveholders in the Ante Bellum South." *Agricultural History* 15(Apr.):54–71.

______. 1960. *Economic Aspects of Southern Sectionalism.* New York: Russell and Russell.

Scarborough, William K. 1966. *The Overseer: Plantation Management in the Old South.* Baton Rouge: Louisiana State Univ. Press.

Schaefer, Donald F. 1983. "The Effect of the Crop Year Upon Relative Productivity in the Antebellum South." *Journal of Economic History* 43(Dec.):851–66.

______. 1985. "A Statistical Profile of Frontier and New South Migration: 1850–1860." *Agricultural History* 59(Oct.):563–78.

______. 1987. "Agricultural Displacement at the Southern Frontier: A Test of the Phillips Hypothesis." Ms., Washington State University.

Schaefer, Donald F., and Mark D. Schmitz. 1979. "The Relative Efficiency of Slave Agriculture: A Comment." *American Economic Review* 69(Mar.):208–12.

______. 1982. "Efficiency of Antebellum Southern Agriculture: A Covariance Approach." *Southern Economic Journal* 49(July):88–98.

______. 1985. "The Parker-Gallman Sample and Wealth Distributions for the Antebellum South: A Comment." *Explorations in Economic History* 22(Apr.):220–26.

Schmitz, Mark D., and Donald F. Schaefer. 1978. "Slavery, Freedom, and the Elasticity of Substitution." *Explorations in Economic History* 15(July):327–37.

Schweikart, Larry. 1987a. "Southern Banks and Economic Growth in the Antebellum Period: A Reassessment." *Journal of Southern History* 53(Feb.):19–36.

______. 1987b. *Banking in the American South from the Age of Jackson to Reconstruction.* Baton Rouge: Louisiana State Univ. Press.

Shlomowitz, Ralph. 1978. "The Transition from Slave to Freedman: Labor Arrangements in Southern Agriculture, 1865–1870." Ph.D. diss., University of Chicago.

Smith, Alfred G., Jr. 1958. *Economic Readjustment of an Old Cotton State: South Carolina, 1820–1860.* Columbia: Univ. of South Carolina Press.

Stampp, Kenneth. 1956. *The Peculiar Institution: Slavery in the Antebellum South.* New York: Vintage Books.

Steckel, Richard. 1983. "The Economic Foundations of East-West Migration." *Explorations in Economic History* 20(Jan.):14–36.

______. 1986a. "A Peculiar Population: The Nutrition, Health, and Mortality of American Slaves from Childhood to Maturity." *Journal of Economic History* 46(Sept.):721–42.

______. 1986b. "A Dreadful Childhood: The Excess Mortality of American Slaves." *Social Science History* 10(Winter):427–66.

Sutch, Richard. 1975. "The Breeding of Slaves for Sale and the Western Expansion of Slavery, 1850–1860." In *Race and Slavery in the Western Hemisphere: Quantitative Studies,* ed. Stanley L. Engerman and Eugene D. Genovese, 173–210. Princeton: Princeton Univ. Press.

______. 1976. "The Care and Feeding of Slaves." In *Reckoning with Slavery: A Critical Study in the Quantitative History of American Negro Slavery,* ed. Paul A. David, Herbert G. Gutman, Richard Sutch, Peter Temin, and Gavin Wright, 231–301. New York: Oxford Univ. Press.

Taaffe, Edward J., Richard L. Morrill, and Peter R. Gould. 1963. "Transport Expansion in Underdeveloped Countries." *Geographical Review* 53(Oct.):503–29.

Terril, Tom E. 1976. "Eager Hands: Labor for Southern Textiles, 1850–1860." *Journal of Economic History* 36(Mar.):84–99.

Terzi, Andrea. 1986–87. "The Independence of Finance from Savings: A Flow of Funds Interpretation." *Journal of Post Keynesian Economics* 9(Winter):188–97.

Thompson, Edgar T. 1975. *Plantation Societies, Race Relations, and the South: The Regimentation of Populations.* Durham: Duke Univ. Press.

Turner, J. A. 1865. *The Cotton Planter's Manual.* New York: Orange Judd.

U.S. Census Office, Tenth Census [1880], Eugene W. Hilgard, Special Agent. 1884. *Report on Cotton Production in the United States.* 2 vols. Washington D.C.: Government Printing Office.

U.S. Department of Agriculture. 1896. Office of Experiment Stations, *The Cotton Plant.* Washington, D.C.: Government Printing Office.

U.S. Department of Agriculture, James R. Covert. 1912. "Seedtime and Harvest: Cereals, Flax, Cotton, and Tobacco." *Bureau of Statistics Bulletin,* no. 85. Washington, D.C.: Government Printing Office.

U.S. Department of Agriculture, H. R. Cates. 1917. "Farm Practices in the Cultivation of Cotton." *Bulletin* no. 511. Washington, D.C.: Government Printing Office.

Vance, James E., Jr. 1970. *The Merchant's World: The Geography of Wholesaling.* Englewood Cliffs: Prentice-Hall.

Vance, Rupert B. 1932. *Human Geography of the South: A Study in Regional Resources and Human Adequacy.* Chapel Hill: Univ. of North Carolina Press.

Wade, Richard C. 1959. *The Urban Frontier: Pioneer Life in Early Pittsburgh, Cincinnati, Lexington, Louisville, and St. Louis.* Chicago: Univ. of Chicago Press.

Wallenstein, Peter. 1987. *From Slave South to New South: Public Policy in Nineteenth Century Georgia.* Chapel Hill: Univ. of North Carolina Press.

Watkins, James L. 1969. *King Cotton.* New York: Negro Univ. Press.

Watkins, Melville H. 1963. "A Staple Theory of Economic Growth." *The Canadian Journal of Economics and Political Science* 29(May):141–58.

Watson, Harry L. 1985. "Conflict and Collaboration: Yeomen, Slaveholders, and Politics in the Antebellum South." *Social History* 10(Oct.):273–98.

Wayne, Michael. 1983. *The Reshaping of Plantation Society: The Natchez District, 1860–1880.* Baton Rouge: Louisiana State Univ. Press.

Weiman, David F. 1984. "Petty Commodity Production in the Cotton South: Upcountry Farmers in the Georgia Cotton Economy, 1840 to 1880." Ph.D. diss., Stanford University.

______. 1985. "The Economic Emancipation of the Non-Slaveholding Class: Upcountry Farmers in the Georgia Cotton Economy." *Journal of Economic History* 45(Mar.):71–93.

______. 1987. "Farmers and the Market in Antebellum America: A View from the Georgia Upcountry." *Journal of Economic History* 47(Sept.):627–47.

______. 1988a. "Urban Growth on the Periphery of the Cotton Belt: Atlanta, 1847–1860." *Journal of Economic History* 48(June):259–72.

______. 1988b. "Families, Farms, and Rural Society in Antebellum America." In *Agrarian Organization in the Century of Industrialization, Research in Economic History,* ed. George Grantham and Carol S. Leonard. Supple., vol. 11 (forthcoming).

Whatley, Warren. 1987. "Southern Agrarian Labor Contracts as Impediments to Cotton Mechanization." *Journal of Economic History* 47(Mar.):45–70.

Wiener, Jonathan M. 1978. *Social Origins of the New South: Alabama 1860–1885.* Baton Rouge: Louisiana State Univ. Press.

Woodman, Harold D. 1968. *King Cotton and His Retainers: Financing and Marketing the Cotton Crop of the South, 1800–1925.* Lexington: Univ. of Kentucky Press.

Wright, Gavin. 1970. "'Economic Democracy' and the Concentration of Agricultural Wealth in the Cotton South, 1850–1860." *Agricultural History* 44(Jan.):62–93.

______. 1971. "An Econometric Study of Cotton Production and Trade, 1830–1860." *Review of Economics and Statistics* 53(May):111–20.

______. 1978. *The Political Economy of the Cotton South.* New York: W. W. Norton.

______. 1979a. "The Efficiency of Slavery: Another Interpretation." *American Economic Review* 69(Mar.):219–26.

______. 1979b. "Cheap Labor and Southern Textiles before 1880." *Journal of Economic History* 39(Sept.):665–80.

______. 1986. *Old South, New South: Revolutions in the Southern Economy Since the Civil War.* New York: Basic Books.

______. 1987. "Capitalism and Slavery: A Lesson from the Mainland." *Journal of Interdisciplinary History* 17(Spring):851–70.

Wright, Gavin, and Howard Kunreuther. 1975. "Cotton, Corn and Risk in the Nineteenth Century." *Journal of Economic History* 35(Sept.):526–51.

Yang, Donghyu. 1984. "Notes on the Wealth Distribution in the United States, 1860: A New Look at Two Manuscript Samples." *Explorations in Economic History* 21(Jan.):88–102.

______. 1985. "The Parker-Gallman Sample and Wealth Distributions for the Antebellum South: A Reply." *Explorations in Economic History* 22(Apr.):227–32.

CHAPTER 7

# Growth and Development in the Antebellum South: Old Debates and New Directions

RICHARD H. STECKEL

In many ways the agenda for cliometric research on the economy of the antebellum South was established before the middle of the twentieth century. The charges and countercharges of the prewar debate over slavery and abolition left a residue of ideas condemning the peculiar institution. Hinton Rowan Helper (1857), for example, argued that inefficiencies inherent in slavery retarded southern economic growth while Frederick Law Olmsted (1861) maintained that slave labor was less productive than free labor and that investments in slaves were generally unprofitable. By the twentieth century themes of growth retardation, inefficiency, nonviability, unprofitability, and harshness of slave life often appeared in works on antebellum southern history (see, for example, Phillips 1905, 1918; Ramsdell 1929; Flanders 1933; Gray 1933; Sydnor 1933; Russell 1939; Stampp 1956).[1] Thus when Conrad and Meyer (1958) published their now famous paper ushering in the new economic history, they confronted widely held views and ways of thinking about slavery and the southern economy.

This chapter reviews the cliometric literature on the antebellum southern economy, emphasizing developments during the past two decades and suggesting new directions for research. Work conducted in the decade or so following the paper by Conrad and Meyer largely settled the profitability issue and the question of whether economic forces alone would have ended the slave system, thereby dropping these subjects from contention (see, for example, Yasuba 1961; Evans 1962; Sutch 1965; Bergstrom 1971).[2] Thereafter research turned toward the productivity of slave labor and economies of scale, slavery and southern economic performance, interregional trade and plantation self-sufficiency, the yeoman experience, and the demography of the southern population. This analysis discusses developments in all these areas but highlights demographic research.

## THE ECONOMY OF THE PLANTATION AND THE REGION

The efficiency of slave labor has been a contentious issue in the recent work on slavery. Robert Fogel and Stanley Engerman (1971, 1974) crystallized thought and debate by expressing efficiency in terms of a geometric index of total factor productivity, which is a ratio of output to a geometric average of inputs.[3] Using data from the manuscript schedules of the U.S. Census of Agriculture, they reported that slave farms were about 40 percent more efficient than northern farms and 28 per cent more efficient than southern free farms in 1860, and that efficiency of slave farms increased with the number of slaves up to 50 and then declined slightly. Fogel and Engerman suggested that specific features of slavery, such as intensive use of labor and superior management, explained the results. Critics offered alternative explanations, citing temporarily inflated prices for cotton in 1860, a large cotton crop in 1860, poor measurement of the land input, a longer work year for slaves, and regional differences in crop mix (David and Temin 1974, 1979; Haskell 1974, 1979; Wright 1975, 1979; Schaefer and Schmitz 1979). Fogel and Engerman (1977, 1980) tested the implications of these criticisms and in my judgment successfully defended their claim of greater efficiency on medium and large plantations, though one may quibble with procedural details and the precise magnitude of the efficiency advantage.[4] Drawing upon the work of Jacob Metzer (1975) and others, Fogel and Engerman also made a plausible case for the importance of the gang system to efficiency but have been unable to measure its contribution relative to other factors affiliated with size such as managerial skills, child-care practices, discipline or force, and complementarities in production.

The thesis that slavery substantially retarded southern economic growth suffered a major setback with the publication of regional income estimates for 1840 to 1860 by Richard Easterlin (1961; later revised by Engerman 1967, 1971), which showed that southern per capita income was about two-thirds of that in the North while the rate of growth was slightly higher in the South. The unexpectedly small size of the gap and especially the growth rate advantage for the South suggested that the peculiar institution may have had little adverse effect on southern economic performance. Subsequent research addressed the adequacy of the income estimates, special conditions of the late antebellum economy that may have affected measured growth, and the mechanisms through which slavery could have retarded growth. Gerald Gunderson (1973) argues that underreporting in the 1840 census of agriculture exaggerates southern performance from 1840 to 1860, but Easterlin's (1975) revised estimates for 1840, 1850, and 1860 indicated that the ratio of northern to southern income changed little in each decade. Roger Ransom and Richard Sutch (1977, 265–66) maintain that Easterlin's figures underestimate the northern advantage and that region's growth rate, but findings by Raymond Cohn (1981) suggest that this bias was modest. Gavin Wright (1975; 1978) emphasizes that the late antebellum cotton boom improved the measured performance of the South, but Ralph Anderson and Robert Gallman (1977) also note that the period from the depression of the late 1830s and early 1840s to 1860 encompassed a Kuznets boom that benefited the North. In sum, the available evidence indicates that per capita income was moderately lower in the South compared with the North but growth rates were roughly equal during the last two decades of the antebellum period.

There is little agreement on models that should be used to investigate the counterfactual of whether southern economic performance in the nineteenth century would have improved without slavery. Recent additions to this literature suggest, however, that the peculiar institution retarded long-run industrial development. Heywood Fleisig (1976) argues that imperfect labor markets in the rural North limited farm size and therefore profit levels, which directed entrepreneurs into urban areas and stimulated industrial growth, while in the South slavery eliminated the labor constraint on large-scale farming operations, agriculture retained entrepreneurial talent, and industrialization was retarded. Anderson and Gallman (1977) explore the implications of Eugene Genovese's (1965, 49) dictum that slavery required all hands to be occupied at all times. In this setting slaves became a form of fixed capital that planters kept busy during lay-by periods through diversification of farm operations. Thus plantations had a high ratio of labor to capital compared with the northern farms, were largely self-sufficient, and had little need of a transportation system and distribution centers (cities and towns) that would have furthered commercial and industrial development. Fred Bateman and Thomas Weiss (1981) argue that planters were reluctant to invest in manufacturing despite high rates of return, possibly because they preferred to avoid risks, were uninformed about the benefits of diversification, or felt threatened economically, socially, and politically by industrialization. Wright (1986) emphasizes that because slave property was portable planters had little stake in community life and little to gain from local improvements in roads, marketing facilities, and education. In this view emancipation was an economic revolution that led to new investment strategies for local development.

The issue of southern self-sufficiency in food production is a byproduct of hypotheses of growth that stress the contribution of interregional trade in widening the market for manufactured goods (see, for example, North 1961) and of claims that southern agriculture lacked the productivity to feed its population. Estimates of food supplies and demands have shown, however, that with the possible exception of certain coastal areas the South was largely independent of western food supplies (Gallman 1970; Lindstrom 1970; Battalio and Kagel 1970; Hutchinson and Williamson 1971). Mark Schmitz (1978) addressed the widespread belief that planters in southern Louisiana imported much of their food from the West. His analysis of manuscript schedules of agriculture indicates that this rural sector could have been self-sufficient if planters exchanged their excess grain for meat produced by small, local farmers.

Views of antebellum southern farmers have been dominated by two opposing schools of thought. U. B. Phillips (1918) and Lewis Cecil Gray (1933) argued that large planters controlled social and economic life in the antebellum South and pushed yeoman farmers into a subordinate position, while members of the Owsley school (see, for example, Owsley 1949 and Clark 1942) emphasized the numerical majority and prosperity of small farmers under democratic institutions. Although recent research tends to favor the planter-dominance position, this work also demonstrates oversimplifications and misconceptions inherent in both schools of thought. Wright (1970) identified several biases and shortcomings in the traditional works and placed the new round of clinometric research on firm methodological foundations. Wright's work on the cotton South showed that the distribution of wealth (as measured by a Gini coefficient and by shares of wealth held by the rich) was more unequal in the South than the North, and that from 1850 to 1860 there was a shift of slaveownership away from small toward larger farmers, resulting in greater concentration of wealth.[5] Studies of other regions and cropping patterns within the South reveal a pattern of inequality similar to that in the cotton South (Neimi 1977; Schaefer 1978; Winters 1987a).[6] Although several studies show that wealth was highly concentrated among the slaveowner class, yeoman were more prosperous and heterogeneous than depicted by the planter-dominance thesis. Small farmers of the plantation belt devoted much of their resources to cotton production, and thereby partook of the cotton boom of the late 1840s and the 1850s, while their counterparts in the upcountry were substantially self-sufficient until there were railroads and other investments in commercial infrastructure beginning in the 1850s. These late-antebellum changes overcame geographic isolation, broke down a culture that emphasized social and political autonomy, and promoted capital gains that were an important first step away from safety-first agriculture and toward market integration (Hilliard 1972; Wright 1978; Hahn 1983; Weiman 1985, 1987; Ford 1986).

## HEALTH

Mortality rates and other ways of measuring health are widely regarded as sensitive indices of the living standards of a population. Given the limited economic information available for the antebellum period, researchers have sought data on health measures to help clarify the southern experience of economic development. In addition, slave mortality rates are important ingredients in profitability calculations.

Although debate over slave health has a long tradition that dates from the abolitionist movement, the most comprehensive investigation of the subject occurred in *Time on the Cross* (Fogel and Engerman 1974) and its aftermath (see, for example, Sutch 1975a). Participants in the modern debate usually agreed that while instances of malnutrition occurred, slaves were reasonably well fed as a group and discussion focused instead on variety in the diet and vitamin and mineral deficiencies. Fogel and Engerman placed slave mortality rates near those of the free population.

The new findings establish the diversity of conditions among slaves. Children's health was comparable to that in the poorest populations ever studied while working-age slaves were remarkably well-off. These conclusions rest on plantation records of births and deaths, measurements of stature recorded on manifests of slaves shipped in the coastwise trade, and an approach that combines these data through the concept of net nutrition (Steckel 1979a, 1979b, 1986a, 1986b, 1986c, 1987). Study of human growth and development makes clear that diet, disease, and work cannot be evaluated in isolation (see, for example, Tanner 1978). Instead, these factors are components of a package that interact to determine a person's level of net nutrition (or nutritional status). Work, disease, and body maintenance place claims on the diet and the net result influences growth and stature. Thus instances of growth retardation and illness and mortality tend to occur together.

Recognizing that decisions were made in an environment of poor medical knowledge and that many deaths would have occurred despite the best intentions and the best care that was available, explanations have been sought for the excess mortality of slaves (Steckel 1986c). Plantation records that were adjusted for underenumeration of deaths occurring soon after birth indicate that slave mortality rates were approximately 350 per thousand among infants and 200 per thousand at ages 1 to 4. These childhood rates were roughly double those of the entire free population in the United States, but the rates of slave and free adults were approximately equal. Mortality rates of slaves and whites are discussed further at the end of this section.

The average slave child fell below the first centile of height standards that characterize Europeans, Americans of European descent, and Afro-Americans who are well-nourished, but the average slave adult exceeded the 27th percentile. Comparisons of average heights in studies of growth from childhood to maturity show that slave children were among the smallest ever measured while growth recovery during and after adolescence was remarkable if not unprecedented.

Study of plantation records suggests that prenatal conditions were poor. A majority of slave newborns probably weighed less than 2500 grams (5.5 pounds), and perhaps two-thirds of the infant losses occurred within the first month of birth (Steckel 1986a, 1986c). Poor prenatal health probably elevated the incidence of sudden infant death among slaves (Savitt 1978; Johnson 1981). Seasonal patterns of mortality help to identify possible causes of the poor health of newborns. About three-quarters of the neonatal losses took place during the six months of February to April and September to November, and stillbirths were concentrated in the late autumn and to a lesser extent in February to April. Nutritional deprivation near the time of conception elevates stillbirths whereas neonatal deaths are caused largely by deprivation near conception or near delivery. The high incidence of stillbirths during the late autumn, and the concentration of neonatal deaths in February to April is consistent with acute nutritional deprivation during the late winter and early spring. Although diet, disease, and work are ingredients in net nutrition, disease is an unlikely cause for the concentration of losses because the sickly season occurred during the late summer and early autumn. Similarly, the nutritional value of the diet was probably at a minimum during the late spring, which is inconsistent with the lack of stillbirths beginning in January and the lack of neonatal deaths during May and June. The implausibility of disease and diet as causes points to hard work as a major source of excess losses before the end of the first month. This hypothesis is supported by the fact that the plowing and planting season of late winter and early spring was particularly strenuous for adults and by evidence from plantation work records and instructions to overseers, which indicates that pregnant women had little relief from work during peak seasons in the demand for labor (Campbell 1984).

Although poor prenatal care and low birth weights contributed to high mortality rates beyond the first month, a poor diet and infections then entered the picture. The meager evidence available on feeding practices suggests that slave infants may have received breast milk for nine months to one year, whereas whites probably continued for more than one year (Steckel 1986c). Slave women usually resumed work within three to five weeks after delivery, and while the mothers were in the fields the young children remained in the nursery, at least on the larger plantations, under the care of older women and their assistants who were older children. Initially the new mothers returned to the nursery two to three times per day for breastfeeding, but within three months after delivery productivity, as measured by the number of pounds of cotton picked per day, reached normal levels. At this time one or more of the daytime breast-feedings may have been eliminated, and infants probably received starchy paps and gruels, often contaminated or fed using contaminated utensils. Owners who recorded births and deaths and who noted causes of infant death frequently mentioned diarrhea or "cholera infantim" as it was called in the medical literature of the era.

Young children who survived the hazardous period of infancy encountered a poor diet and diseases that were often related to poor nutrition (Kiple and King 1981). The child's diet emphasized hominy and fat, and owners and medical practitioners often cited whooping cough, diarrhea, measles, worms, and pneumonia as causes of death.

By ages 8 to 12 work entered the equation of slave health. Other things being equal, work would have placed a claim on the diet that retarded growth. Yet it was at the ages that work usually began, initially as a light activity, that some catch-up growth occurred (Steckel 1986b). Improving health accompanied the growth recovery and mortality rates declined after early childhood, reaching 54 per thousand at ages 5 to 10, 37 per thousand at ages 10 to 14 and 35 per thousand at ages 15 to 19. Other things must not have been equal. As workers, slaves received regular allocations of meat (about one-half pound of pork per day) and other foods. Slaves may have supplemented the allocations by raising garden produce, pigs, and fowl, and by fishing and hunting. In addition, through experience slaves may have become more efficient at their work, thereby leaving more nutrition from a giver food supply for growth. The strong growth recovery as teenagers and workers reinforces the view that the diet was at least adequate, if not exceptional, for the disease load and the tasks performed by slaves. Adolescents may have encountered nutritional deficiencies but they were not widespread or severe enough to impair the substantial recovery of average growth.

The peculiar pattern of food allocation by age may have been achieved partly at the expense of the slave family. Workers ate breakfast and lunch in the fields, and children were often fed earlier and separate from adults in the evening. Therefore parents may have had little time to spend with their children on a regular basis. After emancipation, former slaves who gravitated toward nuclear family norms may have been disadvantaged, through lack of experience, in training young children.

The decision by slaveowners to exclude or reduce meat from the diets of children can be viewed as an investment decision (Steckel 1986b). A better diet during childhood would have produced slaves who were bigger and more productive when they entered the labor force, but the extra rations were costly. Reasonable estimates of the costs of the rations and the benefits of the extra labor suggest that it was unprofitable for slaveowners to feed meat to children.

Although the health of slave children was poor by standards of the nineteenth century, one must distinguish effects specific to the region versus the institution of slavery. It has been argued, for example, that a relatively harsh disease environment in the South was more likely to have exposed residents to diseases, such as malaria and gastrointestinal infections, that caused high death rates. How much of the excess mortality of slave children could be attributed to factors that were special to the South? Some information on this question is available from a national sample of nearly 1,600 families matched from the 1860 to the 1850 census manuscript schedules of population (Steckel 1988b). Although family members could have left the household for reasons other than death, the presence or absence in 1860 of young children and the wife who were recorded in these families in 1850 can be studied as an approximate guide to mortality conditions. Regressions of dichotomous variable representing failure to survive to 1860 on age, wealth, literacy, ethnicity, regions, and number of children show that losses were systematically higher for southern women and may have been higher for some southern children. Women's loss rates were 16.3 percent in the South ($t = 2.51$) compared with 10.0 percent in the Northeast. Among children aged 1 to 4 the expected losses were 16.3 percent in the South and 13.3 percent in the Northeast but the difference was not statistically significant ($t = 1.26$). If the South is divided into two subregions, coastal states from North Carolina to Louisiana and the remaining states (except Texas, which was placed in the frontier), expected losses were higher in the coastal states (18.7 percent; $t = 1.94$) but not the remaining states of the South compared with the Northeast. Because the estimated mortality rates of slave children were approximately double those for the entire free population, the household data suggest that a modest portion of the excess mortality of slave children could be attributed to regional factors such as the disease environment.

The matched sample of households also indicates that the number of young siblings had a significant and powerful adverse effect on the survival of children aged 1 to 4. The expected loss rate of children aged 1 to 4 rose from 10.3 percent to 17.4 percent if the number of children below age 10 in the family increased from 1 to 4. While it is possible that parents who had several children sent some of them away to work or to live elsewhere, this response to crowding is an unlikely explanation for the adverse effect of the number of children on losses of those aged 1 to 4 because the number of children under age 10 had no systematic effect on the survival of those aged 5 to 7. Instead, one could argue that additional children helped replenish the reservoir of childhood diseases, thereby increasing the chances of exposure. If this hypothesis is correct, then concentrations of children in the care facilities of plantations probably contributed to the excess mortality of slave children.

## FERTILITY

Birth rates influenced economic development through the rate of growth of the labor force and by affecting the age distribution of the population, and therefore the share of the population that was in the labor force. In addition, family size in part determined the labor force participation rate of women, and birth rates indirectly influenced migration rates to the West and to urban areas.

The central issues in this research have been the long-term decline and regional differences in fertility. These patterns have been identified and studied using indirect measures of fertility, such as a child-woman ratio, because a comprehensive vital registration system did not exist to record births. Although mortality, migration, and the age distribution of women affect this measure, it is highly correlated with direct measures such as the crude birth rate and total fertility. Among the white population in the entire country the number of children under 10 per thousand women aged 16 to 44 declined by nearly 30 percent from 1800 to 1860 and, depending upon the specific comparisons, was typically 10 to 50 percent higher in the western compared with eastern states (Yasuba 1962, 62). Geographic patterns within the South tended to follow those in the country as a whole. In that region the number of children under 10 per thousand women aged 15 to 49 declined from 1,687 to 1,303 and the average value of the ratio in each census year from 1800 to 1860 and 19.4 percent higher in the states of the West compared with the East.[7] Patterns of slave fertility are discussed at the end of this section.

Factors affiliated with urbanization and industrialization have been suggested as explanations for the time trend in fertility, but less than 20 percent of the population lived in urban areas as late as 1860. Forster and Tucker (1972) estimate that the population shift from rural to urban areas explained about 10 to 15 percent of the antebellum fertility decline. High birth rates, the abundance of farmland, and the importance of agriculture have led to land-availability models of fertility (see, for example, Yasuba 1962; Easterlin 1971; Schapiro 1982). According to this framework, areas with lower population densities (fewer people per square mile) can support more children. If this hypothesis is correct, then low density in the West and the overall increases in density over time could explain both the regional differences and the time trend in fertility. High correlations between density and child-woman ratios at the aggregate level favor this argument, but the approach lacks a clear mechanism of operation at the individual level and the hypothesis has done poorly in tests using household data (McInnis 1977). Moreover, one would expect to observe consistently high fertility on the frontier, where density was low, but in this region child-woman ratios declined over time by the same order of magnitude as in nonfrontier areas (Steckel 1985, 132–33).

Given that marriage and family limitation are often subject to different socioeconomic influences, the search for explanations has tried to measure the extent to which couples may have controlled births. Little direct information is available on this subject before the twentieth century, but indirect approaches suggest that control within marriage probably had little impact on birth rates in the antebellum South. One indirect approach utilizes an equation that defines a synthetic total fertility rate, R:

$$R = [((L-F)/S) + 1]\beta, \tag{1}$$

where L is average age at last birth, F is average at first birth, S is the average child-spacing interval, and $\beta$ is the proportion of women who eventually have children (Steckel 1980). The expression in brackets is the total fertility rate for women who had at least one child, and multiplication by $\beta$ converts the measure to a total fertility rate for all women. The measure is a "synthetic" as opposed to a true total fertility rate because the true underlying fertility schedule is unknown. For this reason the components of R are approximated using cross-section data available from the manuscript schedules of population. Equation (1) is useful for understanding the possible causes of the fertility decline because R can be predicted reliably from the child-woman ratio (which is known from census data beginning in 1800) and because there are limits on plausible values of the components. L, for example, was approximately 40 years, S was seldom less than about 2 years, and $\beta$ rarely exceeded 0.95 in natural fertility populations of the past. Estimates of the components for the South in 1860 are $L = 38.8$, $F = 22.7$, $S = 2.76$, and $\beta = 0.828$ (Steckel 1980, 338). The findings that L was nearly 40 and that S was not much above 2 years in 1860 (after a 23 percent decline in fertility from 1800) suggests that few southern women reduced family size by deliberately spacing births or by averting births near the end of their child-bearing span. Instead, it is likely that delays or avoidance of marriage, which would register as a rise in F and a decline in $\beta$, account for a majority of the southern fertility decline.

The suggestion that birth control within marriage played little role in the antebellum fertility decline is reinforced by evidence from genealogies and from fertility patterns reported by older women in the censuses of 1900 and 1910. The genealogies show that marital age-specific fertility rates changed little until birth cohorts of the mid-1800s and that increases in the age at marriage and especially in the percentage who never married largely explain the antebellum fertility decline (Wahl 1986, 397–403).[8] Unfortunately, the southern genealogy sample is small and few details are available for the region alone. Analysis of the fertility histories of older women reported by the census indicates, however, that the practice of family limitation began in the Northeast in the early 1800s, spread to the Midwest, and became important in the South by the end of the nineteenth century (David and Sanderson 1986, 309–11).

If control within marriage was unimportant in the antebellum South, then what caused the decline in marriage rates and why was fertility higher in the West? Four possible explanatory factors have emerged in recent work. One emphasizes the contribution of education to fertility decisions. According to this line of thought, education may promote awareness of methods of family limitation, increase the diffusion of middle-class values, and raise the costs of rearing children (Schultz 1973; Vinovskis 1981; Caldwell 1982). The proposed relationship between literacy or schooling and fertility has mixed support in historical studies (Graff 1979), however, and the proposed mechanism applies more to control of births within marriage than avoidance of marriage.[9] In a variant of the land-availability model, Easterlin (1976) suggests that farmers' goals to provide land for their children were constrained in old relative to new areas by slow rates of growth in their capital. The aging of agricultural areas could explain the long-term decline in fertility, but this approach also operates through family limitation, which does not fit well with the available evidence for the antebellum South. A third hypothesis emphasizes the effect of the distribution of wealth on fertility. This approach is related to the land-availability models but embraces all assets rather than land alone. Moreover, an increase in wealth may increase income, which increases the demand for children, while also increasing the price of children such that the net outcome on fertility is ambiguous. Jenny Wahl (1986), for example, reports that the number of children per family declined as wealth increased at low levels but then followed an inverted-U pattern as wealth increased further. In a study of southern households in the 1860 census manuscript schedules of population, Richard Steckel (1980) finds the chances that a woman had a child increased substantially (from 29 to 54 percent) as family wealth increased from zero to higher levels, which suggests that a small critical mass of wealth was necessary for family formation. Therefore, growing inequality of wealth, which has been observed for the antebellum period (Williamson and Lindert 1980), could have contributed to a decline in marriage. However, differences in the share of families that had zero wealth explain only 7 percent of the East-West difference in fertility within the South. A fourth hypothesis of fertility patterns emphasizes the old-age security motive for raising children (Sundstrom and David 1988). According to this argument, children expected to receive an inheritance or other transfers in exchange for old-age care, but labor market opportunities for children improved with economic growth, which increased the bargaining power of children in their relationship with parents. The value of children as care providers diminished and parents responded by reducing fertility. Although state-level data are broadly consistent with this approach, one should test the theory against individual experiences that may have been recorded in contracts or written understandings between parents and children. Southerners did not have a penchant for written records, however, and a better setting for empirical work may be found in the North or in Europe.

The time trend of antebellum fertility was downward for both slaves and southern whites, but slaves had a regional pattern opposite that of whites. From 1820 (the first census years for which detailed age information is available for slaves) to 1860 the number of slave children under 10 per thousand women aged 15 to 49 declined by 11.1 percent (Steckel 1982, 241). The average value of the ratio, however, was 4.5 percent higher in the East compared with the West in the census years from 1820 to 1860. Although the child-woman ratios were similar for slaves and southern whites (equalling 1,318 and 1,303, respectively, in 1860), the high ratio in the East combined with beliefs that returns on crop production were low in that region have fueled speculation that techniques of animal husbandry were used in the East to breed slaves for sale in the West. However, evidence at the aggregate level contradicts the widespread existence of this activity; slave prices more than doubled between the early 1840s and 1860 while child-woman ratios declined (Phillips 1918, 371). Farms identified in the manuscript schedules of the 1860 census as having high ratios of women to men and relatively large numbers of children have been cited as evidence for the selective existence of breeding (Sutch 1975b). Although it is difficult to prove or disprove the argument of selectivity, account books recording these operations have never been found and slave sales contributed relatively little to farm income (Fogel and Engerman 1974, 78–86). Moreover, the costs of forcing slaves to cohabit against their will, in the form of shirking field work, could easily have outweighed the monetary benefits of additional children (Steckel 1985, 203–6). The high mortality rates of infants and children also contradict the idea that owners invested heavily in pronatal policies. Nevertheless, owners may have provided indirect incentives, such as housing, that increased childbearing.

If slaveowners provided no more than indirect inducements for family formation, then slaves were largely free to seek partners on the farm providing they were not close relatives. Which relatives may have been eligible is not known in general, but Herbert Gutman's (1976, 88) study of birth and inventory lists indicates that slaves usually did not marry someone as close as a first cousin. If the sex ratio on the plantation was skewed, owners could have purchased slaves in hopes that additional unions would result. But allowing marriage to a slave who lived on a different farm was the only practical solution to the problem of finding an acceptable partner, especially on small units. According to testimony given by witnesses on behalf of former slaves who were seeking a Civil War pension, approximately 50 percent of the slaves lived under these arrangements (Steckel 1985, 227).

Estimates of the components of the synthetic total fertility rate (R in equation 1) for slaves and southern whites provide additional insights into determinants of slave fertility (Steckel 1985). According to slave lists in probate records and a sample of manuscript schedules of population in 1860, age at first birth was about 2 years lower for slaves (20.6 versus 22.7). the lower age for slaves may be explained by the subsidy (housing, food, and clothing) that owners gave for family formation. In contrast, free individuals generally lacked this subsidy and had to amass sufficient wealth to support a family. Yet the age of 20.6 years was not so low for slaves as to suggest maximum use of reproductive potential. An estimate of age at menarche based on stature indicates that most slave women abstained for at least 1 to 2 years after they were capable of having children (Trussell and Steckel 1978). Estimates of $\beta$ were below potential for both slaves (0.858) and southern whites (0.828). The possible role of the wealth distribution on marriage was noted earlier. Slaves did not face the costs of establishing new households, as did whites, but practices against marrying close relatives and geographic restrictions to nearby farms did create difficulties in finding a suitable partner. Estimates of L are approximately equal for slaves (38.7) and southern whites (38.8). Therefore, the child-woman ratios were nearly equal for slaves and whites, despite lower values of F and $\beta$ for slaves, because higher mortality rates lengthened the spacing interval between surviving slave children.

## MIGRATION

Interregional flows, mobility rates, and the distance and direction of movement have been central concerns in migration research. Debate over interregional flows has centered on claims that economic stagnation within the South and an aversion to slavery drove the free population to the North. Peter McClelland and Richard Zeckhauser (1982, 7) argue that an important South-to-North flow existed. According to their census-survivor estimates the South and the West combined lost "some 200,000 white persons in the twenty years before Lincoln's election." However, their results contrast with estimates of interregional flows based on a sample of nearly 1,600 households matched from the 1860 to the 1850 census manuscript schedules of population (Steckel 1989). According to the matched sample only 1.9 percent of those who resided in the South in 1850 lived in the North in 1860. A small inflow from the North reduced the net outflow to that region to 0.9 percent. These figures are small compared with the net flows from the Northeast of 10.0 percent to the North Central States, of 2.5 percent to the West, and the net departure rate of 13.2 percent from the North Central states to the West. By placing Kentucky in the North and Texas in the West, McClelland and Zeckhauser increase the measured rate of outflow from the South. Use of their definitions with Steckel's matched sample raises the estimated net outflow from the South to the North from 0.9 to 1.4

percent. Even with the McClelland-Zeckhauser definitions of regions, however, the net outflow from the South to the North was less than 12 percent of the net interregional east-west flow within the North.

Mobility rates and the distance and direction of migration are important for the operation of labor markets. Wright (1986) has noted that the isolation of southern labor markets observed in the first half of the twentieth century has roots in the antebellum period. According to the matched sample of rural households this isolation cannot be attributed to a low rate of mobility. About 34.2 percent of the families who lived in the South in 1850 lived in a different county in 1860 compared with 16.7 percent in the Northeast, 45.5 percent in the North Central states, and 30.0 percent in the West (Steckel 1989). Southerners changed counties more often than residents of any region except the North Central states, a result that holds after controlling for family characteristics such as age, wealth, and occupation of the head. Southerners were different, however, in their propensity for short moves and their high incidence of reverse flow. Residents of the Northeast traveled the farthest, averaging 852 miles, while those from the North Central states relocated a mean distance of 701 miles and southerners averaged only 289 miles.[10] Geographic features in part explain the long distances for the northern states. In a general east-west flow northeasterners moved long distances to reach newly settled parts of the Midwest and the West. A second explanation notes that only 1.4 percent of southern households left for the West compared with 14.3 percent in the North Central states and 2.65 percent in the Northeast. Steckel (1983) argues that crop and region-specific human capital influenced migration paths. Many rural southerners had acquired skills required to raise cotton or tobacco, for example, and the yields on these investments were low in the West, which discouraged movement to that region.

Southerners also had a high incidence of local, scattered moves (Steckel 1989). Only 20.2 percent of the moves in the North but 46.2 percent of those in the South were under 75 miles. Moreover, the ratio of primary (east to west) to reverse (west to east) flow was 8.7 in the Northeast, 10.8 in the North Central states, and only 3.6 in the South. These figures and the findings of a high probability of movement suggest that labor market information flowed well but over a relatively small area within the South. One hypothesis for this pattern involves a social system in which family ties, friendships, and word of mouth may have played an major role in the communication of opportunities (Owsley

1949). Recent studies of the upcountry tend to confirm the importance of kinship networks as opposed to the market in mediating economic relationships (Hahn 1983). A second explanation centers on the lack of regional integration in product markets, which reduced information flows important for the efficient operation of labor markets. This explanation is consistent with David Weiman's (1987) report that low investments in transportation and communication and in processing and storage facilities substantially limited commercial activity in the Georgia upcountry.

The interregional slave trade is an enduring subject in southern history. The westward expansion of southern agriculture makes clear that the volume of the trade was substantial. Fogel and Engerman (1974) estimate that 835,000 slaves were moved from the exporting to the importing states from 1790 to 1860. The rate of movement reached a peak during the 1830s when approximately 10 percent of the total slave population moved across state lines. There is little doubt that separations of family members occurred as a result of this trade and that the possibility of separation must have caused anxiety among slaves, particularly when a plantation was moved or dissolved. Fogel and Engerman (1974) argued that 84 percent of the slaves involved in the westward movement migrated with their owners, but Gutman and Richard Sutch (1976) rejected the methods underlying their conclusion and stressed instead that a slave had a high probability of being sold at least once during his lifetime. Carstensen and Goodman (1977) note the sensitivity of the Fogel-Engerman results to linearity assumptions.

Whatever its extent, shipping patterns found on manifests of the coastwise trade suggest that separations were more likely to have occurred among residents of the upper states of the South (Steckel 1990). Among slaves sent to Gulf Coast states (Alabama, Mississippi, Louisiana, and Texas), those from the lower as opposed to the upper Atlantic coast states had an age distribution closer to the entire slave population. About 30 percent of the slaves originating from South Carolina and Georgia were aged 30 or above but less than 10 percent of those from Maryland or Virginia fell into this age category. These data suggest that east-west as opposed to northeast-southwest migration was more likely to have preserved family ties. Investments in latitude-specific skills, noted above, may explain the contrasting age distributions. Older slaves accustomed to tobacco and mixed agriculture in the border states had to be retrained for work in the cotton fields of the Southwest, an investment that had a low rate of return. Because older slaves from South Carolina and Georgia tended to raise cotton they did not require retraining and could be moved profitably.

## CONCLUSION

Economic historians have made great strides in understanding the slave economy. Research in the past two decades has overturned the image of slaves as lazy and inept, established slaveowners as rational capitalists, demonstrated that southerners were largely independent of western food supplies, and shown that slave workers were well nourished while young children had extraordinarily poor health. Progress on these and other topics has raised new issues and suggested additional directions for research. Certain results on regional differences in the efficiency of slave labor await explanation and the process of quantifying the sources of higher measured efficiency on larger plantations has just begun. The substantial volume of slave manifests lodged in the National Archives has been only partially exploited for insights into the interregional slave trade and into regional and temporal patterns of slave health. The consequences of poor net nutrition for young slaves have been largely unexamined because studies of the subject emphasized the diet alone and perceptions of food intake were reasonably favorable. The investigation of probate records, which depict the occupations, value, and fertility of slaves, has merely scratched the surface of the available data. Designers of models to study the effect of slavery on southern economic development should look to international comparisons for insights. In addition, the rigor and interest that has characterized study of efficiency, productivity, and self-sufficiency in the Cotton South could be extended further to areas that produced tobacco, rice, or sugar.

Cliometricians have advanced knowledge of the southern free population by measuring the distribution of wealth and charting economic mobility, by exploring the operation of southern labor markets, and by learning of the unimportance of control within marriage to fertility. Yet, in many ways the yeomen and free blacks of the antebellum South have been employed as silent actors who played supporting roles for the study of slavery. Their use as a backdrop has served a useful purpose but these groups are worthy of more research in their own right. In this regard, the

Parker-Gallman sample has been valuable for study of slaves and other groups but its emphasis on farms in the Cotton South limits its value for these other purposes. A large sample of manuscript census records of population and agriculture should be collected for the rest of the South. The combined data base could be used to study wealth, migration, fertility, costs of farm making, production for the market, and profitability—topics investigated in works such as *To Their Own Soil* by Jeremy Atack and Fred Bateman (1987). Further research is also warranted on the southern manufacturing sector. The reports by Bateman and Weiss of high rates of return in industry suggest a need for research on the sensitivity of results to techniques of estimation, on information flows within the South, and on aspirations and constraints of plantation owners.

My last suggestion stresses the benefits and feasibility of research based on longitudinal data. Lack of methods of matching records and the high costs of data collection have discouraged this approach despite its recognized benefits. Backward linkage of manuscript schedules with census data and other records is now possible at a remarkably low cost (Steckel 1988a). The linked data would provide valuable insights into migration, wealth mobility, occupational change, and the responses of farmers to changing economic circumstances. Additional linked data would advance research on causes of localism in southern labor markets, the extent and sources of local or regional differences in health within the South, and the determinants of regional and temporal patterns of fertility.

## NOTES

1. Views on these issues lacked consensus. Gray, for example, held few of the traditional beliefs, rejecting notions of unprofitability and inefficiency of slave labor, but maintained that slavery channeled investment away from industry and commerce. Stampp (1956) emphasized the harshness of slave life while accepting profitability and viability. For a discussion of the evolution of thought see Fogel and Engerman (1974), Chaps. 5–6 and vol. 2, app. C.

2. See Butlin (1971) for dissenting views on profitability and viability. In addition, concepts of viability have been debated; Genovese and Genovese (1979), for example, suggest that the crucial issue is not economic endurance but the political security of slaveholders.

3. The index, G, can be expressed as $G = Q/(L^{\alpha}K^{\beta}T^{\gamma})$ where Q is output, L is the labor input, K is the capital input, T is the land input, and $\alpha$, $\beta$, $\gamma$ are the shares in the

value of output of labor, capital, and land, respectively. Fogel and Engerman measure relative total factor productivity by the ratio of values of G for the South and the North.

4. But see Schaefer and Schmitz (1982) and Schaefer (1983) for another point of view. A recent piece by Elizabeth Field (1988) tends to support the Fogel-Engerman position.

5. Donghyu Yang's (1984) analysis indicates that exclusions of poorer households in the Parker-Gallman sample substantially understates wealth concentration, biasing the Gini coefficient downward by approximately 10 percentage points compared with estimates for northern free farmers.

6. However, Campbell and Lowe (1977) find relatively less inequality but much evidence that slaveowners wielded considerable political power in Texas.

7. The East includes Delaware, the District of Columbia, Georgia, Maryland, North Carolina, South Carolina, and Virginia. The West consists of Kentucky, Tennessee, and Mississippi; Louisiana and Missouri (beginning in 1810); Alabama and Arkansas (beginning in 1820); Florida (beginning in 1830); and Texas (beginning in 1850).

8. This finding contrasts with Sanderson's (1979) report that decreases in marital fertility explain two-thirds to three-quarters of the antebellum fertility decline for the country as a whole. As noted below, however, David and Sanderson agree that control began relatively late in the South.

9. Prospective couples may have viewed delay or avoidance of marriage as a form of control over the number of births.

10. The distance for the South roughly agrees with the figure of 228 miles reported by Donald Schaefer (1985) for the antebellum Southwest.

## REFERENCES AND FURTHER READINGS


Anderson, R. V., and R. E. Gallman. "Slaves as Fixed Capital: Slave Labor and Southern Economic Development." *Journal of American History* 54:24–46.

Atack, J., and F. Bateman. 1987. *To Their Own Soil: Agriculture in the Antebellum North.* Ames: Iowa State Univ. Press.

Bateman, F., and T. Weiss. 1981. *A Deplorable Scarcity: The Failure of Industrialization in the Slave Economy.* Chapel Hill: Univ. of North Carolina Press.

Battalio, R. C., and J. Kagel. 1970. "The Structure of Antebellum Southern Agriculture: South Carolina, A Case Study." *Agricultural History* 44:25–37.

Bergstrom, T. 1971. "On the Existence and Optimality of Competitive Equilibrium for a Slave Economy." *Review of Economic Studies* 38:23–36.

Butlin, N. G. 1971. *Ante-bellum Slavery—Critique of a Debate.* Canberra: Department of Economic History, Australian National Univ.

Caldwell, J. C. 1982. *Theory of Fertility Decline.* New York: Academic Press.

Campbell, J. 1984. "Work, Pregnancy, and Infant Mortality among Southern Slaves." *Journal of Interdisciplinary History* 14:793–812.

Campbell, R. B., and R. G. Lowe. 1977. *Wealth and Power in Antebellum Texas.* College Station: Texas A&M Univ. Press.

Carstensen, F. V., and S. E. Goodman. 1977. "Trouble on the Auction Block: Interregional Slave Sales and the Reliability of a Linear Equation." *Journal of Interdisciplinary History* 8:315–18.

Clark, B. H. 1942. *The Tennessee Yeoman, 1840–1860.* Nashville: Vanderbilt Univ. Press.

Cohn, R. L. 1981. "Antebellum Regional Incomes: Another Look." *Explorations in Economic History* 18:330–46.

Conrad, A. H., and J. R. Meyer. 1958. "The Economics of Slavery in the Ante Bellum South." *Journal of Political Economy* 66:95–130.

David, P. A. and P. Temin. 1974. "Slavery: The Progressive Institution?" *Journal of Economic History* 34:739–83.

______. 1979. "Explaining the Relative Efficiency of Slave Agriculture in the Antebellum South: Comment." *American Economic Review* 69:213–18.

David, P. A., and W. C. Sanderson. 1986. "Rudimentary Contraceptive Methods and the American Transition to Marital Fertility Control, 1855–1915." In *Long-Term Factors in American Economic Growth,* ed. S. L. Engerman and R. E. Gallman, 307–79. Chicago: Univ. of Chicago Press.

Easterlin, R. A. 1961. "Regional Income Trends, 1840–1950." In *American Economic History,* ed. S. Harris, 525–48. New York: McGraw-Hill.

______. 1971. "Does Human Fertility Adjust to the Environment?" *American Economic Review: Papers and Proceedings* 61:399–407.

______. 1975. "Farm Production and Income in Old and New Areas at Mid-Century." In *Essays in Nineteenth Century Economic History: The Old Northwest,* ed. D. C. Klingaman and R. K. Vedder, 77–117. Athens: Ohio Univ. Press.

______. 1976. "Population Change and Farm Settlement in the Northern United States." *Journal of Economic History* 36:45–75.

Engerman, S. L. 1967. "The Effects of Slavery Upon the Southern Economy: A Review of the Recent Debate." *Explorations in Entrepreneurial History* 4:71–147.

______. 1971. "Some Economic Factors in Southern Backwardness in the Nineteenth Century." In *Essays in Regional Economics,* ed. J. F. Kain and J. R. Meyer, 279–306. Cambridge: Harvard Univ. Press.

______. 1975. "A Reconsideration of Southern Economic Growth, 1770–1860." *Agricultural History* 49:343–61.

Evans, R., Jr. 1962. "The Economics of American Negro Slavery." In *Aspects of Labor Economics,* ed. Universities-National Bureau Committee for Economic Research, 185–243. Princeton: Princeton Univ. Press.

Field, E. B. 1988. "The Relative Efficiency of Slavery Revisited: A Translog Production Function Approach." *American Economic Review* 78:543–49.

Flanders, R. B. 1933. *Plantation Slavery in Georgia.* Chapel Hill: Univ. of North Carolina Press.

Fleisig, H. 1976. "Slavery, the Supply of Agricultural Labor, and the Industrialization of the South." *Journal of Economic History* 36:572–97.

Fogel, R. W., and S. L. Engerman. 1971. "The Relative Efficiency of Slavery: A Comparison of Northern and Southern Agriculture in 1860." *Explorations in Economic History* 8:353–67.

______. 1974. *Time on the Cross: The Economics of American Negro Slavery.* 2 vols. Boston: Little, Brown.

______. 1977. "Explaining the Relative Efficiency of Slave Agriculture in the Antebellum South." *American Economic Review* 67:275–96.

______. 1980. "Explaining the Relative Efficiency of Slave Agriculture in the Antebellum South: Reply." *American Economic Review* 70:672–90.

Ford, L. K. 1986. "Yeoman Farmers in the South Carolina Upcountry: Changing Production Patterns in the Late Antebellum Era." *Agricultural History* 60:17–37.

Forster, C., and G. S. L. Tucker. 1972. *Economic Opportunity and White American Fertility Ratios, 1800–1860.* New Haven: Yale Univ. Press.

Gallman, R. E. 1970. "Self-Sufficiency of the Cotton Economy of the Antebellum South." *Agricultural History* 44:5–23.

______. 1975. "Southern Ante-Bellum Income Reconsidered." *Explorations in Economic History* 12:89–99.

______. 1979. "Slavery and Southern Economic Growth." *Southern Economic Journal* 45:1007–22.

Genovese, E. D. 1965. *The Political Economy of Slavery.* New York: Pantheon Books.

Genovese, E. D., and Fox-Genovese, E. 1979. "The Slave Economies in Political Perspective." *Journal of American History* 66:7–23.

Graff, H. J. 1979. "Literacy, Education, and Fertility, Past and Present: A Critical Review." *Population and Development Review* 5:105–40.

Gray, L. C. 1933. *History of Agriculture in the Southern United States to 1860.* 2 vols. Washington, D.C.: Carnegie Institution.

Gunderson, G. 1973. "Southern Ante-bellum Income Reconsidered." *Explorations in Economic History* 10:151–76.

Gutman, H. G. 1976. *The Black Family in Slavery and Freedom, 1750–1925.* New York: Pantheon Books.

Gutman, H. G., and R. Sutch. 1976. "The Slave Family: Protected Agent of Capitalist Masters or Victim of the Slave Trade?" In *Reckoning with Slavery,* ed. P. A. David, H. E. Gutman, R. Sutch, P. Temin, and G. Wright, 94–133. New York: Oxford Univ. Press.

Hahn, S. 1983. *The Roots of Southern Populism: Yeoman Farmers and the Transformation of the Georgia upcountry, 1850–1890.* New York: Oxford Univ. Press.

Haskell, T. L. 1974. "Were Slaves More Efficient?: Some Doubts About *Time on the Cross.*" *New York Review of books,* Sept. 19, 1974, 38–42.

______. 1979. "Explaining the Relative Efficiency of Slave Labor in the Antebellum South: A Reply to Fogel-Engerman." *American Economic Review* 69:206–7.

Helper, H. R. 1857. *The Impending Crisis of the South: How to Meet It.* Cambridge: Belknap Press.

Hilliard, S. B. 1972. *Hog Meat and Hoecake: Food Supply in the Old South, 1840–1860.* Carbondale: Southern Illinois Univ. Press.

Hutchinson, W. K., and S. H. Williamson. 1971. "The Self-Sufficiency of the Antebellum South: Estimates of Food Supply." *Journal of Economic History* 31:591–612.

Johnson, M. P. 1981. "Smothered Slave Infants: Were Slave Mothers at Fault?" *Journal of Southern History* 47:493–520.

Kiple, D. F., and V. H. King. 1981. *Another Dimension to the Black Diaspora: Diet, Disease, and Racism.* Cambridge: Cambridge Univ. Press.

Lindstrom, D. 1970. "Southern Dependence on Interregional Grain Supplies: A Review of the Trade Flows, 1840–1860." *Agricultural History* 44:101–13.

McClelland, P. D., and R. J. Zeckhauser. *Demographic Dimensions of the New Republic: American Interregional Migration, Vital Statistics and Manumissions, 1800–1860.* Cambridge: Cambridge Univ. Press.

McInnis, R. M. 1977. "Childbearing and Land Availability: Some Evidence from Individual Household Data." In *Population Patterns in the Past,* ed. R. D. Lee, 201–27. New York: Academic Press.

Margo, R. A., and R. H. Steckel. "The Height of American Slaves: New Evidence on Slave Nutrition and Health." *Social Science History* 6:516–38.

Metzer, J. 1975. "Rational Management, Modern Business Practices, and Economies of

Scale in Ante-Bellum Southern Plantations." *Explorations in Economic History* 12:123–50.

Niemi, A. W., Jr. 1977. "Inequality in the Distribution of Slave Wealth: The Cotton South and Other Southern Agricultural Regions." *Journal of Economic History* 37:747–53.

North, D. C. 1961. *The Economic Growth of the United States, 1790–1860.* Englewood Cliffs, N.J.: Prentice-Hall.

Olmsted, F. L. 1861. *The Cotton Kingdom.* New York: Mason.

Owsley, F. L. 1949. *Plain Folk of the Old South.* Baton Rouge: Louisiana State Univ. Press.

Phillips, U. B. 1905. "The Economic Cost of Slaveholding in the Cotton Belt." *Political Science Quarterly* 20:798–816.

______. 1918. *American Negro Slavery.* New York: D. Appleton.

Ramsdell, C. W. 1929. "The Natural Limits of Slavery Expansion." *Mississippi Valley Historical Review* 16:151–71.

Ransom, R. L., and R. Sutch. 1977. *One Kind of Freedom: The Economic Consequences of Emancipation.* Cambridge: Cambridge Univ. Press.

Russell, R. R. 1938. "The General Effects of Slavery Upon Southern Economic Progress." *Journal of Southern History* 4:34–54.

Sanderson, W. C. 1979. "Quantitative Aspects of Marriage, Fertility, and Family Limitation in Nineteenth Century America: Another Application of the Coale Specifications." *Demography* 16:339–58.

Savitt, J. L. 1978. *Medicine and Slavery: The Diseases and Health Care of Black in Antebellum Virginia.* Champaign: Univ. of Illinois Press.

Schaefer, D. F. 1978. "Yeoman Farmers and Economic Democracy: A Study of Wealth and Economic Mobility in the Western Tobacco Region, 1850 to 1860." *Explorations in Economic History* 15:421–37.

______. 1983. "The Effect of the 1859 Crop Year Upon Relative Productivity in the Antebellum Cotton South." *Journal of Economic History* 43:851–65.

______. 1985. "A Statistical Profile of Frontier and New South Migration: 1850–1860." *Agricultural History* 59:563–78.

Schaefer, D. F., and M. D. Schmitz. "The Relative Efficiency of Slave Agriculture: A Comment." *American Economic Review* 69:208–12.

______. 1982. "Efficiency in Antebellum Southern Agriculture: A Covariance Approach." *Southern Economic Journal* 49:88–98.

Schapiro, M. O. 1982. "Land Availability and Fertility in the United States, 1760–1870." *Journal of Economic History* 42:577–600.

Schmitz, M. D. 1978. "Farm Interdependence in the Antebellum Sugar Sector." *Agricultural History* 52:93–103.

Schultz, T. W. 1973. "The Value of Children: An Economic Perspective." *Journal of Political Economy* 81:S2–S13.

Stampp, K. M. 1956. *The Peculiar Institution: Slavery in the Antebellum South.* New York: Alfred A. Knopf.

Steckel, R. H. 1979a. "Slave Mortality: Analysis of Evidence from Plantation Records." *Social Science History* 3:86–114.

______. 1979. "Slave Height Profiles from Coastwise Manifests." *Explorations in Economic History* 16:363–80.

______. 1980. "Antebellum Southern White Fertility: A Demographic and Economic Analysis." *Journal of Economic History* 40:331–50.

______. 1982. "The Fertility of American Slaves." *Research in Economic History* 7:239–86.

______. 1983. "The Economic Foundations of East-West Migration during the 19th Century." *Explorations in Economic History* 20:14–36.

______. 1985. *The Economics of U.S. Slave and Southern White Fertility.* New York: Garland Press.

______. 1986a. "Birth Weights and Infant Mortality among American Slaves." *Explorations in Economic History* 23:173–98.

______. 1986b. "A Peculiar Population: The Nutrition, Health, and Mortality of American Slaves from Childhood to Maturity. *Journal of Economic History* 46:721–41.

______. 1986c. "A Dreadful Childhood: The Excess Mortality of American Slaves." *Social Science History* 10:427–65.

______. 1987 "Growth Depression and Recovery: The Remarkable Case of American Slaves." *Annals of Human Biology* 14:111–32.

______. 1988a. "Census Matching and Migration: A Research Strategy." *Historical Methods* 21:52–60.

______. 1988b. "The Health and Mortality of Women and Children, 1850–1860." *Journal of Economic History* 48:333–45.

______. 1989. "Household Migration and Rural Settlement in the United States, 1850–1860." *Explorations in Economic History* 26 (forthcoming).

______. 1990. "The Slavery Period and Its Influence on Family Change in the United States." In *Changing Family Structures and Life Sources in Less Developed Countries,* ed. E. Berquo and P. Xenos. Forthcoming. Oxford: Oxford Univ. Press.

Sundstrom, W. A., and P. A. David. 1988. "Old-Age Security Motives, Labor Markets, and Farm Family Fertility in Antebellum America." *Explorations in Economic History* 25:164–97.

Sutch, R. 1965. "The Profitability of Ante Bellum Slavery—Revisited." *Southern Economic Journal* 31:365–77.

______. 1975a. "The Treatment Received by American Slaves: A Critical Review of the Evidence Presented in *Time on the Cross.*" *Explorations in Economic History* 12:335–438.

______. 1975b. "The Breeding of Slaves for Sale and the Westward Expansion of Slavery, 1850–1860." In *Race and Slavery in the Western Hemisphere: Quantitative Studies,* ed. S. L. Engerman and E. D. Genovese, 173–210. Princeton: Princeton Univ. Press.

Sydnor, C. S. 1933. *Slavery in Mississippi.* New York: Appleton-Century.

Tanner, J. M. 1978. *Fetus Into Man: Physical Growth from Conception to Maturity.* Cambridge: Harvard Univ. Press.

Trussell, J., and R. H. Steckel. 1978. "The Age of Slaves at Menarche and Their First Birth." *Journal of Interdisciplinary History* 8:477–505.

Vinovskis, M. A. 1981. *Fertility in Massachusetts from the Revolution to the Civil War.* New York: Academic Press.

Wahl, J. B. 1986. "New Results on the Decline in Household Fertility in the United States from 1750 to 1900." In *Long-Term Factors in American Economic Growth,* ed. S. L. Engerman and R. E. Gallman, 391–437. Chicago: Univ. of Chicago Press.

Weiman, D. F. 1985. "The Economic Emancipation of the Non-Slaveholding Class: upcountry Farmers in the Georgia Cotton Economy." *Journal of Economic History* 45:71–93.

______. 1987. "Farmers and the Market in Antebellum America: A View from the Georgia upcountry." *Journal of Economic History* 47:627–47.

Williamson, J. G. and P. H. Lindert. 1980. *American Inequality: A Macroeconomic History.* New York: Academic Press.

Winters, D. L. 1987a. "'Plain Folk' of the Old South Reexamined: Economic Democracy in Tennessee." *Journal of Southern History* 53:565–86.

______. 1987b. "The Agricultural Ladder in Southern Agriculture: Tennessee, 1850–1870." *Agricultural History* 61:36–52.

Wright, G. 1970. "'Economic Democracy' and the Concentration of Agricultural Wealth in the Cotton South, 1850–1860." *Agricultural History* 44:63–93.

______. 1975. "Slavery and the Cotton Boom." *Explorations in Economic History* 12:439–51.

______. 1978. *The Political Economy of the Cotton South.* New York: W. W. Norton.

______. 1979. "The Efficiency of Slavery: Another Interpretation." *American Economic Review* 69:219–26.

______. 1986. *Old South, New South: Revolutions in the Southern Economy Since the Civil War.* New York: Basic Books.

Yang, D. 1984. "Notes on the Wealth Distribution of Farm Households in the United States, 1860: A New Look at Two Manuscript Census Samples." *Explorations in Economic History* 21:88–102.

Yasuba, Y. 1961. "The Profitability and Viability of Plantation Slavery in the United States." *Economic Studies Quarterly* 12:60–67.

______. 1962. *Birth Rates of the White Population in the United States, 1800–1860.* Baltimore: Johns Hopkins Univ. Press.

CHAPTER 8

# The Role of Markets and Institutions in Postbellum Southern Development

JAY R. MANDLE

THE STUDY OF THE SOUTH'S postbellum agricultural experience has generated intense controversy within a voluminous literature. The reason for the uncharacteristic passion that has intruded itself into this scholarly discourse is easy to identify. The study of the postemancipation South necessarily involves the question of race. As such it raises issues of great sensitivity to contemporary America. This sensitivity finds expression in both the methods of study and rhetoric employed.

With freedom, the former slaves possessed neither formal education nor substantial material assets. Furthermore, the economy in which they were participants had been severely damaged by the recent war. In such a context, and in the absence of a substantial program of economic assistance to the ex-slaves, it is no surprise to learn that they became the lowest income group in a region that itself was the least economically advanced in the nation.

There has been very little discussion in the economic history literature of the consequences of the failure to endow the freed persons with material or financial resources. Such an endowment would have improved their potential for successful participation in the southern economy. It might therefore have resulted in an entirely different pattern of economic development for the region. Of course, it is also possible that the freed slaves would have failed to use their bequest profitably and as a result might have lost control of their new-found wealth. If so, the path of southern development might not have been very different from what ultimately occurred. But the point is that none of this has been seriously investigated. The configuration of possibilities that were open to the South after the Civil War lies uncharted. What is needed is a comparison of what actually happened during Reconstruction with a systematic assessment of what the likely results would have been of providing, for example, forty acres and a mule to the former slave population. Only then would it be possible to compare what occurred during Reconstruction with the alternatives that were available to the region.

Rather than deal with such a hypothetical alternative, the historical literature has proceeded as if the propertyless condition of the former slaves was the only possible historical option available to the postbellum South. It has then gone on to attempt to assess the extent to which this impoverished population benefited from its insertion as a group of free persons into the southern economy. Primary attention has been drawn to the methods by which the ex-slaves, initially under the tutelage of the Freedman's Bureau, supplied labor to the postbellum cotton plantations. In this regard the issue addressed centers on how sharecropping influenced black income levels and wealth accumulation. Connected to all of this is, in addition, concern with how the role of the black population in the region's economy influenced the course of the South's economic development.

Precisely because little governmental economic intervention occurred to assist the black population in the South, its experience has come to be seen as a test of the efficacy of unimpeded markets in promoting equity. Market theorists have long argued that unregulated markets are efficient vehicles for achieving advances in human welfare. A finding that despite all of the handicaps that blacks admittedly encountered in the aftermath of slavery, the freed persons nonetheless achieved satisfactory economic advances would provide important support for the more general argument that markets should not be interfered with. The contrary, of course, is also the case. Less than satisfactory advances would seem to provide justification for positive intervention by government to achieve welfare gains. Much, therefore, is at stake ideologically in the discussion of the postbellum southern economy. For not only can it shed light on the historical sources of contemporary black poverty in the United States, it also provides insight into the strategies that might be adopted to overcome that poverty.

Robert Higgs is the most prominent of those scholars who have emphasized the beneficent consequences of the South's markets. In his book *Competition and Coercion,* Higgs points to the gains achieved by blacks as a result of their postemancipation ability to migrate voluntarily and to accumulate both human and physical capital. According to him, increased literacy allowed the black labor force to experience upward occupational mobility. With the latter emerged a class of landowning black farmers and business persons. Higgs argues that "writers who have portrayed the economic conditions of the blacks in the late nineteenth century or early twentieth centuries as having improved only slightly are surely mistaken" (Higgs 1977, 123).

Higgs does acknowledge that even more economic achievement could have been accomplished. What Higgs does, however, is argue that the obstacles that impeded feasible black economic advance were noneconomic in nature. In his argument, markets are largely absolved from responsibility for the failures that were experienced. To Higgs "the fundamental problem was the failure to achieve free and equal participation in politics. . . ." This political failure meant, on the one hand, that blacks in the South were unable to influence patterns of public expenditures. Particularly important in this regard, according to Higgs, was the gap existing in expenditures on education between whites and blacks. The inadequate level of funding for black education was responsible for the "sending of masses of young blacks into the labor market unprepared for any but the most unskilled occupations." On the other hand, the exclusion of blacks from government influence permitted the passage and enforcement of discriminatory laws. The effect of such codified discrimination was to constrain aggressive entrepreneurial behavior by blacks and to shut off promising new areas of economic activity from black participation. In this way black economic success was limited. Overall, Higgs's judgment is that the exclusion of blacks from politics was the critical factor in producing a pace of black economic advance below the level that could have been achieved (Higgs 1977, 123, 144).

Before the adequacy of black economic advance can be defended, however, its extent must be quantified. This Higgs does in a 1982 article in which he estimated black wealth accumulation in Georgia before World War I (Higgs 1982; Margo 1984; Higgs 1984). Higgs summarizes his view by declaring that "during the half century following emancipation the Georgia freedmen and their descendants—and other southern blacks as well—rapidly accumulated wealth." He writes that "over the entire half century before World War I, the blacks made tremendous relative gains on the whites in property" (Higgs 1982, 735). Higgs's choice of language strongly suggests he believes that, though black wealth accumulation was slower than it could have been, it nonetheless progressed at a satisfactory rate.

Closer examination of Higgs's data, however, raises serious questions concerning the validity of Higgs's optimistic interpretation. Table 8.1 reproduces Higgs's wealth estimates. It turns out that though black per capita wealth did, as Higgs reports, more than double between 1880 and 1900, the absolute increase that occurred was only \$8.62. During these same years that growth of white per capita wealth was \$100.83. Since the white's increase represented less than a doubling, Higgs is technically correct in pointing out the *rate* of black wealth accumulation was in excess of that of whites. Still, the fact remains that the absolute growth of wealth per capita among whites was almost 12 times that of the black population. A similar but less dramatic pattern emerges in the first decade of the twentieth century during which the growth of black wealth per capita (\$9.97) represented a more rapid *rate* of growth than the \$27.55 increase that occurred among the white population.

TABLE 8.1. Assessed Value of Total Property Holdings by Race, Georgia, 1880–1910

| Year | Assessed Value of Black Property Per Capita | Assessed Value of White Property Per Capita |
|---|---|---|
| 1880 | \$8.00 | \$285.40 |
| 1890 | \$17.46 | \$455.19 |
| 1900 | \$16.62 | \$386.23 |
| 1910 | \$26.59 | \$413.78 |

Source: Higgs (1982, 720).

Higgs's celebration of the growth of black wealth is therefore critically dependent upon the fact the black population at Emancipation was "practically penniless" (Higgs 1982, 728). Because of their low level of wealth at the beginning of the period almost any growth in black property holding would have produced a large percentage increase and rapid rate of expansion. These large percentage increases should not obscure the very limited growth in black wealth and the continued gap between black and white per capita wealth in this period. Dire poverty remained the lot of a very large majority of the black southern population in the years before World War I.[1] The limited growth in wealth that had occurred did not promise to alleviate this condition any time soon.

Indeed, it is precisely the large percentage but small absolute increases in wealth that suggests the limited extent to which markets can achieve a fair distribution of wealth from an initial condition of gross inequality. This is because markets act at the margin. Resources can be shifted by markets from one group to another only incrementally, and only to the extent that those with wealth are willing or are compelled to bring that wealth to the market. Those with substantial resources typically need not place a large share of their wealth at risk. Thus the extent to which markets typically effectuate a substantial shift in the distribution of wealth is strictly limited. A substantial transfer of resources requires a large, once-and-for-all shift—precisely the kind that markets rarely facilitate.

In this regard, Steven DeCanio's econometric analysis provides empirical support for the view that in the absence of a land redistribution program, a substantial closing of the black-white wealth differential was not to be achieved. After reviewing the results of his model of wealth accumulation, DeCanio concludes that "emancipation without property condemned the blacks to a lengthy period of economic disadvantage" (DeCanio 1981, 104). According to him, "even if all markets had operated perfectly and no discrimination had been practiced against the freedmen either in wage payments or in their access to occupations, this initial gap in tangible capital would have produced by itself most of the gap in income between blacks and whites throughout the late 19th and early 20th centuries." Markets, in short, could not have done the job of achieving racial justice.

Robert Higgs's view (1977, 127) that "a half century was not long enough for the blacks to close the economic gap separating them from the whites, even under the most favorable circumstances" is simply untrue. A dramatic increase in black wealth could have been accomplished virtually overnight. A redistribution of land away from the former slaveowners to the former slaves would have endowed the southern black population with the material resources necessary to participate more nearly as equals with whites in the region's economy. A movement toward racial justice would thus have been accomplished.

While markets could not have effectuated a substantial transfer of financial and material resources, Joseph Reid has ingeniously argued that they were, nonetheless, active agents of progressive social change. Reid's hypothesis rests on the assumption that the "freedmen were woefully unprepared to farm on their own," because "slaves did not learn many managerial and marketing skills." Therefore, the former slaves could not have filled the role of farmer-manager effectively. What was required by the freed persons, according to this view, was training in those areas in which they were deficient. According to Reid, sharecropping provided just such training. For in this tenantry system "the offer to landlords of shares of the crops by these new black and white tenants promised to secure landlords' assistance and attention and thereby to make tenants' prosperity quick and sure." As seen by Reid, sharecropping, the typical form of tenantry in the postbellum South, was in the economic interest of both owner and tenant. According to him not only is it the case that generally "knowledgeable landlords will want share tenants," it also it true that "somewhat unknowledgeable but ambitious farm workers will be share tenants." This form of tenantry was adopted as a product of negotiations between landlords and farm laborers in which "the price of farming know-how was set in the market place just as were the prices of land, labor and capital" (Reid 1981, 36, 47, 43; Reid 1973).

Intriguing though it is, Reid's hypothesis faces an intractable objection. There simply is no evidence whatever that the former slaves preferred sharecropping to an owner-occupier status. On the contrary, evidence is overwhelming that the former slaves saw in "forty acres and a mule" the road to economic emancipation. The actions taken by Sherman and his troops (particularly a federal order issued in January 1865 creating a reservation exclusively for black use) may have been the foundation of the expectation, held by many freedmen, that "the benevolent government which had freed them would provide forty acres and a mule." The belief in an imminent land transfer was strong. As late as September 1868 a Republican mass meeting in Georgia issued the slogan, "Division Now; don't wait until Christmas" (Oukre 1978, 184).

It is possible to argue, as Reid perhaps would, that the former slaves were economically irrational in their preference for an owner-occupier status. But market theory makes no allowance for such misperceptions; it assumes that economic agents know their own interests. As a result, Reid seems to be guilty of violating the norms of economic reasoning in believing that black aspiration to landowning status was irrational. For the fact remains that former slaves did not see sharecropping as their economic status of choice. No amount of theoretical reasoning can overcome that historical fact.

Reid's view of the symbiotic relationship existing between the managing planter and the learning cropper has not received much support in the literature. There is instead a general consensus among scholars that sharecropping represented a means of accommodating the *conflicting* interests of planters and former slaves. The former sought continued supervisory control, while the latter sought escape from the gang labor associated with slavery. Sharecropping achieved both. Planters still maintained a high degree of control over output levels and methods of cultivation, but croppers were able to work in household units with the autonomy associated with discrete and defined units of production (Hughes 1983, 247).

Both Higgs and Reid believe that the tragedy for southern blacks lay in the realm of politics, not economics. According to Higgs, whites in the South succeeded in achieving "political supremacy" and used the supremacy "to exclude the blacks from effective participation in political affairs." As a result black advances in education and training were impaired. This same political exclusion "placed the personal and property rights of the blacks in serious jeopardy." Conceding that there was some marketplace discrimination, Higgs nonetheless argues that "the fountainhead of effective discrimination . . . lay in the governments of the Southern states, counties and cities where the monopoly of racial politics allowed the hostile whites to treat blacks as they pleased" (Higgs 1977, 133, 134). Reid argues in the same way. For him post-1900 racism and poverty did not derive their strength from economic relationships. Instead they "came immediately and directly to blacks from the discontinuous manipulation of the law by the sheriff. . . ." Indeed, what happened in the South, according to Reid, was a negative evolution from "new hope in 1865 to no hope in 1900" as the positive impact of markets declined under the weight of an increased monopoly of political power by whites (Reid 1981, 55).

The problem with the conclusion advanced by Reid and Higgs is that it is obtained exclusively through a deductive process of reasoning. Neither, in fact, investigates the nonpolitical sources of black oppression nor does either offer a framework in which noneconomic variables could be systematically investigated. At no point does either provide an empirical analysis of how the politics of discrimination were practiced, or the strength and intensity of the economic consequences that resulted from political discrimination. Their method is to assume that a market economy always functions to promote equity. The fact of continuous racial inequality therefore has to be assigned to noneconomic factors.

It is precisely this conclusion Gerald David Jaynes (1986) challenges when he questions whether the labor markets existing in the South after the Civil War were free. In the end, Jaynes answers this question in the negative. If Jaynes is right, then at least the question of assigning responsibility for black deprivation would have to be reajudicated. The assumption, made by Reid and Higgs, that the problem must have been political because it could not have been economic, would no longer be tenable.

In arguing that free labor markets were not operative in the South, Jaynes draws a distinction between free and competitive markets. He points out that the existence of the second does not imply the presence of the first. One obvious example not cited by Jaynes is the case of slavery. Slave prices were competitively determined in the market for chattel, but by no means were the slaves themselves free to choose their employment. Jaynes himself cites the case of indentured immigrants. Contracts accepted by indentured servants might have been competitively determined, but writes Jaynes, "the servant is not free and often realizes significant income losses due to missed opportunities over the duration of servitude" (Jaynes 1986, 302).

The same was true, argues Jaynes, with regard to black labor under sharecropping in the South. A sharecropper, argues Jaynes, was not endowed with the attributes of a free worker—"the power to select a day-to-day work rhythm of its own choice and . . . to abandon the employ of any planter at any time during the crop season." The prospect of such a situation would have brought "trepidation" to the planters, according to Jaynes, and spurred their search for a framework that would drastically narrow the range of options available to the black labor force (Jaynes 1986, 303; Woodman 1979, 324).[2]

Various elements were brought together to create a condition which, according to Jaynes, "confined the southern laborer within a system that bred exploitation." These included "the long pay"—delay of compensation until the end of the crop year—as well as the region's lien laws vesting in the planter or merchant control of the crop and the share tenant's property, the South's antienticement legislation making it illegal for employers to hire workers who had broken an employment contract, and stringent vagrancy laws allowing states to become important sources of convict labor in potential competition with croppers (Jaynes 1986, 314, 303–9).

Jaynes concedes that in the face of the kind of debt bondage he describes, surreptitious flight was possible for the tenants. But in a rebuttal to Higgs, Jaynes denies that such a possibility of escape symbolized the existence of market freedom. An escaping laborer would be forced to relinquish earnings not yet received. In addition, theorizes Jaynes, such flight would necessitate movement out of the immediate area in order to escape the enforcement of antienticement legislation. Jaynes writes that "with children to feed, little or no money, and the knowledge that the local labor market was closed to them, enticement laws must have been severely effective." In such a setting "the laborer taking to the woods during the night would hardly have recognized the blessing of the competitive forces to which Higgs' position pays so much homage" (Jaynes 1986, 308).

A difficulty with Jaynes's presentation, however, is that he makes no effort actually to measure the extent and representativeness of the market circumstances he describes. Instead of such a quantitative effort, Jaynes employs qualitative information, principally field reports found in the archives of the Freedmen's Bureau. Though this kind of evidence is sufficient to establish the plausibility of his hypothesis, it is not adequate for a full testing of the existence of exploitative market relations. To establish the extent to which the South's markets was in fact exploitative, Jaynes would have to estimate the difference that would have existed in labor's supply behavior under a simple rental relationship compared to that of sharecropping. This Jaynes makes no effort to do.

Despite this shortcoming, Jaynes has made a valuable contribution in establishing that the claims made by Higgs concerning the efficacy of the market are excessive. That a market was present in the South is not in and of itself sufficient to demonstrate it produced a tendency toward equity. But similarly, Jaynes has not demonstrated the extent to which the market "imperfections" he points to influenced black economic achievement.

In a book published like Jaynes's in 1986, Gavin Wright provides at least a partial refutation of one of Jaynes's principal claims. In *Old South, New South* Wright rejects the argument that "labor-tying" contracts—Jaynes's "long wage"—represent involuntary servitude (Wright 1986, 93). Wright cites a high turnover rate among share tenants as evidence that croppers were able to take advantage of market opportunities. Further, he points out that when croppers changed employers their movement typically was geographically circumscribed, extending over only a short distance, not the long-distance flight postulated by Jaynes.

That the evidence seems to show short-distance moves were frequent casts doubt on the strength of Jaynes's position.

According to Wright, though labor markets were operative in the South, the region's problem was that these markets were shut off from those of the rest of the country. This isolation was reinforced by the fact that the region lacked its own indigenous technological tradition and its expenditures on education were far below those of the rest of the nation. In combination these meant that the South emerged as "a low wage region in a high wage country." As a result, according to Wright, the South "was not in a position to make use of the best products emerging from the American machine tool industry [which were] designed for the high wage setting" (Wright 1986, 50, 79).

Under the circumstances, Wright observes, what the South required to develop was an inflow of both capital and the kind of people who could reverse the region's technical and educational backwardness. The problem, according to Wright, was that southern leaders "could not take vigorous steps to open communications and interactions with the outside world without loss of labor." Wright's view is that "the isolation of the [South's] labor market was the basis of the isolation of the capital market" and that these in combination ensured southern poverty (Wright 1986, 78).

This thesis is extremely helpful. With it, Wright underlines and offers an explanation for the South's experience of a development pattern separate and distinct from that of the rest of the country. Furthermore his hypothesis that the foundation of this difference lay in the contrasting status of labor in the southern social setting compared to the rest of the country is plausible and ably defended. In presenting this position, Wright avoids both the excessively optimistic interpretation of the functioning of the South's markets for labor and also the unsubstantiated hypotheses advanced by Jaynes.

As useful as his hypothesis is, however, Wright does not pursue in detail how the contrasting status of southern and northern labor produced divergent development paths. That the South "lacked a strong indigenous technological tradition" is noted (Wright 1986, 79), but the reasons for this absence are not provided. The same is true with regard to the region's lack of investment in education. Wright attributes this lack to rural planters' inability to obtain the benefits of such outlays because raising the educational level of the southern labor force might have increased the rate of emigration from the region. But he does not explain why a leadership that could have benefited from such educational expenditures did not emerge in the South as it did elsewhere in the country. Wright, in short, does not explain why the South remained trapped in a self-reinforcing pattern of poverty while other regions of the United States embarked on a technologically driven growth path.

The problem here is that Wright has touched on the essentials of the development process—technological progressivity and the educational attainment of the labor force—but these factors are not rooted in the analytic framework he employs. In Wright's case, as in most neoclassical economic theorizing, the level of technological development is assessed on an ad hoc basis. It is not incorporated in the model used by Wright to make causal inferences. This omission is very costly. Excluding the issue of the level of technological development from systematic consideration means that a central contributor to the development process is neglected. Further, important feedbacks emanating from the limited technological development are also missed. For example, limited technological progress may have resulted in only a minimal pressure to upgrade educational performance and expenditures. Too much of importance, in short, is omitted if the causes of the level and rate of change technology are not part of the model employed to explain southern distinctiveness.

It is this task which the plantation economy school of thought has addressed. The effort in this approach is to identify the distinctive nature of the South's social structure and, on that basis, to account for the region's lagging behind the rest of the country in all of the principle indicators of economic development. The thesis that a plantation economy persisted in the South after the Civil War can be traced to the work of the historian of the South, Ulrich B. Phillips. Phillips, writing in 1910, underlined the continuity in the southern experience, asserting that the plantation system "was less dependent upon slavery than slavery was upon it." It was, according to Phillips, the plantation system that formed "the industrial and social frame of government in the black-belt communities." Slavery, in contrast, was merely "a code of written laws enacted for the furtherance of that system's purposes." It was true that with the ending of slavery, plantation communities "have had to provide new laws and customs for the adjustment of employers and employees." But the transition had been successfully completed, especially in the Black Belt, and the plantation system continued as "the most common tether" linking the black and white races together (Phillips 1986, 245, 267). It is this plantation system that Edgar T. Thompson argues was the "institutional fault line" separating the South from the rest of the country. Thompson writes that "whatever is 'different', whatever is special, about the South appears to go back to the plantation and to the system of institutions which have grown up around it" (Thompson 1975, 86 54–5).

Central to the successful functioning of large-scale, labor-intensive, low-labor productivity agricultural cultivation—the defining characteristics of plantation agriculture—was the entrapment in the South of black labor. Only with labor's mobility constrained could this form of agriculture have remained commercially viable. Such constraints were required to ensure that a sufficiently large number of low-cost workers was available to the producing units. Thus labor mobility was the enemy of plantation agriculture. In the antebellum period this enemy was held in check by slavery. After emancipation the viability of plantation agriculture depended upon other means of limiting the mobility of the South's—particularly the black—labor force (Mandle 1978, 1–15).

It is ironic, therefore, that the most important source of black entrapment was to be found outside the South. The decisive obstacle to black mobility at least until World War I, was that blacks were unsuccessful in securing employment in the North. This lack of success underlies both the separate labor markets identified by Gavin Wright, and the fact that plantation agriculture remained viable well into the twentieth century.

The obvious hypothesis implicit in this failure was that Northern employers discriminated—they preferred to hire European immigrants rather than black immigrants from the South. Support for this view is provided by the "great migration" of blacks from the South that occurred when international migration from Europe was shut down during World War I (Mandle 1978, 71–83). The outpouring from the South suggests that the prior failure to relocate may not have been the result of a lack of interest on the part of the black labor force (Higgs 1971, 121–22). When finally the Europeans became unavailable, employers had no choice but to hire black laborers from the South who in turn readily made themselves available.

Before World War I, in short, black labor was compelled to work in the South. Widely denied access to land and discriminated against in industrial employment in that region, large numbers of black workers found that employment on the region's plantations was their only realistic option (Thomas 1973, 330–33; Mandle 1978). They found work on the region's plantations through the vehicle of sharecropping. In that sense it is true, as Reid and Higgs have insisted, that market relations were present. Almost certainly, however, the croppers' participation in these markets was not free in the sense that Jaynes uses the term.

Beyond discrimination in hiring in the North, black labor mobility was impeded within the South itself. The delayed form of compensation embodied in sharecropping was one source of this immobility. It virtually compelled croppers to remain with the same employer for a full year since compensation was received only once—at the harvest. A second limitation on mobility derived from sharecropping as a credit relationship. Planters and merchants extended material and supplies to tenants until the harvest occurred (Ransom and Sutch 1977, 120–25). Thus they had much at risk economically. Consequently they were reluctant to come to an agreement with an individual they did not know at least by reputation. Croppers would have to demonstrate their credit worthiness to be able to secure employment on a new plantation. This is why most moves by sharecroppers were of short distances. It was difficult for black farmers to move to a new location if their reputation—both as persons who knew their place in the social order and who were good credit risks—did not precede them (Wright 1986, 92, 98). Black agricultural laborers therefore found the geographic scope for their potential relocation severely circumscribed. As Davis has put it, "the average black family moved from place to place, from landlord to landlord, and even from supplier to supplier without ever leaving the neighborhood" (Davis 1982, 179).

The consequence of this situation was to constrain southern economic development. With limited alternatives, labor supplied itself at low wages to planters. Work in the low-labor productivity cultivation of cotton ensured that these workers would receive poverty-level incomes. The planters, in turn, chose methods of production reflecting the relative prices of inputs and utilized labor-intensive techniques of cultivation. This bias away from the use of capital and the technology embodied in new tools and equipment meant that southern and especially cotton productivity was low and grew less rapidly than productivity elsewhere and in other comparable crops (Mandle 1978, 52–70).

Though equipment such as cotton seed planters, fertilizer spreaders, stalk cutters, and improved plows and harrows had become available, they largely went unutilized in the South. According to Fite, "except for plowing which was done with mule and horse power, cotton growing was mainly a hand operation" (Fite 1984, 25). It is likely that this bias away from the use of tools and equipment in production was the source of the continuing lack of a progressive technological tradition in the region noted by Wright. Absence of demand may explain the weak development of a capital goods sector in the region, the sector that would have been the wellspring of a tradition of technological advance if one had emerged (Ferleger 1985). It also may well be the reason that the South lagged behind the rest of the country in its expenditures on education. Actually, an educated or skilled labor force was not required for the kind of cultivation carried out in the South. In short, the important role of plantation agriculture in the South and that form of cultivation's technological backwardness, remained the continuing source of regional underdevelopment.

But, as emphasized by Higgs and Reid, the difficulties blacks encountered in the South were noneconomic as well as economic. In this regard plantation agriculture made itself felt as well. It not only was an important source of the South's economic underdevelopment, it also was a major factor in the persistence of the region's racism. Planter authority was virtually unchallenged by legal authorities in this period. The anthropological and sociological literature on postbellum plantation life is in unanimous agreement on this point. Edgar T. Thompson, for example, writes that the "southern plantation has always had the character of a coercive institution," and Morton Rubin agrees that the power of the planter allowed for control "to a degree far exceeding comparable institutions in a democratic society" (Thompson 1975, 98; Rubin 1951, 27).

This planter authority was almost always cast in racial terms. Phillips in 1910 reflected this view when he commented that since Southern blacks were "deficient in industrial talents and economic motives," it was desirable that they, "the inefficient and unstable element be provided with firm control and skilled management." Racism thus provided the ideological defense of the system of coercion by which planters gained access to inexpensive labor. In this way, as Thompson affirms, "capital and labor relations on the plantations are commonly at the same time race relations" (Phillips 1968, 249; Thompson 1975, 217).

Extraregional and regional limitations on employment choice, combined with the power structure embodied in the plantations themselves and the strength of the racial color bar, all conspired to dictate an exploited role for the southern black labor force. They also were decisive in turning the plantation-dominated states away from sustained technological innovation, the hallmark of the development process. The South was different from the rest of the country in many of the ways suggested by Wright. But the hypothesis offered here is that those differences found their origins in the continuation, beyond the ending of slavery, of plantation production and life. Finally, it was only under the pressure of the demand for labor during World War II that the black labor force was able completely to escape the domination of plantation agriculture and southern culture. The plantation economy's persistence thus established the context of the race relations with which we live today in the United States.

It is true that labor markets were operative in the South after the Civil War and that black sharecroppers were participants in those markets. But those labor markets were not salient in determining the role the black population played in the United States in this period. What was basic was that within the South, land was not provided to the former slaves, and that outside of the South, employment opportunities were generally not made available. These two constraints were fundamentally responsible for blacks remaining southern and poor during the half-century after the Civil War. Only in the context of these constraints did the southern labor market function. Whatever income-raising effects it might have been responsible for were swamped by the consequences of the geographic and occupational restrictions blacks encountered.

The lesson to be drawn from an examination of this experience is obvious. Market theory and market analysis are important. But markets are not all there is to society. They too, like other social institutions, must be seen in context. What this means is that it is not sufficient to treat noneconomic institutions as residuals, undeserving of analysis in their own right. All of the important institutions in society must be assessed, particularly as they interact with each other. But as we have seen, unfortunately, this is the kind of analysis economists tend to eschew.

In the case of blacks in the South after the Civil War, it is the relative unimportance of the labor market that must be emphasized. Blacks were able to choose among employers, but largely only in agriculture and in geographically circumscribed regions. A limited labor market was thus operative. But blacks remained in the region's plantation economy not because it was their voluntary choice in a market to do so. That role was an imposition, the consequence of the denial of opportunities imposed within the South and elsewhere. Those restrictions helped to define the region's economy and the role blacks played in that system.

The study of the black experience in the South after the Civil War requires that attention be paid both to the way markets functioned and the restrictions that impeded their functioning. To date, however, most attention has been directed to the first, with the second relatively neglected. The lost opportunities that the failure of land reform caused, and the costs associated with labor immobility are thus two areas requiring further analysis before a thorough appraisal of the post emancipation experience of the black labor force can be produced.

## NOTES

1. Poverty, too, was the fate of most of the one in four white southern farmers who were sharecroppers (Wright 1986, 107). Gilbert Fite cites a report of a Senate committee indicating that low incomes prevented "farmers from any enjoyment of the luxuries and elegance of life and prevents the education of their children" (Fite 1984, 34). Much work remains to be done in comparing the economics of black and white sharecropping. Differences in location, size, and output mix require close investigation. In his recent book Steven Hahn touches on the subject, but agrees "scholarly treatments of white yeoman, tenants and sharecroppers in the postwar South are still few and far between. . . ." (Hahn 1983, 160–65, 324).

2. Jonathan Wiener shares Jaynes's view that the black labor force was unfree (Wiener 1978, 69). Barbara Jeanne Fields provides a critique of this point of view from a Marxist perspective (Fields 1985, 7–28).

## REFERENCES AND FURTHER READINGS

Billings, Dwight B. Jr. 1979. *Planters and the Making of a "New South": Class, Politics and Development in North Carolina, 1865–1900.* Chapel Hill: Univ. of North Carolina Press.

Daniel, Pete. 1972. *The Shadow of Slavery: Peonage in the South, 1901–1969.* Urbana: Univ. of Illinois Press.

Davis, Ronald L. F. 1982. *Good and Faithful Labor: From Slavery to Sharecropping in Natchez District, 1860–1890.* Westport, Conn.: Greenwood Press.

DeCanio, Stephen J. 1981. "Accumulation and Discrimination in the Postbellum South." In *Market Institutions and Economic Progress in the New South, 1865–1900,* ed. Gary M. Walton and James F. Shepherd. New York: Academic Press.

DuBois, W. E. Burghardt. 1935. *Black Reconstruction: An Essay Toward a History of the Part which Black Folks Played in the Attempt to Reconstruct Democracy in America, 1860–1880.* New York: Harcourt, Brace.

Ferleger, Louis. 1985. "Capital Goods and Southern Economic Development." *Journal of Economic History* 45:411–17.

Fields, Barbara Jeanne. 1985. "The Advent of Capitalist Agriculture: The New South in a

Bourgeois World." In *Essays on the Postbellum Southern Economy,* ed. Thavolia Glymph and John J. Kushma. Arlington: Univ. of Texas.

Fite, Gilbert C. 1984. *Cotton Fields No More: Southern Agriculture 1865–1900.* Lexington: Univ. of Kentucky Press.

Foner, Eric. 1988. *Reconstruction: America's Unfinished Revolution, 1863–1877.* New York: Harper and Row.

Hahn, Steven. 1983. *The Roots of Southern Populism.* New York: Oxford Univ. Press.

Higgs, Robert. 1971. *The Transformation of the American Economy, 1865–1914: An Essay in Interpretation.* New York: John Wiley.

______. 1977. *Competition and Coercion: Blacks in the American Economy.* New York: Cambridge Univ. Press.

______. 1982. "Accumulation of Property by Southern Blacks before World War I." *American Economic Review* 72, no. 4:725–37.

______. 1984. "Accumulation of Property before World War I: Reply." *American Economic Review,* 74, no. 4:777–81.

Hughes, Jonathan. 1983. *American Economic History.* Glenview Ill: Scott, Foresman.

Jaynes, Gerald David. 1986. *Branches Without Roots: Genesis of the Black Working Class in the American South, 1862–1882.* New York: Oxford Univ. Press.

Mandle, Jay R. 1978. *The Roots of Black Poverty: The Southern Plantation Economy After the Civil War.* Durham: Duke Univ. Press.

______. 1983. "Sharecropping and the Plantation Economy in the United States South." In *Sharecropping and Sharecroppers,* ed. T. J. Byres. London: Frank Cass.

Margo, Robert A. 1984. "Accumulation of Property by Southern Blacks before World War I: Comment." *American Economic Review* 74, no. 4:768–76.

Oukre, Claude F. 1978. *Forty Acres and a Mule: The Freedmen's Bureau and Black Land Ownership.* Baton Rouge: Louisiana State Univ. Press.

Parker, William N. "The South in the National Economy, 1865–1970." *Southern Economic Journal* 46 (Apr.):1019–48.

Phillips, Ulrich Bonnell. 1968. *The Slave Economy of the Old South: Selected Essays in Economic and Social History.* Ed. Eugene D. Genovese. Baton Rouge: Louisiana State Univ. Press.

Ransom, Roger L., and Richard Sutch. 1977. *One Kind of Freedom: The Economic Consequences of Emancipation.* New York: Cambridge Press.

Reid, Joseph D. Jr. 1973. "Sharecropping as an Understandable Market Response: The Postbellum South," *Journal of Economic History* 33, no. 1:106–30.

______. 1981. "White Land, Black Labor and Agricultural Stagnation: The Causes and Effects of Sharecropping in the Postbellum South." In *Market Institutions and Economic Progress,* ed. Gary Walton and James F. Shepherd. New York: Academic Press.

Rubin, Morton. 1951. *Plantation County.* Chapel Hill: Univ. of North Carolina Press.

Thomas, Brinley. 1973. *Migration and Economic Growth: A Study of Great Britain and the Atlantic Economy.* Cambridge: Cambridge Univ. Press.

Thompson, Edgar T. 1975. *Plantation Societies, Race Relations and the South: The Regimentation of Populations.* Durham: Duke Univ. Press.

Walton, Gary, and James F. Shepherd, eds. 1981. *Market Institutions and Economic Progress in the New South, 1865–1900.* New York: Academic Press.

Wiener, Jonathan M. 1978. *Social Origins of the New South: Alabama 1860–1885.* Baton Rouge: Louisiana State Univ. Press.

Woodman, Harold. 1979. "Southern Agriculture and the Law." *Agricultural History* 53:319–37.

______. 1985. "The Reconstruction of the Cotton Plantation in the New South." In *Essays on the Postbellum Southern Economy,* ed. Thavolia Glymph and John J. Kushma. Arlington: Univ. of Texas.

______. 1987. "Economic Reconstruction and the Rise of the New South, 1865–1900." In *Interpreting Southern History: Historiographic Essays in Honor of Sanford W. Higginbotham,* ed. John B. Boles and Evelyn Thomas Nolen. Baton Rouge: Louisiana State Univ. Press.

Woodward, C. Vann. 1951. *Origins of the New South, 1877–1913.* Baton Rouge: Louisiana State Univ. Press..

Wright, Gavin. 1986. *Old South, New South: Revolutions in the Southern Economy Since the Civil War.* New York: Basic Books.

CHAPTER 9

# *Issues in Postbellum Southern Agriculture*

LEE J. ALSTON

THERE IS AND PROBABLY ALWAYS WILL BE considerable controversy about agriculture in the postbellum South. In large part this controversy reflects the rich cast of characters involved: Yankee soldiers running the South, country merchants coercing tenants into debt peonage, ex-slaves facing discrimination, plantation landlords regaining political power, and independent yeoman farmers forced into producing for the market. To make sense of this drama it is necessary to know what was grown where and to understand the more complex issues of labor contracts, credit arrangements, and racial discrimination. Additional issues underlying the drama are the importance of plantations and the reasons Southern income lagged behind Northern income.

## CROPS

In the postbellum period corn was the leading southern crop in terms of acreage, but cotton was by far the dominant cash crop. Moreover, cotton's relative importance increased over the nineteenth century, largely at the expense of feed crops and livestock. In terms of output per capita of rural population, corn and hog production declined between 1860 and 1880 in all the southern states. Corn production per capita fell 29 percent while hog production fell 41 percent (Wright 1986, 35). Meanwhile cotton production increased dramatically: from 3.8 million bales in 1860 to 6.6 million bales in 1880 to 11.6 million bales in 1910 (U.S. Bureau of Census 1975, 517–18). Cotton remained king in the postbellum period because production increased where it was grown in 1860—principally South Carolina, Georgia, Alabama, Mississippi, Louisiana, and Arkansas—and because the domain of cotton increased, spreading east, north, and west.[1]

What prompted the South's increased emphasis on cotton? Simply put, this was the profit-maximizing crop to grow (McGuire and Higgs 1977; DeCanio 1974; Wright 1986). Of the four developments favoring a shift into cotton, an increase in the relative price of cotton and a dramatic fall in the price of fertilizer were perhaps the most important factors. Additionally, upland farmers shifted into more cotton because railroads now served their areas and enabled them to forego self-sufficiency and to produce for the market. And finally, country store owners preferred cotton over other crops as collateral for credit.

Cotton was not the only cash crop grown in the South. Tobacco, rice, and sugar had their domains. Before 1890 tobacco was grown chiefly in Kentucky, Virginia, and Tennessee. After 1890 many farmers in North Carolina turned to the specialized production of bright flu-cured tobacco, the demand for which was increased by the expanding cigarette industry.[2]

Before the Civil War most rice was produced in the coastal lowlands of Georgia and South Carolina. Although some farmers continued to produce rice in the southeast until about 1900, the dominant producers of rice emerged in southwestern Louisiana, southeastern Texas, and parts of Arkansas after irrigation techniques were perfected (Fite 1984, 11). The new rice farmers ran the most modernized farms in the postbellum South.

Sugarcane production required specialized climatic conditions found mainly in Louisiana. The production process also required considerable capital in mills and irrigation ditches, which were largely destroyed in the Civil War. As a result, it took Louisiana sugar farmers some time to regain antebellum production levels.

As transportation costs fell during the postbellum period, some farmers, particularly on the eastern seaboard, turned their efforts to producing fruits and vegetables for southern and northern urban areas. The production of specialty crops required greater care in handling and greater attention to marketing than did the main cash crops. Many farmers shied away from the production of fruits and vegetables because of the high risk associated with uncertain marketing opportunities and a perishable output.

## CONTRACT FORM[3]

Emancipation threw labor relations into total disarray. Given the initial conditions, it is a wonder that any cotton was produced in the years immediately following the Civil War.[4] Plantation owners and ex-slaves were unaccustomed to dealing with one another in contractual relationships. Despite emancipation, plantation owners expected subservience and obedience from their ex-slaves. However, many ex-slaves were eager to test their new-found freedom and expected land and capital—forty acres and a mule—as compensation for their past enslavement. A third actor, the Freedman's Bureau composed of Yankee soldiers, oversaw race and labor relations and had the power of martial law. Clearly a disequilibrium market situation prevailed, and the participants in the early years were engaged in the process of search. Participants tried different forms of contracts, but some type of share contracting was the most prevalent even in the first year following the Civil War (Shlomowitz 1984, 592). In the earliest share arrangements, all workers on the plantation shared the proceeds. The collective share contract suffered from the free-rider problem and gave way to "squads." The squad was a subset of the plantation work force, usually no more than ten hands, and it was self-selecting (Shlomowitz 1984). Because the group was small and self-selecting, one might expect that the free-rider problem was minimized. Nevertheless, disputes about fair shares for all the participants, including women and children, doomed the squad system. The squad system, as distinct from family units, prevailed for a decade after the Civil War. By the late 1870s the sharecropping arrangement based on the family unit had become the most common form of contract—though by no means the exclusive one. In general terms, though with numerous variations, planters and workers settled on three forms of contract: fixed wages, shares, and fixed rent.

What factors determined the mix of contracts? One might suspect that planters would initially favor wage payments with workers organized in "gangs" as they had been under slavery. Some did, but gang labor, although successful in terms of productivity under slavery (Fogel and Engerman 1974), rested in part on the ability of planters to induce work effort through physical punishment (Fenoaltea 1984). Freedmen would no longer tolerate such "slave conditions." Even without physical punishment, ex-slaves found the close supervision necessitated by wages irksome at best. Many workers would work for wages only if they were paid a premium over what they would make as sharecroppers.

Planters too had an incentive to move away from wage contracts. Workers with a share contract had a greater incentive to work without costly supervision. Also, in the immediate postwar period it was difficult for planters to ensure the availability of sufficient harvest labor. It appears that share contracts provided a greater incentive than wage contracts for workers to stay through the harvest.

In the early postbellum years an additional incentive for share contracts was the paucity of credit (Jaynes 1985). Planters had difficulty coming up with money to pay workers during the year and consequently favored paying workers out of future output. Given the risk associated with planters paying wages, workers felt they too might fare better by being paid out of output. This explains why workers demanded a wage premium to work on wages.

Among the three principal forms of contract, the mix varied over time. Table 9.1 presents figures on the percentage distribution of farm operators by tenure between 1880 and 1910. The category of owners includes farms or parts of farms that utilize wage laborers. Over the thirty years indicated, owner-operated farms gave way to tenancy arrangements. Fixed-rent arrangements gained ground relative to share arrangements between 1880 to 1900 and lost ground between 1900 to 1910.

TABLE 9.1. Percentage Distribution of Southern Farm Operators by Tenure, 1880–1910

| | 1880 | 1890 | 1900 | 1910 |
|---|---|---|---|---|
| Owners | 63.8 | 61.5 | 53.0 | 50.4 |
| Tenants | 36.2 | 38.5 | 47.0 | 49.6 |
| Rent | 11.8 | 13.5 | 17.5 | 14.6 |
| Share | 24.4 | 25.0 | 29.5 | 33.0 |

Source: Reid (1979b, 40).

Rent and share tenants may not sum to tenants because of omission of others and unspecified tenants.

Before examining the determinants of contractual mix, it is necessary to understand the economic differences among the prevailing contract forms. First, price and yield risk borne by the contracting parties varies. Whoever is the residual claimant to output bears the risk from changing prices or yields. Wage workers, who are paid a fixed amount per unit of time, do not bear any such risk. Share contracts spread the risk between landlords and workers according to their shares in the crop.

Fixed-rent contracts place all the risk on the worker if the value of the crop is sufficient to pay the rent and any advances; in practice landlords faced some risk because tenants frequently did not have sufficient collateral to pay the rent when disaster wiped out the crop. Though the three types of contracts vary with respect to the amount of risk borne by the contracting parties, it is not immediately apparent how workers and landlords rearranged the mix of contracts in the face of increased risk. Robert Higgs (1973) has tested the hypothesis that landlords are more willing to take risks than tenants because of better access to capital markets or greater wealth, and therefore higher risk should be associated with fewer fixed-rent contracts relative to share or wage contracts. Higgs found that increased risk was indeed negatively correlated with the use of fixed-rent contracts in 1910.

Second, as discussed above, different contract forms give different incentives to honor the full term of the contract. In premechanized cotton production, the peak demand for labor occurred at harvest time. Consequently, high wages for cotton picking might lure some workers away from their present contractual obligations. But not all workers face the same costs of leaving. Costs depend on how much one would forgo by leaving. For a wage worker, ignoring reputational effects, the cost is simply the current wage. For a sharecropper and a fixed renter, the opportunity cost of leaving is how much one would clear after harvest net of advances. In theory, wages throughout the season could be withheld. If so, then the question is: Under which contract form do workers receive the most remuneration? The answer is that on average tenants make more than sharecroppers, and sharecroppers more than wage workers. As a result, if a landlord is worried about insufficient labor at harvest, he will try to negotiate more fixed-rent and share contracts.

It is not always equally damaging if labor leaves before the harvest. The damage depends on how difficult it is to replace labor, which in turn depends on unemployment and access to a large pool of labor. Immediately after the Civil War the problem of securing labor through harvest seemed particularly great. This is not surprising considering that emancipation brought about a reduction in labor services, particularly of women and children. One would expect that wage contracting should be more prevalent close to urban areas, where employers have a larger pool of labor to draw from, and when unemployment increases.[5]

A third economic difference among the contract forms is that they give rise to different incentives to take on-the-job leisure, that is, to shirk. When effort is closely linked to reward, as is the case for fixed renters, there is little incentive to shirk. Alternatively, when workers are paid according to a proxy for output (that is, per hour), they have little incentive to work without supervision. Sharecroppers, because they receive a fraction of their marginal product, have greater incentive to labor than wage workers but less incentive than fixed renters. Contract form is in part a response to high labor-supervision costs. Attributes of the landlord and worker determine supervision costs and influence contract form. Given that an owner has only a fixed amount of time, supervision costs rise with the size of the farm. To reduce supervision costs, planters with greater acreage negotiated relatively more tenant contracts (Higgs 1974).

The marginal cost of supervising workers depends on the human and physical capital supplied by landlords and workers. If landlords are supplying physical capital (for example, workstock) then they have an incentive to monitor its use to prevent depreciation. When a landlord is present to monitor workstock, the marginal cost of supervising work effort is negligible. Similarly if a landlord has to be present to provide know-how, the extra cost of monitoring effort is low. This hypothesis explains in part why young workers tend to be wage hands and why older, more experienced workers who own their workstock tend to be fixed renters.[6]

## COUNTRY STORES

The Civil War destroyed the factor system of rural finance on which antebellum planters relied. Factors (financial intermediaries between investors and plantation owners) specialized in providing large loans to plantation owners. Credit needs changed with emancipation. Now tenants were the demanders of credit. Most tenants had little wealth and needed capital for farm operations (such as fertilizer and seeds) and personal consumption. Tenants and some owners incurred debts at the beginning of the crop year and settled accounts after harvest when the crop was sold. Small country merchants filled the credit vacuum left by the disappearance of factors. The number of country stores in the Cotton South increased dramatically: 18.5 percent per year from 1870 to 1875, 4.9 percent per year between 1875 and 1880, and 7.8 percent per year between 1880 and 1885 (Goldin 1979, 17). This expansion suggests that country merchants were well suited to sell goods and furnish credit. Credit to tenants was risky business, and merchants were in a position to know the personal characteristics of tenants and minimize the risk of lending.

In general, lending is not done without some form of collateral. Credit to tenants was no exception. What was somewhat unusual was that loans were made to people with virtually no assets. The most valuable (marketable) asset that tenants possessed was their share of the crop, yet to be harvested. To allow a future crop to serve as collateral, the legislatures of southern states passed crop lien laws in the early postbellum years. A crop lien was a note given to a storekeeper or landlord that gave the holder, in event of default, the right to have the crop of the debtor sold for repayment of the debt. Merchants accepted crop liens and in exchange gave advances. They did not want to extend credit beyond the value of the lien, but the value of the lien depended on luck and how much effort a worker put into the crop. Thus lien holders (merchants) had an incentive to monitor the work effort of those to whom they advanced credit. The value of the lien also depended on a worker's skill, the size and fertility of acreage farmed, and the number of dependents (wife and children) who could help out during peak labor times.

So far I have described the relationship between merchants and debtors as one of mutually beneficial exchange. Without credit, wealthless workers would not have been able to advance from wage worker to tenant. Yet controversy surrounds the role of the merchant. At issue are whether crop lien holders coerced tenants into growing "too much" cotton, charged "extortionate" interest rates, and kept their tenants in perpetual debt.

Contemporaries railed against the practice of merchants insisting on cotton production before accepting a crop lien, and yet on average cotton was the most profitable crop for southerners to grow. If merchants possessed monopoly power, and if cotton was not the most profitable crop, then merchants were not exploiting all profit opportunities. This is because many items in tenants' consumption bundles (for example, tobacco, tea, sugar, and coffee) could not be grown on many farms. Tenants had to purchase these commodities at the country store. To the extent that these goods had a low price elasticity of demand, store owners would want to maximize the income of tenants by having them produce the profit–maximizing crop mix and then extract the profits by charging monopoly prices on the goods that tenants purchased. Moreover, if merchants coerced sharecroppers into producing a crop mix that did not maximize profits, then merchants were directly cutting into the profits of the landlords who got a share of the output either directly or in the form of the fixed rent (Lebergott 1984, 265). If the crop mix was not optimal for the sharecropper, then it was also not optimal for the landlord.

The notion that landlords could have been so taxed by merchants does not seem plausible. Landlords had considerably more political power and financial resources than did merchants. In any struggle over who would be able to tax whom, landowners would seem to have had the upper hand. Consistent with this view is the fact that landlords in the cotton states were given the first claim to a tenant's crops (Wiener 1978). That is, landlords received rent before merchants could claim payment for advances. Perhaps what angered workers was not the loss of self-sufficiency that came from growing more cotton, for they were never really self-sufficient in foodstuffs, but the loss of choice over their production and marketing decisions (Wright 1986, 113–14).

It is legend that indebted Southern sharecroppers and tenants paid extortionate interest on their debts. How high were interest rates, and did competition put a ceiling on interest rates? Ransom and Sutch (1977) maintain that merchants on average charged 60 percent on an annual basis for credit. Ransom and Sutch estimate their implicit interest rate by calculating the difference in prices between goods purchased on credit and goods purchased for cash and then adjusting the difference to an annual rate. Prices for goods purchased on credit averaged 30 percent more than cash prices. Since credit was extended from May until harvest, the period of the loan was approximately six months. The 30 percent markup for six months converts into an annual rate of 60 percent. However, the 30 percent markup does not necessarily imply that merchants earned 60 percent. Implicit in the 60 percent is that merchants earned 60 percent on their money for the half of the year that tenants did not receive credit (Temin 1979, 62). Temin notes that the total annual return of merchants is a mixture of the 30 percent they received for six months and what they could receive at a financial institution for the remaining six months. Temin suggests that 5 percent was a reasonable short–term market rate and hence suggests that the gross return was closer to 35 percent. From this gross return, bad loans need to be deducted. Some loans were never repaid while other loans were extended, usually at an annual rate of 8 percent (Higgs 1977, 56).

Other evidence suggests that although loan rates were high, they were not a result of monopoly power but rather of risk. For one to have monopoly power, there need to be barriers to entry. There appear to have been few constraints to entry.[7]

No legal constraints precluded entry. The net worth of merchant stores was between $5000 and $10,000 (Ransom and Sutch 1977, 138). This relatively small amount suggests that if country stores were extremely profitable, the profits would be competed away. Actually there was considerable entry as indicated earlier. If merchants had monopoly power, then it is implicit that many farmers had no access to a store before the arrival of new entrants (Goldin 1979, 17). Surely most farmers had some outlet from which they purchased goods.

Suppose one acknowledges that merchants had some monopoly power. How large were monopoly profits? Lebergott estimates that monopoly profits from credit ranged between $8 and $18 per loan (Lebergott 1984, 264). Lebergott derived his estimates by taking the difference between the interest rate charged by merchants (60 percent) and a justifiable interest rate (15 percent to 28 percent), times the average lien ($18), times the length of loan (one-half year).[8] Given that the average income of sharecroppers was $184, it appears that either merchants were benevolent monopolists or simply did not have much monopoly power (Lebergott 1984, 262).

Distinct from the argument about high interest rates is the allegation that merchants held their tenants in debt peonage. There is a strong and weak variant of the argument. The strong view is that tenants were held in debt perpetually. They could not escape because their income after harvest was not sufficient to pay their advances (Wiener 1988, 979). The weak variant is that most tenants made enough to pay out after the harvest but did not have sufficient wealth to get through the next year without borrowing from the same merchant (Ransom and Sutch, 1977).

The strong variant has a theoretical weakness. Holding someone in perpetual debt is not desirable for a lender because it implies giving away real resources. To understand why, answer the question: Would you be willing to loan me $1000 if you never expected to get paid back? Of course, if the debt is fictitious, then a lender might gain if the debt reduces the mobility of a valued tenant (Higgs 1977, 58). But if the loan was real, merchants were quite wary of extending too much credit for fear that tenants would abandon the farm without notice. It seems reasonable to assume that merchants tried to limit credit to an amount they expected tenants would repay. How much credit a tenant could secure depended on his reputation, the state of his current crop, and how many dependents he had to help with the harvest.

Despite the best efforts of merchants in monitoring their debtors, some failed to pay out at the end of the year. Unfortunately there is little systematic evidence on the degree of indebtedness or changes in indebtedness.[9] The data that exist indicate that for Georgia, indebtedness of farmers to merchants fell between 1883 and 1887: the average indebtedness in 1887 was between 74 percent and 80 percent of the 1883 level (Fishback 1987, 4–5). Perhaps of more interest is what factors determined the changes in indebtedness. Fishback (1989) estimated the impact of various factors on the indebtedness of farmers in one year compared to the previous year. The results are consistent with the view that good farm conditions increase the likelihood of paying out and that increased purchases from merchants increase the likelihood of debt. The results did not support the view that farmers lived in perpetual indebtedness: indebtedness the previous year had little effect on indebtedness in the current year. The view that seems to be emerging is that farmers were in and out of debt depending on the prices and yields of crops, cotton in particular.

Although on average merchants did not hold tenants in perpetual debt peonage, no doubt there were some farmers who were defrauded through fictitious debts and cowed into remaining with the same creditor. But it seems that the crop lien system was simply the vehicle for fraud and not the reason. Ignorance on the part of tenants and uncertain (or an absence of) legal rights provided the underlying basis for fraud.

## BLACKS IN AGRICULTURE

> We have only to reflect for a moment upon the situation in which these people found themselves when liberated. Consider their ignorance, their poverty, their destitution, and their absolute dependence upon the very class by which they had been held in bondage for centuries, a class whose every sentiment was averse to their freedom, and we shall be prepared to marvel that they have, under the circumstances, done so well. (Frederick Douglass, 1880. Quoted in Higgs [1977, 37].)

Judging the economic condition of blacks in agriculture is a lot like determining whether the glass is half full or half empty. In an absolute sense, poverty abounded throughout the postbellum period (Fite 1984, Chap. 1). Hand-to-mouth existence was the norm for many black farmers, as it was for many white sharecroppers and tenants as well. A farmer's (black or white) wealth position was largely determined by where he stood on the agricultural ladder. Table 9.2 gives the percentage distribution of farm operators by tenure and race. Several comparisons are noteworthy. First, from no ownership in 1865, 20 percent of black farm operators owned the land they tilled by 1880. Relative to whites, black operators were a third as likely to be owners, three times more likely to be fixed renters, and twice as likely to be sharecroppers. Clearly in 1880 blacks tended to be on the lower rungs of the agricultural ladder relative to whites. Between 1880 and 1900 there was an absolute and relative improvement in the farm status of blacks. The percentage of operators represented by black sharecroppers fell while the percentage of fixed-rent tenants and owners increased. Whites did not match the gains by blacks: there was little change in the mix of classifications between 1880 and 1900.

The next ten years, 1900–1910, witnessed a decline in conditions for blacks. The percentage of black share tenants increased while the percentage of fixed renters and owners fell. For whites there was a small decline in status. That by 1900 a quarter of black operators achieved ownership is impressive, particularly in light of the difficulty in accumulating sufficient wealth to purchase land and the discriminatory obstacles faced by blacks. To own land, blacks had to be acceptable to the white community. They had to know their "place" and not be too aggressive.

TABLE 9.2. Percentage Distribution of Southern Farm Operators, 1880–1910

| | 1880 | 1900 | 1910 |
|---|---|---|---|
| Owners | | | |
| Black | 20 | 25 | 25 |
| White | 66 | 64 | 61 |
| Fixed and Other Renters | | | |
| Black | 26 | 37 | 29 |
| White | 8 | 10 | 10 |
| Share Renters and Croppers | | | |
| Black | 54 | 38 | 46 |
| White | 27 | 26 | 29 |

Sources: The figures for 1880 come from Reid (1979b, 40). The figures for 1900 and 1910 come from Reid (1979a, 286).

Because of rounding the figures may not sum to 100%.

In addition to farm ownership there are other indicators of black progress, in absolute terms and relative to whites. Robert Higgs estimates that for Georgia, black real wealth grew by almost 9 percent per year from 1875 to 1892, stagnated during the remainder of the nineties, and then grew 6 percent per year from 1902 to 1915.[10] Relative to whites, black income per capita grew faster from the Civil War to the depression of the 1890s, slower from 1892 to 1902 and at the same rate or faster from 1902 to 1920.

Changes in mortality also indicate that the standard of living of blacks improved over the postbellum period. Declines in mortality rates in this period most likely resulted from changes in housing and diet because there was little advance in medical science or reduction in the incidence of infectious diseases. In the 1850s the black expectation of life at birth was approximately thirty-one or thirty-two years. Health deteriorated during and immediately following the Civil War. For the later postbellum period recent estimates indicate that life expectancy increased, reaching forty to forty-two years by the mid-1890s.

Despite the absolute and relative gains in income over the postbellum period, black income remained considerably below white income. We do not have estimates for the rural South, but for the United States as a whole black income relative to white stood at about 35 percent in 1900.

What factors can account for large differences in income between the blacks and whites? There is little evidence that blacks and whites received different agricultural wages or tenancy terms (Wright 1986, 96; Higgs 1977, 54, 65). However, it is clear that humanitarian concerns did not prompt the payment of equal wages. Consider the statement of Colonel Samuel Thomas in 1865.

> [T]hey [the Southern whites] are yet unable to conceive of the negro as possessing any rights at all. Men who are honorable in their dealings with their white neighbors will cheat a negro without feeling a single twinge of their honor. To kill a negro they do not deem murder; to debauch a negro woman they do not think fornication; to take the property away from a negro they do not consider robbery. The people boast that when they get freedmen affairs in their own hands, to use their own classic expression, "the niggers will catch hell." The reason for all this is simple and manifest. The whites esteem the blacks their property by natural right, and however much they may admit that the individual relations of masters

> and slaves have been destroyed by the war and by the President's emancipation proclamation, they still have an ingrained feeling that the blacks at large belong to the whites at large, and whenever opportunity serves they treat the colored people just as their profit, caprice, or passion may dictate. (Quoted in Higgs 1977, 9).

Attitudes probably changed little by the turn of the century. Despite a taste for discrimination, it appears as if a taste for greed was stronger. Competition among landlords bid up wages for blacks. Landlords tried to form cartels to prevent competition, but evidence suggests that the cartels were not effective. The repeated proposals throughout the postbellum period for cooperation would not have been necessary had cooperation existed.

Part of the difference in income can be explained by differences in initial conditions. For blacks to have attained income equality with whites by World War I, black income per capita needed to increase almost elevenfold (Higgs 1977, 126). It is important to keep in mind that white incomes were also growing over the period.

Another reason why white incomes exceeded black incomes stems from the role of property (DeCanio 1979, 1984). At the national level property generated about a third of national income in the late nineteenth century. At emancipation blacks had virtually no property so they could not close the income gap until differences in property income were eliminated. As we can see in Table 2, by 1900 blacks in southern agriculture had begun to accumulate property but these figures understate the white/black difference in ownership. Blacks owned smaller and less fertile farms than whites. For example, in a sample of Georgia counties in 1902 the average acreage of white proprietorships was 264.8 acres compared to 64.3 acres for black proprietorships (Wright 1986, 105).

No doubt the lack of initial property ownership is important in accounting for income differences, but so too are the discriminatory forces faced by blacks in the postbellum South. It was noted earlier that blacks and whites tended to receive comparable wages for the same work, but they often did not have access to the same work. Blacks in the postbellum South were overwhelmingly in agriculture. Gavin Wright (1986) argues that the lack of blacks in the textile industry was not the result of outright discrimination but rather was a historical legacy: in the postbellum period whites were employed first in textiles. To the extent that whites passed on skills to their children, blacks could only be employed in textiles if they were willing to work for less than whites. But blacks had no incentive to work for less than whites because the going wage in textile mills was set by the agricultural wage.

With respect to other occupations there is some evidence of exclusion. Craft unions in the South and particularly in the North did not encourage black membership (Higgs 1977, 85–86). Of course any occupations in which blacks would oversee whites were out of the question. Some jobs were simply "white man's" jobs.

Blacks had considerable reason to pursue an agricultural career given the socio-legal conditions. Educational expenditures were grossly unequal beginning with emancipation and worsening over time as blacks lost political power after 1890 (Margo 1982, 11). Gavin Wright suggests that low expenditures on education for blacks and whites was a result of decision makers' (rural planters and mill owners) inability to capture the returns from education. Planters and mill owners feared that education would prompt migration and thereby raise labor costs. In 1910 per pupil southern expenditures as a percent of the U.S. average ranged from 24 percent in South Carolina to 65 percent in Texas (Wright 1986, 80). Considering that within the South per pupil expenditures on whites were as great as five times the expenditures on blacks, the amount of education for blacks was pitiful indeed (Margo 1985, Table I.1).[11]

Perhaps more than anything else, uncertainty over property rights reduced the opportunity for blacks to acquire skills and possessions. Blacks feared reprisals should they transgress the norms of race etiquette, one of which was never appearing uppity. Given the savagery with which whites loosed their wrath on blacks—namely, beatings, lynchings, rapes—the fear of white reprisal was well warranted.

Since the southern legal authorities allowed and at times participated in crimes against blacks, blacks had every incentive to seek for themselves a powerful patron, one who could protect them from hostilities. White plantation owners fulfilled the role of "protector" (Alston and Ferrie 1989). In return for protection, blacks had to perform as "good and faithful labor." Good and faithful meant working hard over several years. Plantation owners, through their political power, fostered an intimidating, discriminatory state and then gave protection to increase work intensity and reduce labor turnover. Protection took other forms as well, including old-age assistance and medical care. Poorer whites also entered into "paternalistic" relationships, but in general they had a lower demand for protection than blacks. The feedback effect on blacks of a discriminatory legal and judicial system was to adopt a risk-averse strategy: stay in agriculture and manage as best one could.

## WHY THE SOUTH LAGGED BEHIND

Available evidence indicates that real wages for unskilled farm labor were greater in the North than the South dating back to 1866. Moreover, the difference increased over time. A comparison between Mississippi (a high-wage southern state) with Ohio (a low-wage northern state) reveals that the absolute difference in wages was about $0.10 a day in 1865, $0.20 in 1880, $0.50 in 1902, and $0.80 in 1914 (Wright 1986, 66). These differences are significant because absolute real wages per day in Mississippi ranged from $0.70 to $1.20 between the Civil War and World War I (Wright 1986, 66). Thus the cost of staying in the South increased over the postbellum period. Was southern agricultural labor bound? It would appear not, given the considerable migration within the South. Labor moved from the low-wage Southeast to the high-wage Southwest in large numbers throughout the postbellum period. What then prevented labor from moving North?

Limited localized knowledge about seeds, crops, and climate is one part of the answer. People tend to feel physically and psychologically comfortable in a similar climate. Agricultural skills and knowledge that are well suited to one region may be useless in another. Moreover, a move from the South to the North most likely meant abandoning agriculture for an urban job. Since migration flows North never really got started in the antebellum period, this further stunted postbellum migration because a prime source of information about jobs is letters from former migrants whom one trusts.

The southern labor market could have remained separate but equal.[12] Instead of labor flowing out to equalize wages, capital could have flowed in to utilize cheap labor, eventually equalizing wage rates. This did not happen in the South. Some capital did flow in, but southern industry never really took off because northern capitalists brought with them northern technology unsuited to the resource endowments of the South (Wright 1986). The South needed its own technology. But for technology one needs education, and the South invested few resources in education. The South as a whole might have benefitted from increased educational expenditures, but the rural elite who controlled the purse strings worried about the impact of education on labor costs.

As a result of racism, low educational expenditures, inappropriate technology, and little out-migration, the South at the end of the postbellum period remained an isolated low-wage region within the U.S. economy.

At the turn of the twentieth century more than a million black and white southern farmers were mired in what seemed to be inescapable poverty. These farm families lived out their lives in a hand-to-mouth existence. Their homes, their diets, their clothes, and their meager expenditures for items other than food, fuel, and shelter all testified to perpetual hard times. Hundreds of thousands of farmers were unable to accumulate enough capital to own a farm, to earn sufficient money to provide a decent house and furnishings, or to buy the amenities that might make life a little more pleasant. In an economy where money was essential to acquire the industrial products being turned out by the nation's expanding factories, poor southern farmers were left in the economic and social backwash. Although both white and black farm families developed institutions that met some of their social and cultural needs, life was too often dull and monotonous. (Fite 1984, 47).

Life was not dull and monotonous for everyone. One suspects that the life of plantation families could be quite jolly indeed. The high life of the rural elite was made possible by the impoverished conditions elsewhere. If the proximate causes of southern poverty were low education, racism, and little out-migration, the root of southern poverty was most likely the political power of the rural elite. The connections between the economic incentive of the southern elite to maintain a dependent low–wage labor force and the ways in which they accomplished their goal is a relatively unexplored area of research.

What was the incentive of the rural elite to maintain a dependent low–wage labor force? By a dependent low–wage labor force I mean the bottom half of the paternalistic relationship already described between landlord and tenant.

Landlords offered to compensate workers with paternalism to increase work intensity, reduce mobility, and instill in workers a longer time horizon (Alston and Ferrie 1989). Southern agriculture in postbellum times is typified by high supervision costs of labor because southern landlords, unlike their northern counterparts, had a large work force spread over vast acreages. By giving their workers paternalistic goods that could be seen as goodwill gestures, landlords might legitimate their position in the social hierarchy and also might cause workers to feel grateful and work harder. Perhaps more importantly, paternalism would increase the cost to workers of being detected shirking by raising the cost of losing their jobs. Paternalism was not something a worker could readily replicate with another employer or purchase in the market. It was something that employers gave only to their "good and faithful" laborers. Thus it not only increased work effort but also reduced job mobility. This was important to landlords because before advances in science the only way to ascertain and control soil fertility was through learning by doing. The learning by doing was embedded in tenants who had farmed the same plot over a number of years. Paternalism also represents a commitment on both sides to a long-term relationship. The advantage to employers is that workers under a paternalistic relationship have an incentive to labor in the long-run interests of the landlord: rather than maximizing short-run yields they would maximize the present value of the stream of yields over the expected relationship. This was accomplished by the appropriate application of fertilizers and maintaining capital assets such as ditches and fences.

The above rationale for the provision of paternalism helps us to explain why, in general, tenants and sharecroppers were recipients of paternalism while wage workers were not. To a certain extent direct supervision, contract mix, and paternalism are substitute means for encouraging work effort. Wage workers are directly supervised to increase work effort, and the provision of paternalism may not further increase work effort to warrant its costs. Sharecroppers are less frequently supervised, and paternalism functions to increase work effort and reduce turnover. Tenants on plantations are the most likely to receive paternalism, which reduces the incentive to change employers. The provision of paternalism rested on the ability of landlords to provide goods that markets or governments did not supply, and the landlords maintained a monopoly on this ability through their disproportionate political power. Before examining the results of the political power of the plantation elite, I will attempt to assess the importance of plantations in southern agriculture.

Plantations dominated the production of cotton in the antebellum years. In the postbellum years it is difficult to determine the importance of plantations because the census did not enumerate farms according to ownership or management but only by operator. That is, the census considered all sharecroppers as farmers operating a farm even though the farm was part of a large management ownership unit. If what one means by a plantation is that a large tract of land was worked with gang labor, then it is correct to say that plantation-style agriculture died out after the Civil War. But if one is interested in large ownership units that employed large work forces under central management, then it is a mistake to suppose plantations were unimportant in the postbellum period. A recent case study of the Natchez district in Mississippi indicates that antebellum plantation owners were able to hold on to their land in the postbellum period (Wayne 1983). Land was neither confiscated by the victorious North nor seized by creditors for payment of debts. In some instances to avoid seizure landowners transferred land titles to their unencumbered spouses.

How representative was the Natchez district? My guess is that Natchez was not atypical, and in postbellum times plantations continued to dominate rural agriculture in some areas of the South. Plantations declined in relative importance over the postbellum period because of the increased production for market by upcountry farmers. Nevertheless, in absolute terms, at the end of the postbellum period plantations were important enough for the Census Bureau in 1910 to undertake a special survey of their prevalence in the South. The Census defined a plantation as "a contiguous tract of land of considerable area under the general supervision or control of a single individual or firm, all or part of such tract being divided into at least five smaller tracts which are leased out to tenants" (U.S. Bureau of the Census 1916, quoted in Whatley [1987]). The Census never published the county data but presented a map showing 325 plantation counties. The plantation belt, typified by fertile soil and a large black population, ran in an arc from Texas to Virginia and represented more than 30 percent of improved acreage in Alabama, Arkansas, Georgia, Louisiana, Mississippi, and South Carolina (Whatley 1987, 54). In the fertile Mississippi Delta plantations dominated cotton cultivation.

For our purposes the point is not simply that plantations survived and thrived as agricultural production units but that the plantation owners and their allies (bankers and merchants) dominated political decisions (Harvard 1972, 9–10). The disfranchisement of blacks and legal segregation can be viewed as the political means of maintaining a cheap and reliable labor force. Similarly, the low levels of educational expenditures reduced the likelihood of rural out-migration and contributed to the impoverishment of rural areas, which again enhanced the paternalistic arrangements of the planters.

The political power of the rural elite operated at all levels of government—local, state, and federal. After Reconstruction local and state governments maintained a policy platform of relatively low expenditures on education, welfare, and old-age assistance. At the national level, in the 1930s and 1940s when social legislation was proposed that would have affected the labor costs of the rural elite, southerner legislators moved quickly to exempt farm labor from the provisions of the legislation (Alston and Ferrie 1985). Examples include social security legislation and minimum wages. Rather than market arrangements, such as sharecropping and crop liens, causing the South to lag behind the rest of the country, it appears as if the culprit was the political power of the rural Southern elite.

## NOTES

For comments I thank Mary Alston, Robert Higgs, and Price Fishback. This chapter was written while on leave in the Department of Economic History in the Research School of Social Sciences at the Australian National University. I am grateful for their support.

1. For maps depicting the increase in cotton production, see U.S. Department of Agriculture (1918, 17).

2. Nanny May Tilley (1948) gives the authoritative account of the tobacco industry in North Carolina. For a recent study focusing on the economic and social relations in an old tobacco region see Shifflett (1984). Anyone searching for a topic in southern history is advised to analyze the labor and capital relations in tobacco cultivation.

3. The following account draws heavily on Alston and Higgs (1982).

4. Lebergott (1984, 249) maintains that the Southern production of cotton following emancipation declined far less than sugar production following emancipation in two British sugar colonies.

5. In the South a large shift into wage contracts occurred during the Great Depression.

6. Perhaps the most dramatic change in enforcement and supervision costs came with the adoption of the tractor and cotton picker in the twentieth century. The tractor standardized the marginal productivity of labor. That is, it controlled the quality of plowing and thereby reduced the need to monitor labor (Alston 1981). The cotton picker reduced the need for downstream harvest labor and enabled planters to use more wage contracts (Whatley 1985, 1987).

Gavin Wright (1986) explains contract form largely through sorting: young, unmarried workers sought wage contracts in a large geographic area whereas married workers sought more permanent employment as sharecroppers and tenants. The need to secure credit through one's reputation limited widespread geographic mobility for married workers. Wright's interpretation is consistent with the explanation presented here but cannot account for differences in contractual mix across demographically similar places or times.

7. For an alternative view, see Ransom and Sutch (1979) who argue that social constraints precluded entry.

8. The lower band of 15% was taken from Ransom and Sutch (1977) who specified 15 percent as justifiable. Lebergott considered 28 percent, which was the rate rural banks charged white farmers, as a "low upper bound" (Lebergott 1984, 265).

9. Ideally we would like evidence on indebtedness and turnover. Even if a tenant was in debt to one merchant of the landlord, another merchant or landlord might be willing to pay the debt to acquire the labor services of the tenant.

10. The following paragraphs on health and wealth of blacks are based on data found in Higgs (forthcoming).

11. Robert Margo (1986) using data from 1930 to 1940 argues that even if school inputs had been equal for blacks and whites, educational achievement of blacks would have lagged behind whites because of relatively low black incomes and wealth and a high rate of illiteracy.

12. The separateness of southern labor markets is the unifying theme in Wright (1986).

## REFERENCES AND FURTHER READINGS

Alston, Lee J. 1981. "Tenure Choice in Southern Agriculture." *Explorations in Economic History* 18(July):211–32.

______. 1986a. "Race Etiquette in the South: The Role of Tenancy." *Research in Economic History* 10:193–205.

______. 1986b. "Resisting the Welfare State: Southern Opposition to the Farm Security Administration." In *The Emergence of the Modern Political Economy,* ed. Robert Higgs. Greenwich, Conn.: JAI Press.

Alston, Lee J., and Joseph P. Ferrie. 1985. "Labor Costs, Paternalism and Loyalty in Southern Agriculture: A Constraint on the Growth of the Welfare State." *Journal of Economic History* 45(Mar.):95–117.

______. 1989. "Social Control and Labor Relations in the American South before the Mechanization of the Cotton Harvest in the 1950s." *Journal of Institutional and Theoretical Economics* (Mar.):133–57.

Alston, Lee J., and Robert Higgs. 1982. "Contractual Mix in Southern Agriculture: Facts, Hypothesis and Tests." *Journal of Economic History* 42(June):327–53.

Alston, Lee J., Samar K. Datta, and Jeffrey Nugent. 1984. "Tenancy Choice in a Competitive Framework with Transactions Costs." *Journal of Political Economy* 45(Mar.):95–117.

Davis, Ronald L. F. 1982. *Good and Faithful Labor: From Slavery to Sharecropping in the Natchez District, 1860–1890.* Westport, Conn.: Greenwood Press.

DeCanio, Stephen J. 1974. *Agriculture in the Postbellum South: The Economics of Production and Supply.* Cambridge: MIT Press.

______. 1979. "Accumulation and Discrimination in the Postbellum South." *Explorations in Economic History* 16(Apr.): 182–206.

Fenoaltea, Stefano. 1984. "Slavery and Supervision in Comparative Perspective." *Journal of Economic History* 44(Sept.):635–68.

Fishback, Price V. 1989. "Debt Peonage in Postbellum Georgia."*Explorations in Economic History* 26(Apr.):219–36.

Fite, Gilbert. 1984. *Cotton Fields No More: Southern Agriculture 1865–1980.* Lexington: Univ. Press of Kentucky.

Fogel, Robert William, and Stanley R. Engerman. 1974. *Time on the Cross: The Economics of American Negro Slavery.* Boston: Little, Brown.

Goldin, Claudia. 1979. "'N' Kinds of Freedom." *Explorations in Economic History* 16(Jan.):8–30.

Havard, William C., ed. 1972. *The Changing Politics of the South.* Baton Rouge: Louisiana State Univ. Press.

Higgs, Robert. 1973. "Race, Tenure, and Resource Allocation in Southern Agriculture, 1910." *Journal of Economic History* 33(Mar.):149–69.

______. 1974. "Patterns of Farm Rental in the Georgia Cotton Belt, 1880–1900." *Journal of Economic History* 34(June):468–82.

______. 1977. *Competition and Coercion: Blacks in the American Economy, 1865–1914.* New York: Cambridge Univ. Press.

______. (Forthcoming.) "Black Progress and the Persistence of Racial Economic Inequalities, 1865–1940." In *Blacks and Whites in the U.S. Labor Market: Analyses of Racial Inequality,* ed. William A. Darity, Jr., and Steven Shulman. Middletown: Wesleyan Univ. Press.

Jaynes, Gerald David. 1985. *Branches Without Roots: Genesis of the Black Working Class.* New York: Oxford Univ. Press.

Lebergott, Stanley. 1984. *The Americans: An Economic Record.* New York: W. W. Norton.

McQuire, Robert, and Robert Higgs. 1977. "Cotton, Corn and Risk in the Nineteenth Century: Another View." *Explorations in Economic History* 14(Apr.):183–95.

Mandle, Jay. 1978. *The Roots of Black Poverty.* Durham, N.C.: Duke Univ. Press.

Margo, Robert A. 1982. "Race Differences in Public School Expenditures: Disfranchisement and School Finance in Louisiana, 1890–1910." *Social Science History* 6(Winter):9–33.

______. 1985. *Disfranchisement, School Finance, and the Economics of Segregated Schools in the U.S. South, 1890–1910.* New York: Garland Press.

______. 1986. "Educational Achievement in Segregated School Systems; The Effects of 'Separate-but-Equal.' " *American Economic Review* 76(Sept.):794–801.

Ransom, Roger, and Richard Sutch. 1977. *One Kind of Freedom: The Economic Consequences of Emancipation.* New York: Cambridge Univ. Press.

Reid, Joseph. 1973. "Sharecropping as an Understandable Market Response." *Journal of Economic History* 33(Mar.):106–30.

______. 1975. "Sharecropping in History and Theory." *Agricultural History* 49(Apr.):426–40.

______. 1976. "Sharecropping and Agricultural Uncertainty." *Economic Development and Cultural Change* 24(Apr.):549–76.

______. 1979a. "Sharecropping in American History." In *Risk, Uncertainty, and Agricultural Development,* ed. James Roumasset et al., 283–319. New York: Agricultural Development Council.

______. 1979b. "White Land, Black Labor and Agricultural Stagnation: The Causes and Effects of Sharecropping in the Postbellum South." *Explorations in Economic History* 16(Jan.):31.55.

Shifflett, Crandall A. 1984. *Patronage and Poverty in the Tobacco South: Louisa County, Virginia, 1860–1900.*

Schlomowitz, Ralph. 1979. "The Origins of Sharecropping." *Agricultural History* 53(July):557–75.

______. 1984. "'Bound or Free'? Black Labor in Cotton and Sugarcane Farming, 1865–1880." *Journal of Southern History* 50(Nov.):569–96.

Temin, Peter. 1979. "Freedom and Coercion: Notes on the Analysis of Debt Peonage in One Kind of Freedom." *Exploration in Economic History* 16(Jan.):56–63.

Tilley, Nanny May. 1948. *The Bright-Tobacco Industry.* Chapel Hill: Univ. of North Carolina Press.

U.S. Bureau of the Census. 1975. *Historical Statistics of the United States, Colonial Times to 1970.* Bicentennial edition, Part 1. Washington, D.C.: Government Printing Office.

U.S. Department of Agriculture. 1918. *Atlas of American Agriculture.* Part 5, Sec. A. Washington, D.C.: Government Printing Office.

Wayne, Michael. 1983. *The Reshaping of Plantation Society: The Natchez District, 1860–1880.* Baton Rouge: Louisiana State Univ. Press.

Whatley, Warren. 1983. "Labor for the Picking: The New Deal in the South." *Journal of Economic History* 43(Dec.):905–30.

______. 1985. "A History of Mechanization in the Cotton South: The Institutional Hypothesis." *Quarterly Journal of Economics* (Nov.):1191–1215.

______. 1987. "Southern Agrarian Labor Contracts as Impediments to Cotton Mechanization." *Journal of Economic History* 47(Mar.):45–70.

Wiener, Jonathan M. 1978. *Social Origins of the New South.* Baton Rouge: Louisiana State Univ. Press.

Wright, Gavin, 1979. "Freedom and the Southern Economy." *Explorations in Economic History* 16(Jan.):90–108.

______. 1982. "The Strange Career of the New Southern Economic History." *Reviews in American History* 10(Dec.):164–80.

______. 1986. *Old South, New South: Revolutions in the Southern Economy Since the Civil War.* New York: Basic Books.

______. 1987. "The Economic Revolution in the American South." *Journal of Economic Perspectives* (Summer):161–78.

CHAPTER 10

# *Slavery, Emancipation, and the Capitalist Transformation of Southern Agriculture, 1850–1910*

JOSEPH P. REIDY

BETWEEN 1865 AND 1900 the most important general trend in southern agriculture was the development of capitalist social relations.[1] Raw materials–producing areas in other parts of the world also felt developmental pressures of a similar nature simultaneously, though the position of the South within the national economy and reunified national state made it different from Europe's colonial dependencies. The process echoed changes characteristic of the early history of capitalism in England. Both transformations involved the creation of a "free" laboring class, liberated from direct domination of the kind exercised, for instance, by slave masters over their slave property, but also separated from customary rights to productive property, particularly land, from which subsistence could be derived. Thus freed, these incipient proletarians had to exchange their ability to labor for remuneration with which to obtain the necessities of life (Marx 1967; Dobb 1947; Hilton 1976; Aston and Philpin 1985). In the emancipation process, southern slaves became such a free laboring class. Freed from their masters by the victorious Union army, they had to, in the vernacular of the time, work or starve.

In any number of particulars the experience of the American South did not exactly parallel that of England, the social heritage of racial slavery being only the most obvious. Whereas the initial transformation took several centuries to accomplish in England, for instance, it was essentially over in one century in the South. Moreover, whereas improving English landlords began to accumulate capital internally early in the transformation process, internal capital accumulation in the South lagged for a variety of reasons, including federal monetary and banking policies and the general decline in commodity prices over the Long Depression of 1873–1898. After 1900, however, the trend was toward internal capital accumulation, outside capital investment, application of science and technology to agricultural "problems," and the qualitative (not just quantitative) exploitation of labor.[2] Under the direction of southern "Progressives," this process moved forward in fits and starts until the 1930s, when New Dealers infused sufficient capital to allow the mechanization of field operations. The present essay will focus upon the period down to 1910, during which capitalist social relations displaced those of slavery, but before tractors and mechanical harvesters completed the transformation process.

## THE SWAN SONG OF SLAVERY, 1850–1861

Before the Civil War the economic prosperity of the entire nation rested upon the backs of southern slaves. As Douglass C. North (1966) has shown, southern, slave-grown cotton provided nearly all the raw material needed by both European and northern textile mills and the bulk of the fledgling nation's foreign exchange. Other staples—tobacco, rice, and sugar specifically—did not figure as prominently in international finance but did enter into both international and national markets. Yet, despite the legitimate claims of the critics of plantation agriculture that it was wasteful and destructive, the nation could neither deny nor escape its reliance upon slave labor.

Although slaves made up the bulk of the South's commercial agricultural work force, whites constituted nearly two-thirds of the southern population in 1860. Most whites lived in nonplantation areas, where subsistence agriculture rather than production for the market promised the sturdy independence yeoman farmers prized. On occasion the yeomen clashed with Black Belt planters, for despite essential agreement over slavery their views of the ideal society did not coincide. Within the rubric of shared political and social institutions, the grandees and the plain folk coexisted.

Whether producing for use or sale, all farmers transform the surface of the land. Southern planters did so on a grand scale, with generally little regard for long-term effects. As Eugene D. Genovese and Gavin Wright have demonstrated from different perspectives, the social value of slaves far exceeded that of land. Land constituted the means, slavery the end (Genovese 1965; Wright 1986). Economic values reflected that reality, as did southerners' notorious land-use policies. In some areas, however, such as those along the rivers of the South Atlantic and Gulf coasts, planters both oversaw drastic modifications of the physical geography and took great pains to conserve the soil and other natural resources. Sugar and rice thrived in the coastal areas, but successful cultivation required flood control. The labor of keeping rivers within their banks seemed endless. It was physically demanding and often dangerous, with the added risk of exposure to debilitating diseases such as malaria. Attesting to the difficulty of labor in such settings, antebellum Louisiana sugar planters considered their slaves too valuable to risk digging drainage canals; they hired Irishmen instead. And they consistently preferred men over women for plantation labor, thereby producing a demographic imbalance unique in American slavery though typical of other New World sugar-plantation areas (Sitterson 1953; Shugg 1939).

Of all the feats of labor performed in the South, none matched those of rice plantation slaves in coastal South Carolina and Georgia. During the eighteenth century, rice planters abandoned swamp-based rice cultivation in favor of hydraulic cultivation along the banks of rivers affected by the tides. The change involved new technology in which the flow of water was regulated by an elaborate and extensive network of dikes, locks, canals, and drainage ditches. The change required an enormous amount of concentrated and coordinated labor, and doubtless exacted a staggering human toll. Out of the crucible of such change emerged the task system of labor. As a growing number of recent studies has demonstrated, that system provided slaves with access to land to work at their own discretion after they had finished assigned tasks. Although the larger constraints of slavery remained, nowhere else in North America did slaves enjoy such control over their labor and its fruits (Clifton 1978; Armstrong 1980; Berlin 1980; Morgan 1982; Coclanis 1988).

In the cotton plantation area slaves also modified the natural environment. Between 1800 and the 1840s the cotton frontier had spread from upland South Carolina and Georgia across the South as far as Texas. In the process, slaves brought millions of new acres under the plow. but in addition to clearing new land, at times they had to revamp old fields as well. Numerous studies of antebellum agriculture indicate that depressions, particularly the slump of the 1840s, prompted cost-cutting and improved farming techniques, such as contouring and ditching. In season and out, slaves also performed routine maintenance work—especially constructing and maintaining fences and buildings and clearing land and ditches of debris—essential to southern agricultural prosperity (Gray 1933; J. H. Moore 1956; Bonner 1964).

The antebellum southern subsistence economy is at last receiving the attention it deserves, although appreciation of the slaves' role in it is only beginning to emerge. Steven Hahn's (1982) pioneering monograph on the Georgia upcountry, for instance, examines the transformation in the lives of subsistence producers (which began even before the Civil War) and demonstrates the connections between social relations of production and larger manifestations of power. The studies of antebellum southern herdsmen, especially those by Forrest McDonald and Grady McWhiney (1978), focus on another facet of antebellum rural life that combined elements of the subsistence and commercial economies. Although these studies spotlight white yeoman areas, they illuminate processes equally relevant to plantation districts.

Simply put, slaves contributed significantly to the subsistence needs of the plantation South. Slaves produced a variety of food ranging from fruits and vegetables to chickens and hogs and took both plants and animals from the wild. At one level, these products reduced planters' expenses and provided essential vitamins and minerals in the diets of masters no less than slaves. But more important, they lessened the slaves' dependence upon their owners for subsistence. They also provided slaves access to a marketing system based upon barter and even at times upon cash transactions. Recognizing that some slaves had experience with commodity markets must not obscure the fact that they remained chattel property. Such arrangements depict the various forms of the struggle between masters and slaves—the day-to-day reality of the peculiar institution. While masters strove for optimum subserviency, slaves pushed in the other direction. The lessons slaves learned and the ground they won in such contests proved valuable during slavery and invaluable after emancipation.

In sum, before the Civil War slaves supported southern commercial agriculture, which in turn undergirt the national economy. Although they constituted a minority of the whole southern population, they formed a sizeable majority (and virtually the entire work force) in the staple-crop areas. Moreover, nearly every slave over the age of ten worked, and they consistently worked longer hours, with greater intensity, than did other southerners. In addition to producing the South's marketable crops, they provided substantial amounts of their own food. Finally, they cleared new land and performed maintenance labor on old land to the benefit of not just individual masters but southern agriculture as a whole. And they developed an understanding of the value of their labor at wide variance with the views of their masters. Then the Civil War demolished antebellum society as though it were a house of cards.

## REVOLUTION, 1861–1865

The southern economy underwent an industrial revolution during the Civil War. Although it was imposed from above by political authorities and was aimed toward the specific objective of war production (as Emory Thomas has demonstrated), this revolution thoroughly transformed the countryside (Thomas 1971). Confederate war planners ordered the reduction of cotton acreage and channeled manpower and other resources away from agriculture. Impressment officers wielded a heavy hand, especially in subsistence farming areas outside the Black Belts, seizing articles of military usefulness (Escott 1978, Chaps. 4–5; Hahn 1983a, Chap. 3). Throughout the South, fences, buildings, animals, tools, and processing equipment suffered from hard usage, neglect, or damage coincident upon the fighting. At war's end, General Ulysses Grant's proverbial gesture permitting defeated Confederates to keep their horses could do but little to reverse the impoverishment of southern agriculture.

More important than transforming the southern agricultural landscape, the war also revolutionized patterns of social relationships, particularly those between masters and slaves. In countless ways, slaves witnessed the erosion of their masters' power and confidence, especially during the last two war years. Once omnipotent patriarchs found themselves subordinate to military orders, often enforced by mere boys. In the absence of military-age men, slaves became harder to manage—"demoralized" in the contemporary vernacular. The most daring attempted escape, to the Yankees if practicable, to the woods or swamps otherwise. The flight of masters before the Yankee advance produced additional dislocations. In short, as the masters' authority dissipated, slaves found openings and took advantage of them. Slavery did not end either quickly or dramatically; its demise constituted an uneven, halting process with as many stops and reverses as forward movements. But each slave who gained greater distance from his master shifted the balance against the survival of the institution. And that, after all, was the point of all the bloodshed and destruction (Berlin et al. 1985).

Historians have made much of the extent to which wartime policies on the part of the Union army attempted to reestablish staple-crop production and plantation agriculture (Gerteis 1974; Davis 1982). The debate is largely irrelevant in that it focuses upon certain continuities of form while ignoring discontinuities of substance. In the nineteenth-century American context, the distinction between slavery and freedom permeated economic and political discourse. The Unions' commitment to end slavery coupled with the slaves' determination to press the revolution forward meant a radical restructuring of southern labor relations: no longer was the laborer a commodity to be whipped or coddled, bought or sold at the whim of the master. This meant all the difference in the world to the freedpeople—in concrete practical terms and not just abstract philosophical ones.

Inconsistency, not to mention contradiction, among northern policymakers should not be surprising, in that Northern free-labor ideology was itself in turmoil, and that the chaos and confusion of war often confounded the best-laid plans in any event. As the painstaking work of Lawrence N. Powell has demonstrated, the federal experiment of leasing abandoned plantations to northerners foundered, and within a few years of the war most lessees had returned North (Powell 1980). The North's wartime contribution to southern agriculture had less to do with the physical layout of plantations than with the relationship of slave to master. Above all else, the architects of federal policy intended that laborers receive compensation—the amount and, indeed, the form were secondary to the principle itself. This represented revolutionary change.

Northerners' conceptions of "free labor" encompassed the situations of independent proprietors as well as "hirelings." Neither proprietors nor laborers, of course, had masters; but whereas the former owned productive property, the latter generally did not. Laborers traded their ability to work in the labor market for wages. Underlying the system was the belief that the market was free: ideally, laborers in pursuit of the most advantageous terms of employment could bypass or abandon unsuitable situations. And frugal laborers could themselves become proprietors, even employers. It was this putatively free market in labor power that northern policy makers wished to replicate in the South. Slaves had a similar though not identical vision of how northern victory would transform the South's labor system. They, too, believed in the producer's right to the fruits of his toil, but they aspired to the kind enjoyed by the independent proprietor rather than the free, but propertyless, worker (Fields 1983, 1985b; Foner 1983; 1988, Chaps. 3–4).

Within Union lines compensation to black laborers took a wide variety of forms, ranging from comparatively high wages (at times as much as $35 per month) for skilled workers in government employ to rations and shelter for "contrabands" at refugee camps. In addition, freedmen in certain sections of the Union-occupied South leased land from the government and enjoyed virtual proprietors' rights over the products of their labor. In areas of the Mississippi Valley, for instance, black lessees fared extremely well growing cotton at inflated wartime prices. Less spectacularly, blacks in the vicinity of Union military camps throughout the South grew provisions for sale to army quartermasters. Along the coast of South Carolina and Georgia General William T. Sherman's famous order setting aside land for exclusive occupation by blacks promised the evolution of a black peasantry, though the effect of the order proved short lived (Berlin et al. 1990).

Both within and without Union lines, blacks squatted on abandoned land, raising provisions and sometimes cash crops. The autonomy these squatters enjoyed came at a heavy price: Confederates found them easy prey and Union officials broke up their operations in the interest of protecting them and "making them useful to themselves." Self-subsistence did not fit the Yankees' notion of industry. Even within Confederate lines, as slavery disintegrated over the last two war years, some masters and slaves devised novel modes of compensation (often in terms of crop shares) to the slaves. Though by no means universal, such experiments acknowledged the erosion of the masters' authority, and in some cases became the basis for the formal transition to compensated labor following Confederate defeat (Berlin et al. 1990).

The legacy of the war upon southern agriculture is thus a mixed one. The shift away from cotton, which in any event never matched the level indicated by patriotic exhortations or legal proscriptions, proved ephemeral. Production of food crops fell far short of what was needed to sustain both the army and the civilian population. The shift of resources out of the agricultural sector and the willful or accidental damage done to land, buildings, fences, and other appurtenances did long-term damage. In any circumstances southern agriculture would have been hard pressed to recover from the effects of the Civil War. But emancipation forced monumental social changes, too.

## AGRICULTURAL RECONSTRUCTION, 1865–1873

Few debates about the nature of the early postwar years have generated as much heated controversy as the land question. Historians differ widely about the extent of the federal government's commitment to confiscating the land of rebels and distributing it among black families in forty-acre parcels (Du Bois 1935; Oubre 1978; Magdol 1977). Commentators have observed that such a mass redistribution would have been inconsistent with northerners' well-established reverence for private property. True as the observation is, it does not always address the fact that expropriation of the planters' most valuable property—the slaves—involved assets whose commercial value was at least ten times that of the land.[3] It also overlooks the enormous grants of real estate to the railroads, though admittedly that land consisted not of private but of "public" land, that is, the homelands of the western Indians.[4]

Another line of reasoning holds that even had the land been distributed to the freedmen, their inexperience in market transactions coupled with the unscrupulousness of speculators would have stripped many of their allotments, perhaps even before they set foot on the land. And even those black farmers who might have survived such a process would still have faced all the difficulties of falling crop prices, rising expenses, and entanglements with creditors and other "middlemen" that plagued farmers throughout the nation over the last third of the nineteenth century (Belz 1973). According to such a view, land distribution in the end would not have made much difference.

But it would have. For one thing, almost every such analysis presumes commercial rather than subsistence agriculture. Yet the history of slave emancipation in most New World plantation societies indicates a clear pattern of withdrawal from staple production in favor of food crops and of reluctance to become totally enmeshed in the market economy (Rodney 1981). Even had this process of "peasantization" not taken place in the aftermath of land distribution, there is reason to suspect that confiscation would have so shattered the planter class that the speculators who invested in southern land would have had northern rather than southern roots. Agricultural development would inevitably have felt the influence of this northern capital investment with perhaps a more rapid and thorough introduction of wage labor and a program of economic diversification.[5]

In any event, President Andrew Johnson's policy of restoring the land to its antebellum owners dashed the hope of a federally supervised land distribution. A number of new and unanticipated problems quickly surfaced. Northern victory meant an end to property in man and to the masters' special claims upon the slaves' labor power and its fruits. "Free labor" would replace slavery, but given the growing divisions among northerners over the nature of their own economy and polity, there simply was no agreed-upon blueprint for introducing free labor to the South, except by outlawing its supposed opposite, slavery. In the circumstances, former masters and former slaves would play the crucial role in devising the new social relationships, within certain limitations imposed by the federal government. The ensuing struggle over the content of "free labor" undergirt the economic, political, and social reconstruction of the South (Foner 1980).

Both econometricians with neoclassical biases and historians with liberal biases have misinterpreted the nature of this process. The former postulate the instantaneous flourishing of perfectly functioning market mechanisms once slavery was destroyed (Reid 1973; Higgs 1977). The latter see the continuation of various types of coercive, nonmarket mechanisms despite the destruction of slavery (Gerteis 1974; Cohen 1976; Novack 1978). Whereas the most extreme of the former consider sharecropping as "an understandable market mechanism," the most extreme of the latter find little difference between postwar "freedom" and prewar slavery.

Some Marxists, whose work reflects the influence of structural sociology more than historical materialism, have also misunderstood the postbellum South. One school identifies a plantation mode of production—characterized by large estates, powerful landed elites, coerced labor of one variety of another, and the absence of markets in labor power and other commodities—in which the formal differences between slave labor and free labor are largely irrelevant. Another school identifies a sharecropping mode of production—neither slave nor capitalist—as characteristic of the period from the 1860s to the 1930s. Preoccupied with the nonmarket forms of coercion binding the work force and the low level of capital investment in agricultural machinery, this view sees sharecropping as a social formation essentially different from slavery, which it superseded, and capitalist agriculture, which in turn superseded it during the New Deal. The South, in other words, followed a "Prussian road" to capitalist development, wherein a politically dominant planter class directed the process to suit its interests, subordinating both laborers and fledgling industrialists to its program (Mandle 1978; Wiener 1978, 1979).

There is enough of a kernel of truth in all these positions to grant them a degree of plausibility. However, each also contains flaws. To the extent that "markets" exist in reality and not simply in the imagination of theorists, they cannot be understood apart from specific historical contexts. Market behavior must be learned—it is not something in the genetic makeup of homo sapiens. Much of the postwar period involved precisely such a learning process that affected former masters at least as much as former slaves. Even when markets (especially in labor power) can be said to exist, rarely do they function according to the presumed ideal type, that is, without nonmarket forms of coercion. Employers and employees almost never enter into labor agreements as equals. What is more, even in the quintessentially capitalist society, England, blatantly coercive labor laws remained on the books until the second half of the nineteenth century. Nonmarket coercion by itself reveals little about the fundamental nature of social relationships.

Postbellum southern society is best understood as developing along capitalist lines, a point most succinctly made by the work of Barbara Jeanne Fields (Fields 1983, 1985a, 1985b). Though spanning more than a human lifetime, the capitalist transformation of the South occurred with comparative rapidity.[6] The process occurred unevenly, with certain periods—Reconstruction, for instance—registering extreme flux whereas other periods demonstrated much less volatility. Moreover, it affected various areas of the South differently. And both legally sanctioned and extralegal force characterized every phase of development. Everywhere it involved a protracted struggle between former masters and former slaves in which both wielded weapons at their disposal in an attempt to influence the new social order. For a time in the early postwar years federal authorities oversaw this contest—not as exclusive partisans of either side, despite what some historians persist in arguing. After roughly a decade, federal oversight ended, but not the struggle itself.

During the early postwar years, all southerners were preoccupied with reconstructing a new labor system to take the place of slavery. They inched their way across strange terrain. Despite their wishes, former masters lost their bid to resurrect slavery with cosmetic changes, and former slaves lost their bid for a division of the lands. The reservations of the freedpeople notwithstanding, staple production survived. White southerners considered such a system the only way to prevent the freedpeople from lapsing into "barbarism." Northerners—who did not necessarily share the racial presuppositions of the former masters—viewed commercial agriculture as the seedbed in which Yankee virtues would flourish and the means by which the South grow into the image of the North.

Although unable to leave the plantations, most former slaves took determined steps to reorganize the productive process in ways that reflected their aspirations and expectations of freedom. These steps often ran directly counter to those of the former masters. As growing numbers of recent studies have made clear, the planters wished to perpetuate gang labor under overseers. They hoped thereby to replicate as much as possible the control over work processes (and indeed the nonworking lives of their laborers) they had enjoyed under the old order (Ransom and Sutch 1977; Wiener 1978; Davis 1982; Wayne 1983; Jaynes 1986).

Employers quickly learned that free laborers were not slaves. Even under the restrictions imposed by Black Codes, former slaves enjoyed unprecedented control over themselves and their labor. The wholesale appearance of vagrancy laws during 1865 and 1866 attests to this; after all, vagrancy, as such, does not exist in a slave labor system. Moreover, freedpeople could make demands, threatening to leave for greener pastures should their views be ignored. In precisely such an exercise of leverage, women and children deserted the fields for other pursuits, such as managing households and attending school. (Widely misunderstood by both contemporaries and more recent observers, this move aimed primarily to gain control over the disposition of family members' labor power.) Freedpeople who remained in the field worked shorter hours. Roger Ransom and Richard Sutch estimated that this withdrawal of labor represented roughly a one-third reduction in the annual amount of labor performed by blacks in the Cotton South (Ransom and Sutch 1977, 44–46). In similar fashion, southern blacks bargained for rights such as access to garden plots. Not all the negotiations reached resolution: often irreconcilable differences arose over such issues as tending stock and other animals on Sundays and holidays and performing routine maintenance chores and other labor not directly related to the crops in the ground upon which the laborers' compensation depended.

Within the first few years after the Civil War, former slaves throughout the South had made clear their desire to assert greater control over field routines and the allocation of family labor. In their struggle with landowners over the conditions of labor and terms of compensation, they pursued a variety of strategies, depending upon the relative balance of power and the nature of the class conflict inherited from slavery days. These conflicts had different outcomes in the different crop regions of the South. Sharecropping represented the stalemated nature of the struggle in cotton- and tobacco-growing areas. In the rice-growing region of coastal Carolina and Georgia, where there was a stalemate of a different variety, a peasantry resulted. Former slaves got possession of small parcels of land upon which they subsisted themselves by engaging in various productive enterprises and working casually for wages. In the sugar-growing region of southern Louisiana, there was no such stalemate. Former masters rapidly evolved into an agrarian bourgeoisie, determined to extract surplus value from the highly competitive production of sugar by treating labor power as a commodity to be purchased as needed at the cheapest possible rate. At the same time and as part of the same process, former slaves evolved into a rural proletariat, propertyless and hence obliged to work for wages to obtain the necessities of life. Full appreciation of the larger evolution of late-nineteenth southern agriculture requires consideration of developments in each of these separate regions.

Understanding developments in the sugar country must begin with the transformation of the planters (Sitterson 1953). As Roger Shugg suggested almost fifty years ago, there was a "revolution in land titles" accompanied by the phenomenal growth of direct and indirect influence by northerners (Shugg 1939, 248; Woodward 1951, 179). Although some old-style sugar planters acted as Brazilian *senhores de engenho,* retaining the ways of the past, all but the staunchest ideologues began to bend with the winds of change. Intent upon taking advantage of the growing domestic sugar market, they formed alliances with Yankee businessmen and politicians. Both northerners and southerners readily agreed that large production units were the key to success. They forestalled division of the land (for in the productive sense, white yeomen were no better than freedmen), even when that meant insolvent plantations would lie fallow for years awaiting a suitable buyer. Also in the name of bigness, they insisted that each plantation process its own cane into sugar crystals according to the "honorable" practice of slavery times.

New-style planters also pressed their vision of the new order into the realm of plantation organization. In the interest of exerting maximum control, they continued the practice of gang labor under overseers. They even maintained the old barracks system of housing. Although the work was strenuous and conditions less than ideal, planters maintained an adequate work force by paying high wages (as much as 50 percent above what prevailed in the old Cotton Belt) and by actively recruiting laborers from as far away as Virginia.

Unlike former slaves elsewhere in the South, freedpeople on sugar plantations failed to gain control over work routines. As gang laborers, they performed all duties—both crop-related and incidental maintenance tasks—that had been done by antebellum slaves. Despite the lack of control over field labor, the freedpeople maintained access to garden plots. The provision grounds came to embody the failed hopes of fundamentally restructuring plantation agriculture. There freedpeople were masters of their time and the fruits of their labor. The plots, however, could not sustain existence, and access to them in any event was predicated upon plantation employment. Hence land functioned as a means of tying laborers to the plantations through the harvest season, even while relieving some of the objectionable features of the new plantation order.

In the rice area along the South Atlantic coast, however, freedpeople's desire not to carry on as before led in quite a different direction, ultimately to the development of a black peasantry (Heyward 1937; Clifton 1978; Armstrong 1980; Morgan 1982; Foner 1983; Strickland 1983, 1985; Coclanis 1988). It is ironic that the rice planters failed in their objective to resurrect plantation agriculture with free labor even though Louisiana sugar planters succeeded. The capital requirements of both were extensive: on average, the value of buildings and machinery at sugar mills and boiling houses exceeded that of rice mills; at the same time, the elaborate water works on the average rice plantation were more intricate and extensive, not to mention more difficult to maintain, than those on the average sugar estate. But whereas in the sugar country northern capital seems to have played an important role in preserving the antebellum layout of the land and reestablishing cultivation on the principle of large productive units, the same did not hold true for the rice country. Without such outside financial support at their disposal, low-country planters attempted to accumulate capital through the surplus-labor mechanism. They failed miserably.

To begin with, many low-country blacks had enjoyed possession of land during the Civil War. When forced to relinquish it in the fall of 1865, they disputed the return of the former masters; moreover, they resisted demotion from landholders to employees. In protest, many deserted the rice plantations altogether, squatting on nearby pinelands. Firmly establishing themselves as subsistence agriculturists, these freedpeople sought seasonal labor on the plantations, or with the timber, phosphate, and railroad interests intent upon "developing" the coastal region during the late-nineteenth century. And those blacks who remained on the plantations did not passively accept the program of restoration envisioned by the planters.

Laborers insisted that the antebellum task system govern the organization of field work. The day's work of each "hand" would consist of an assigned task, determined by custom according to the nature of the work and the physical capabilities of the worker. Upon completion of the task, the worker would enjoy the remainder of the day as free time. Before the war, such free time could be used productively on land made available by the masters. Then the task system had helped ease some of the special burdens of rice-plantation slavery. But after the war it would serve an entirely different function by providing access to land upon which freed people could pursue their vision of "free labor." They held such access to land as a precondition for their labor in the rice fields.

By 1867 or 1868, a so-called two-day system came to characterize labor arrangements on rice plantations. Laborers worked by tasks in the main plantation fields for two days, in return for which each received a house, fuel, and access to 6 acres of land rent free. By the 1870s the laborers were also drawing wages for their task labor. During the harvest planters hired additional laborers for cash, who also worked by the task but did not enjoy access to land. As noted above, most of the hired hands had small patches away from the plantations, generally in the pinelands.

In cotton- and tobacco-plantation areas, former slaves took a different approach. Both cotton and tobacco grew in a variety of soils and on small plots as well as large. Even subsistence-oriented southerners customarily grew cotton and tobacco for home use. Neither crop required enormous capital outlays either to raise or to prepare for market. Such flexibility provided an opening wedge through which freedpeople could assault the plantation system. Given the scarcity of credit in the postwar South (in no small measure the result of Republican monetary policy), payments in kind served in lieu of cash wages. As Gerald Jaynes and others have demonstrated, a mind-bogglingly diverse array of crop-sharing arrangements emerged, evolving from share wages through a squad system ultimately to family tenancy (Shlomowitz 1979, 1982, 1984; Glymph 1985; Jaynes 1986). At best, these represented tenure systems wherein the cultivators were actually renters (either in cash or in kind); at worst, they were simply another variety of wage labor in which the laborers merely collected their remuneration in a form other than a fixed amount of cash money per unit of time worked. In both, cultivators provided an interest-free loan of labor power to landowners until after the harvest, in a manner typical of such arrangements the world over (Scott 1978). Regardless of specific terms, these dealings implied a mutual interest between laborers and landowners. Freedpeople insisted upon the notion of "partnership," whereby those who shared such interest should also mutually determine crop mixes, cultivation methods, and work routines (Glymph 1985; Woodman 1985; Jaynes 1986).

As recent studies have made clear, southern blacks would not have enjoyed the leverage they did without the advent of Radical Reconstruction. Although they never exercised undisputed control at a local level (much less statewide or beyond) anywhere in the South, they often succeeded in blunting efforts by former masters to use the state as a blatant instrument of coercion. Most important, the near-universal participation of southern black men in Republican party organizations provided a network of solidarity and leadership development indispensable to pressing demands at the workplace. In that sense, the ubiquitous Republican clubs represented the politicized black working class. "Redeemers" well understood the connection between the politics and the economics of postbellum southern agriculture.

While they plotted to regain political control, former masters also aimed to undermine the independence of southern blacks by assaulting the latter's access to the means of subsistence, as Steven Hahn's (1982b, 1983a) pioneering work demonstrates. Landowners began pressing for legal restrictions upon access to the open range for hunting, fishing, wood-gathering, and pasturing animals. By such strictures as well as various extralegal ones, employers attempted to make freedpeople more dependent, even as the latter strove for increased distance and autonomy. Such developments in the Black Belts foreshadowed (and to an extent precipitated) similar disputes slightly later among whites in nonplantation areas.

In short, the South underwent a social revolution, initiated by the defeat of the Confederacy, overseen for a time by the victorious Union army and its subsidiary agency, the Freedmen's Bureau, and calculated to accomplish overnight what had taken several centuries in England—the development of a system of free labor. The planters were unwilling partners in this process, to say the least. Determined to maintain control over production, they locked horns with the freedpeople—equally determined to gain that control for themselves. This struggle raged hottest during the first half-decade after the Civil War. During that time, Radical Reconstruction inaugurated new political relations in the South. Even by the most pessimistic reading, freedpeople gained new leverage in their quest for economic and social justice. The Panic of 1873, however, drastically affected both the ultimate objective and the means of attaining it.

## AGRICULTURAL ADJUSTMENT AMIDST DEPRESSION, 1873–1898

The early years of Reconstruction coincided with the last sustained period of economic prosperity of the nineteenth century. In that sense, the early adjustment from slavery to compensated labor occurred in the best of possible times. The Panic of 1873 changed all that, calling back into question issues that had already been resolved and raising new ones. None of the solutions seemed simple. Landowners devised elaborate plans to increase productivity and cut unit costs of production. As commodity prices fell, rising outputs only aggravated the situation. Meanwhile, the ever-growing number of debtors considered staple cultivation the only means to retire their obligations. The pressures initiated by the Long Depression advanced the development of capitalist social relations in southern agriculture, often despite the best wishes and aspirations of those whose lives were most immediately affected.

The South's major staple crops each reflected the changes wrought by the depression. Take cotton, for instance. Over the last third of the nineteenth century the world price of cotton declined by approximately 60 percent (Hammond 1897, Chaps. 9, 11, App. 1). Meanwhile, American farmers increased their cotton acreage approximately threefold from roughly 8 million to 25 million acres. The size of the crop grew even more dramatically, roughly from 2 million to 10 million bales (Wright

1978, 96; 1986, App. 1). But late nineteenth-century cotton growers functioned in a different world from the prewar one. No longer did cotton revenue support the nation's balance of payments. And although American growers continued to supply the vast majority of the world's raw cotton, their virtual monopoly over supply was a thing of the past.[7] Arguably, the falling price of American cotton had less to do with the increasing size of the crop than with the general decline in commodity prices during the Long Depression and, more specifically, the less-than-anticipated growth in demand for cotton cloth. But to southerners who staked their all on raising cotton, such technical explanations for why they worked longer and harder but had less to show for it made little sense.[8]

The increase in acreage devoted to staples affected nonplantation as well as plantation areas. As the studies of Steven Hahn and others demonstrate, the shift to commercial agriculture in formerly subsistence-oriented regions undermined the yeomen's way of life. Though they may have still farmed the same fields, debt transformed even the most reluctant into cotton farmers, and declining prices kept them growing the staple—and falling deeper into debt. In the process, farmers became tenants and tenants mere laborers (Hahn 1983a: Ford 1983; Weiman 1985).

Within the older Black Belt areas, cotton acreage also increased. Landowners used fertilizers to revitalize worn-out land and to bring marginally productive land on the peripheries of the plantation belt under cultivation. To some extent the patterns of credit contraction and railroad expansion that characterized the postwar South drew farmers into the seemingly endless cycle of cotton and fertilizer, with its companion debt. But the new credit system influenced the spread of cotton cultivation in another sense: the phenomenon of cotton plants growing virtually to the doorsteps of the cultivators. Landlords, tenants, and merchants alike knew that cash grew on cotton plants, though each party sought the cash for different reasons. The end result, however, was the eventual conversion of yards and gardens, not to mention fields previously used for other purposes, to cotton cultivation. Over the last third of the nineteenth century, this process accounted for the addition of thousands of new cotton acres throughout the South (Fite 1984, Chap. 1; Range 1954, Pt. 2; Taylor 1976, Chap. 4; Daniel 1979). Paralleling these developments, the Black Belts themselves expanded onto fresh land, particularly in Arkansas, Louisiana, and Mississippi, thanks in large measure to the creation by Congress of the Mississippi River Levee Commission in 1879 (Brandfon 1967).

For freedpeople in general, the spread of cotton became synonymous with the spread of sharecropping. The outline of that system was clearly discernable by 1870, when former slaves succeeded in establishing the principle of family-sized farming units. At that time cropping held hope of providing a "ladder" to eventual proprietorship. But it gradually became transformed from a variety of tenure (in which the cultivator enjoyed legal rights to the crops) into a variety of compensated labor (in which the laborers held no such rights) (Reidy, forthcoming).

Court decisions proved crucial to that process of stripping croppers of their rights. The first came in 1872 when the Georgia Supreme Court ruled sharecropping equivalent to wage labor and not tenancy. (Not coincidentally, that was the same year Democrats resumed control of the state government, after having overturned the Radical regime the year before.) Courts in the other states made similar rulings over the last part of the nineteenth century. These, in conjunction with assaults upon traditional rights such as the open range, pushed sharecroppers toward the status of an agricultural proletariat (Woodman 1979).

Upon the onset of depression in 1873, the combination of falling prices and tight money dimmed hope that tenancy might serve as a step to proprietorship. By mid-decade furnishing merchants required chattel mortgages obligating borrowers to pledge as collateral not just crops, but animals, tools, and household possessions. Reconstruction-era homestead exemptions were reduced or eliminated, all in the name of creating a sound investment climate for lenders. Under such a system, years of hard work and savings could be nullified in one bad crop year.

The evolution of southern agriculture during the late nineteenth century forced both black and white agriculturists into similar circumstances. Important differences remained, however. Whereas white farmers covered the gamut of owners, tenants, sharecroppers, and wage laborers, the heritage of slavery and racism, coupled with the depressed economy and the oppressive credit structure, kept most blacks out of proprietorship. Real class differences had begun to emerge that operated within as well as across the color line. Southern farmers made various attempts to organize in the face of adversity and regain control over their lives. The earliest organizations, the Grange for instance, tended to reflect the interests of landowners, often those of planters. Both Robert

McMath and Lawrence Goodwin have demonstrated that as more landowners became dispossessed, the organizations took on a more democratic character, representing the interests of tenants, sharecroppers, and even at times farm laborers, as well as proprietors. The Alliance was the premier expression of this multiclass farmers' movement (McMath 1975; Goodwyn 1976; Schwartz 1976; Palmer 1980).

Despite the recent appearance of excellent studies of the Farmers' Alliance, the specific history of the organization's black adjunct, the colored Farmers' Alliance, remains something of an enigma. In approximate figures, the organization claimed a million members representing every southern and a handful of northern states. From its inception, the Colored Alliance had a strong local flavor, in part reflecting local conditions and the appeals of individual organizers. It also suffered from factional disputes, in particular the rival claims to national leadership exercised by white superintendent R. M. Humphrey and a black Alliance organizer, Andrew J. Carothers. The Colored Alliance weathered these difficulties, but the renowned cotton pickers' strike of 1891 brought its demise. The strike exposed fundamental differences between harvest laborers, who demanded $1 per day, and employers (tenants as well as owners), who considered such a wage exorbitant. It began with less than wholehearted support among black Alliancemen and quickly fizzled in the face of stiff opposition by employers (Saloutos 1960, 79–81; Goodwyn 1971, 1976, Chap. 10; Holmes 1975; Taylor 1976, 33–36, 90–91).

This defeat notwithstanding, blacks continued to unite with dissident whites all over the South. Sidestepping the class differences that had made the cotton pickers' strike a disaster, farmers, tenants, and laborers fused their respective political affiliations into coalitions aimed at wresting control from the combination of planters, merchants, financiers, and industrialists who apparently benefitted at the expense of all other southerners. The elites' program of disfranchisement intended to break up these political coalitions. And the new racial ideology served to justify reducing southern blacks to the most debased proletarian status, thoroughly vulnerable before the combined economic and political power of the large landholders and their commercial and financial allies. White small farmers were invited to partake of the spoils of this debasement, but at the price of their political and ideological independence. Thus did the New South bourgeoisie achieve hegemony (Fredrickson 1981; Cell 1982).

Whereas the process of capitalist development in the cotton South thus took a roundabout course over the late nineteenth century, agrarian capitalism developed comparatively quickly in the Louisiana sugar bowl. Predictably, the Long Depression speeded the process of proletarianization begun during the early postwar years. The depression in fact threw sugar producers worldwide into cutthroat competition. England opened its borders to free trade in sugar, and continental European nations offered lavish protection to their sugar-beet farmers and encouraged a scientific as well as a technological revolution in sugar processing. Louisiana planters benefitted from a growing home market, which was rapidly emerging as the premier sugar market in the world. They also enjoyed a certain amount of tariff protection, though on terms dictated by the emerging "sugar trust" of northern refiners rather than themselves. During the 1880s, as the trust gained a virtual monopoly over distribution in the United States, it forced growers to adopt uniform standards of product quality, to refrain from refining to table grade, and to accept low prices. Planters seeking access to the European and North American markets competed feverishly to reduce production costs. World prices plummeted (Sitterson 1953; Eichner 1969; Albert and Graves 1984; Klein and Engerman 1985; Moreno Fraginals 1985; Heitmann 1987).

To the extent that Louisiana planters reduced costs, their success resulted from the infusion of northern men and money. Capital made possible the repair or replacement of mills and boiling apparatuses. Aware of the need to reduce costs—especially after 1873—they sought more efficient grinding mechanisms and vacuum pans, and they switched to coal-fired furnaces. Ironically, as capital transformed sugar manufacturing, the most antiquated methods continued to govern field work, which had changed little over the five centuries that sugar had been grown commercially on plantations. Still, the similarity of work routines must not mask the radical change in social relations that accompanied the transition from slavery to compensated labor.

The depression wreaked havoc with the planters' wage system, wherein the workers were "hired hands" subject to the directions of the planter or his overseer and emphatically not "partners" in the enterprise. Only such a system, planters reasoned, would guarantee the degree of control to themselves and incentive to the workers essential during the harvest. But the depression squeezed the operating funds necessary for regular cash payments. When planters attempted to cut wages, a series of labor disputes ensued. Lasting more than a decade, the conflict further speeded the proletarianization process begun upon emancipation (Hair 1969; Shlomowitz 1979; Ferleger 1982, 1984; Gould 1984).

The first episode occurred early in 1874, following a wage reduction. Striking workers demanded that they be granted a choice between either renting land or earning $20 monthly, paid in cash. Planters eventually broke the strike with the aid of the state militia. In 1880, when strikers threatened to migrate to Kansas unless paid $1 per day, the militia again forced an end to the dispute. Later in the decade workers struck repeatedly. As they abandoned hope for renting land and concentrated their efforts on improving wages and working conditions, they identified the planters' Achilles heel—the harvest. Moreover, in the spirit of the planters, they recognized the importance of outside allies. Supported by the Knights of Labor, sugar workers struck at the start of the 1887 harvest season, aiming to raise wages, eliminate payment in scrip, and improve working conditions. After nearly a month of impatient negotiations and veiled threats, planters called upon the militia, which suppressed the strike and inaugurated a reign of terror. Somewhere between 30 and 300 strikers lost their lives. The so-called Louisiana sugar war broke the effort of sugar workers to organize until into the twentieth century.

Assaulted by intense competition in the international market and strike activity at home, planters sought additional means to cut production costs and undermine labor militancy. (Sugar planters elsewhere, particularly Cuba, reached similar conclusions around the same time.) With a new infusion of northern capital, they undertook the total reorganization of sugar production predicated upon central factories—huge, centrally located, steam-powered mills—to refine the cane of surrounding plantations. The centrals effectively separated sugar production into its two constituent parts: cultivating and processing. The most prosperous planters became, in effect, agribusinessmen, with one foot in the field and the other in the factory. Marginal planters, on the other hand, became mere cane growers, selling their crop to the centrals. Even smallholders (including tenants leasing land from marginal planters) could contract with the centrals to grow cane. Some blacks managed to gain tenancies, but most black sugar workers were reduced to the status of unskilled cane cutters. Planters resorted increasingly to seasonal labor, thereby rendering gardens essentially irrelevant. Moreover, by reserving most factory jobs for whites, the directors of the centrals depressed the skill level of black workers and imposed a racial division upon the sugar work force (Gould 1984). This strategy did not solve all the planters' problems, but it certainly demoralized black workers.

The Panic of 1873 ravaged the rice country no less than the sugar country, as the work of Eric Foner (1983, Chap. 3) makes clear. Like Louisiana sugar planters, Carolina rice planters cut daily wages for harvest workers and began paying in scrip instead of cash. By spring 1876 such conditions had become intolerable, especially for those who lived in the pinelands and worked only seasonally on the plantations. Workers stayed out of the fields in protest. As the strike dragged on, planters lobbied to have the militia intervene on their behalf. They failed, due in part to the large presence of blacks in the militia and to the dependence of the Republican governor upon black voters. Within months, the end of Radical Reconstruction tipped the balance of power from field hands to planters. But that did not resolve all the planters' difficulties.

The new political climate made rice workers more determined than ever to distance themselves from plantation labor. And with land relatively accessible (at least for the squatting), freedpeople had an outlet. They were thus able to frustrate every effort of the employers to achieve absolute control over production, or, as one remarked, to assimilate rice planting "to the industrial establishment of other parts of the world" (Foner 1983, 85). Through a combination of strategies rice workers succeeded where sugar workers failed: with access to land, they became a subsistence-oriented peasantry despite the most pointed exertions of the planters to reduce them to a propertyless and hence dependent proletariat. In the circumstances, planters not laborers became "demoralized." Then natural disasters took their toll. Hurricanes had imperiled rice cultivation from colonial times, but given the planters' inability any longer to command laborers, damage went unrepaired. One by one the descendants of the rice aristocracy gave up planting, selling land piecemeal to workers for homesteads or wholesale to Du Ponts and Carnegies for hunting preserves.[9]

## THE PROLETARIANIZATION OF COTTON AGRICULTURE, 1898–1910

As blacks along the South Atlantic coast developed into a peasantry, and those in Louisiana's sugar bowl into a proletariat, sharecroppers in the upland cotton areas came to resemble the latter more than the former. To be sure, large numbers of black cotton growers were tenants rather than sharecroppers; as late as 1900 there were more tenants than croppers in South Carolina, Florida, Alabama, and Mississippi, and in Georgia the numbers were roughly equal. Although the experience of Ned Cobb was exceptional in many ways, his determination to avoid sharecropping and thereby enjoy greater control over his labor and its fruits typified the aspirations of countless others who left no record (Rosengarten 1974).

Still, southern blacks had a hard time stepping up the tenure "ladder." If neoclassical economic theories held true, the era of prosperity, which began in the late 1890s and except for the recession of 1907–1908 continued through World War I, should have benefitted rural southern blacks. By some crude measures it did, but upon closer inspection these signs point in a different direction. The case of black landholders makes the point.

As numerous observers have pointed out, the absolute number of black landholders in the South increased between 1865 and 1920, when the total peaked. The most dramatic increase occurred between 1890 and 1920, presumably in reflection of rising commodity prices and farm income. What those figures fail to indicate, however, is the falling proportion of farm owners among all black agriculturists over the period.[10] (See Tables 10.1 and 10.2).

Between 1900 and 1910, the proportion of farmer owners rose in the Upper South states of Maryland, Virginia, and Kentucky, where a system of mixed farming prevailed. But in only one state of the plantation South, namely Louisiana, did it rise and there only negligibly. In Florida and Texas the proportion decreased dramatically. Figuratively speaking, black landowners lost ground. Over the same period, the proportion of tenants in every state fell, in most cases more alarmingly than the decline of landowners. The losses in proportions of landowners and tenants were counterbalanced by increased proportions of sharecroppers and wage laborers.

Simply put, at the dawn of the twentieth century the transformation of the postemancipation labor system reached a critical stage, which the changed structure of farm tenancy reflected. These changes were directly related to, in fact, were the necessary outcome of the process of capitalist development that had proceeded unevenly since the Civil War. Yet the intensification of those tendencies from the late 1890s onward must be emphasized.

The new era resulted from the temporary consolidation of class

TABLE 10.1. Percentages of Owners, Tenants, Sharecroppers, and Laborers among All Black Agriculturists by State, 1900

| State | Owners | Tenants | Croppers | Laborers | Whole Number |
|---|---|---|---|---|---|
| | | Mixed Farming States of the Upper South | | | |
| Md. | 12.1 | 2.1 | 7.1 | 78.4 | 26,997 |
| Va. | 25.2 | 6.5 | 10.6 | 57.5 | 105,429 |
| Ky. | 14.6 | 2.1 | 13.4 | 69.7 | 37,085 |
| Tenn. | 10.7 | 12.4 | 15.3 | 61.5 | 88,148 |
| | | | Plantation States | | |
| N.C. | 11.0 | 6.5 | 17.0 | 65.4 | 158,663 |
| S.C. | 7.1 | 16.0 | 9.0 | 67.8 | 265,579 |
| Ga. | 4.3 | 13.2 | 13.8 | 68.6 | 263,816 |
| Fla. | 19.0 | 16.0 | 4.0 | 60.7 | 34,434 |
| Ala. | 5.1 | 20.5 | 8.6 | 65.7 | 274,443 |
| Miss. | 6.4 | 17.5 | 15.4 | 60.7 | 327,543 |
| Ark. | 10.5 | 13.9 | 16.8 | 58.8 | 113,916 |
| La. | 5.0 | 11.4 | 14.7 | 68.8 | 186,456 |
| Tex. | 14.0 | 5.9 | 25.7 | 54.3 | 143,453 |

Source: U.S. Department of Commerce, Bureau of the Census, *Negro Population in the United States, 1790–1915* (Washington, D.C.: Government Printing Office, 1918), 528, 609. The figures represent the combination of "Negro" farm laborers (p. 528) and "Colored" farmers (p. 609). The latter category includes small numbers of non-Negroes who were impossible to disaggregate by tenure categories (see p. 588).

Note: Tenants include cash tenants and share-cash tenants. Croppers include sharecroppers and tenants not specified.

TABLE 10.2. Percentages of Owners, Tenants, Sharecroppers, and Laborers among All Black Agriculturists by State, 1910

| State | Owners | Tenants | Croppers | Laborers | Whole Number |
|---|---|---|---|---|---|
| | | Mixed Farming States of the Upper South | | | |
| Md. | 13.1 | 1.5 | 6.3 | 78.9 | 30,242 |
| Va. | 25.6 | 3.2 | 9.3 | 61.8 | 125,843 |
| Ky. | 16.1 | 1.7 | 14.0 | 68.1 | 36,714 |
| Tenn. | 9.9 | 10.6 | 14.8 | 64.7 | 108,540 |
| | | | Plantation States | | |
| N.C. | 9.6 | 4.9 | 14.8 | 70.6 | 223,407 |
| S.C. | 5.8 | 10.9 | 10.9 | 72.4 | 350,297 |
| Ga. | 4.0 | 12.1 | 14.7 | 69.1 | 397,269 |
| Fla. | 14.0 | 8.5 | 5.6 | 71.7 | 51,958 |
| Ala. | 4.9 | 17.9 | 8.8 | 68.4 | 349,997 |
| Miss. | 5.4 | 12.3 | 17.6 | 64.8 | 467,504 |
| Ark. | 8.4 | 10.6 | 17.3 | 63.8 | 175,485 |
| La. | 5.2 | 4.7 | 16.8 | 73.2 | 204,836 |
| Tex. | 9.9 | 3.2 | 19.4 | 67.5 | 215,060 |

Source: See Table 10.1.

Note: Tenants include cash tenants and share-cash tenants. Croppers include sharecroppers and tenants not specified.

forces, specifically, the planters' emerging perception of themselves as an agrarian bourgeoisie, their defeat of the Populist challenge, and the effective reduction of the freedpeople (and increasing numbers of rural whites, too) to the status of an agricultural proletariat. In short, bourgeois planters gained the upper hand by trapping the work force in commercial farming, restricting alternative employment, and curtailing independent means of subsistence; by systematically defeating every effort at working-class organization; and, when it was feared that other measures were insufficient to the task, by disfranchising the black proletariat and its political allies—blacks in general and certain whites.

The Long Depression was a stern school that produced a new generation of graduates: planters who looked upon agriculture not so much as a way of life as a way of producing returns on investments. As Robert L. Brandfon's (1967, Chaps. 5 and 6) important study of Mississippi makes clear, such planters intended to mine the land for timber and other natural resources as well as cotton and other crops. They struck alliances with other parties interested in development (increasingly, northern capitalists). And, as some articulate spokesmen of the group clearly indicated, they considered subsistence-oriented white yeomen as obstructive to "progress" as black farm workers (Stone 1905; Percy 1941).

As both land values and commodity prices rose, bourgeois planters increased their exploitation of human resources too. The black population working in agriculture increased dramatically, as women and children returned to the fields on a scale unmatched since emancipation. Between 1890 and 1910 the increase in numbers of southern black agriculturists far outstripped population growth rates. Only in Louisiana were the increases of the same proportion; in Georgia, by way of contrast, the black population grew by less than 14 percent while the black farm work force grew by more than 50 percent. (See Table 10.3).

Not content with exacting more from the rural population in a quantitative sense, new-style planters sought more fundamental changes in social relations. They substituted sharecroppers for tenants, and wage laborers for both. By 1910 the combined proportion of croppers and laborers in no southern state was below 70 percent. In Maryland, Kentucky, the two Carolinas, Georgia, Mississippi, Arkansas, and Texas, it was above 80 percent. In Louisiana, it was above 90 percent (see Table 10.2). Tenants came under closer supervision—in their own best interest, planters all insisted—and plowing and picking gangs worked through the

TABLE 10.3. Percentage Increases in Black Agriculturists and the Black Population at Large by State, 1900–1910

| State | Black Agriculturists | Black Population |
|---|---|---|
| Md. | 12.0 | −1.2 |
| Va. | 19.4 | 1.6 |
| Ky. | −1.0 | −8.1 |
| Tenn. | 23.1 | −1.5 |
| N.C. | 40.8 | 11.7 |
| S.C. | 31.9 | 6.8 |
| Ga. | 50.6 | 13.7 |
| Fla. | 50.9 | 33.8 |
| Ala. | 27.5 | 9.8 |
| Miss. | 42.7 | 11.2 |
| Ark. | 54.0 | 20.7 |
| La. | 9.8 | 9.7 |
| Tex. | 49.9 | 11.2 |

Source: Tables 10.1 and 10.2: *Negro Population in the U.S.*, p. 37.

boundaries that had previously separated the tenants' family plots. All signs indicated that planters considered sharecroppers' labor power—and not tenants' crops—the commodity of choice for the twentieth century. Although still short of mechanizing field work, these changes marked a qualitative transformation of productive relations.

## CONCLUSION: THE CAPITALIST TRANSFORMATION OF SOUTHERN AGRICULTURE, 1865–1910

The capitalist transformation of southern agriculture during the late nineteenth century did not occur evenly. Nor could it have. The North's victory in the Civil War began the process, but, except for banning slavery and its most blatant carryovers, the Yankees offered no one blueprint for the new era of freedom. Rather, with varying degrees of oversight, southerners themselves fashioned the new social order to replace the old. Former masters and former slaves had certain expectations, most of which went unfulfilled. At times what resulted was literally a compromise; more often it was simply a modus vivendi, to be renegotiated at the insistence of one party or the other.

Both the pace of change and the content of new social relations differed considerably from one section of the South to another. On the surface, such differences resulted from different crop cultures and different political cultures. But they also resulted from differential dynamics of class formation, which themselves reflected local, regional, national, and international influences. While it is important to understand that there was both unity and diversity within the larger process, it is more important to recognize that the process itself involved some southerners' pursing interests perceived as distinct from—often incompatible with—the interests of other southerners. Thus was the class structure redefined after slavery had been removed as the keystone of the antebellum social order.

The Long Depression of the late nineteenth century speeded the process of class reformation, nowhere more quickly or clearly than in the sugar plantation area of southern Louisiana. By the late 1880s, sugar planters had become an agrarian bourgeoisie with close links to kindred spirits in New Orleans, New York, and Havana. Freedpeople had become a proletariat, lacking control over productive property and working for wages at the lowest skill level. Tenancy never was a viable option until the emergence of central factories, which from the workers' standpoint was too late.

In cotton and tobacco areas, tenancy and sharecropping obscure the lines of capitalist development present from 1865. By their possession of such productive resources as mules and implements, tenants were able to retain certain legal rights to crops and thereby forestall reduction to proletarian status. Sharecroppers did not fare as well as tenants: with fewer resources, they were more vulnerable. Striping them of their rights to the crops, courts in state after state viewed them as the equivalent of wage laborers. That took time, but by the dawn of the twentieth century cotton and tobacco sharecroppers were for all practical purposes proletarians too.

Only in the rice area of the South Atlantic coast did former slaves successfully resist proletarianization. Their success—like that of freedpeople on Caribbean islands such as Jamaica—depended upon gaining access to land, which enabled them to forsake plantation labor unless conditions suited them. To be sure, planters resisted that development, but their ranks thinned steadily over the late nineteenth century. The demise of the coastal rice industry guaranteed the future of the one authentic black peasantry in the United States.

The capitalist development of southern agriculture did not end in 1910. But subsequent changes rested upon the prior separation of the producer from the means of production and the establishment of capitalist relations of production, which was essentially complete by 1910. The New Deal later added a new dimension by superimposing an industrial revolution in the form of tractors and mechanical harvesters.[11] Ironically, the final phase of the proletarianization of the southern black workforce coincided with the South's integration into national labor markets. Insect pests, natural disasters, New Deal programs, and racist violence in the South combined with employment opportunities in the North to prompt a great migration. Thus did the agricultural proletariat become an industrial one.[12]

## NOTES

1. The author wishes to thank Eric Foner and Leslie S. Rowland for comments on an earlier draft on this essay.

2. Increased quantitative exploitation of labor involves expanding the number of workers, lengthening the work day, speeding the pace, or reducing wages in an effort to lower production costs. Increased qualitative exploitation involves altering the social relations between the laborers and the expropriator of the surplus with a corresponding redefinition of rights to the products of labor.

3. The 4.5 million slaves in 1860 represented a total capital value of perhaps $4 billion. The confiscation and distribution of 40 acres of land to each of the approximately 1 million black families would have required 40 million acres; if that land were choice (worth say, $10 per acre) the transfer would have involved assets worth $400 million.

4. The federal government alone distributed some 180 million acres of land to the railroads after the Civil War.

5. Industrial capitalists tended to prefer pure market relations between employer and employee. They presupposed an organization of production in which they enjoyed complete control over both the products of labor and the labor power they purchased with wages. Boston textile manufacture Edward Atkinson, who invested in South Carolina cotton plantations during the war, is a good example (see Rose 1964; Powell 1980, 76–77). Louisiana sugar planters, about whom more will be said below, quickly demonstrated the same preferences.

6. Comparison with the capitalist transformation of England has already been noted. Despite the interpretive attention focused upon nineteenth-century Prussia, that of Russia appears to offer a more relevant and fruitful contemporary comparison to developments in the South (see Lenin 1956; Kolchin 1987).

7. In 1860, 84 percent of the cotton imported to Europe was grown in the United States. During the Civil War—in a pattern reminiscent of the birth of the European beet sugar industry during the Haitian Revolution—the Cotton Supply Association of Great Britain sponsored the development of Indian cotton in an effort to free English manufacturers from their dependence upon southern slaveholders. As late as 1890, however, southern growers still supplied 74 percent of European cotton imports and thus continued to exert a phenomenal degree of control over the supply of one of the key raw materials of world trade. In that same year the United States produced 55.9 percent of the cotton grown in the world; the East Indies (chiefly India) was next with 21.3 percent; Egypt was a distant fourth with 4.9 percent, behind China, 9.5 percent. (See Hammond 1857, Chap. 9, 346.)

8. The South's other major staple crop, tobacco, fared similarly. The area of bright tobacco cultivation, the so-called Old Belt along the Virginia-North Carolina border, re-

mained relatively stable until the 1890s, though the size of the crop expanded slightly. Tobacco prices declined gradually over the Long Depression, with considerable yearly fluctuation until the last decade of the nineteenth century when they hit record lows. Growers increased their acreage with corresponding increases in crops and decreases in prices, especially during the 1890s, when the low price of cotton prompted farmers in the coastal plain of the Carolinas, the so-called New Belt, to forsake cotton in favor of tobacco. By that time, tobacco growers had improved both cultivation and curing methods, the most significant of which involved the introduction of furnaces and flues in curing barns. The development of the smoking tobacco industry (prompted by the development of a successful cigarette rolling machine during the 1880s) promised expanded demand for bright tobacco. The boom came at a price, however, in that the "tobacco trust" soon cast a long shadow over the entire tobacco industry, growers no less than manufacturers (See Tilley 1948, especially Chap. 9; Daniel 1985, Chap. 2.) There is no study of the burley tobacco industry comparable to Tilley's, despite Kentucky's claim of more tobacco acreage than Virginia and North Carolina combined in 1890. Even more glaring is the absence of any work on the transition from slavery to free labor in the burley-tobacco areas of Kentucky and Tennessee. (Burley was the leaf of choice among chewing-tobacco manufacturers because of its superior ability to absorb flavoring.)

9. As the coastal rice industry perished, a new system of rice cultivation flourished in upland regions of Arkansas, Louisiana, and Texas. Modeled after Midwest family farms and predicated upon heavy capital investment in irrigating and harvesting equipment, dry rice culture solved the "labor problem" of wet rice culture essentially by eliminating the laborers. (See Daniel 1985, Chaps. 3, 10, 13.)

10. The common tendency of lumping together as "farmers" persons in all tenure categories creates additional sources of confusion, the most serious of which rests upon the tacit assumption that farmers own the land they farm. For a recent example see Fite 1984, 238.

11. Wiener (1979) and Daniel (1986) attribute more fundamental significance to the changes inaugurated during the New Deal.

12. Two excellent recent treatments of the formation of this industrial proletariat are Trotter (1984) and Gottlieb (1987).

## REFERENCES AND FURTHER READINGS

Scholarly focus upon the historical development of capitalist social relations in agriculture has produced a bumper crop of outstanding recent works. Contributors represent a wide range of disciplines, geographical areas, and historical periods. Yet despite these differences, they have identified certain processes of class formation that tie the more specialized studies together. The result is a body of literature that speaks to broadly theoretical issues while analyzing the concrete historical experiences of particular places at particular times.

Three significant contributions toward generalizing the broad sweep of these processes over a large time frame are Hobsbawm (1962, 1975, and 1987) from an historical perspective, Wallerstein (1974 and 1980) from a sociological perspective, and Wolf (1982) from an anthropological perspective. Much of the best work on the historical process of class formation remains concerned with the "transition" of Europe from feudalism to capitalism analyzed by Marx. Marx's original inquiry has been reinvigorated by the seminal studies of Dobb (1947) and Brenner (1976), and the respective "debates" that ensued (Hilton 1976;

Aston and Philpin 1985). Students of every facet of the nineteenth and twentieth centuries have benefitted enormously from this work.

Those who concentrate upon the American South are no exceptions. Jonathan Wiener (1978), Jay Mandle (1978), and Dwight Billings (1979) addressed historical problems raised by those inquiries, though their theoretical insights emanate more from the work of Barrington Moore (1966) than Dobb. The more recent works of Berlin et al. (1982, 1985, 1990), Fields (1983, 1985a, 1985b), Robinson (1986), and Foner (1988) point in the other, more fruitful, direction. As the works of Genovese (1969) and Foner (1983) also make clear, such an approach opens exciting comparative dimensions with other systems of bound labor, in the Americas and elsewhere, that experienced emancipation during the nineteenth century.

The body of such work grows daily, even as its quality gets richer. Outstanding example include studies of the West Indies by Mintz (1979), Rodney (1981), Holt (1982), Rebecca Scott (1985) and those in Moreno Fraginals, Moya Pons, and Engerman (1985); of Spanish South America in Duncan and Rutledge (1977); and of East Africa by Cooper (1980). Although the potential for comparison with Brazil has yet to be fully realized (Hahn 1983), parallels between the slave societies of the Americas and the serf societies of eastern Europe have been broached by Blum (1978) and systematically examined by the pathbreaking work of Kolchin (1987). Albert and Graves's collection (1984) provides important comparative insight into social relations of production in various sugar-producing areas. Comparisons with contemporary peasant societies also prove fruitful, as evidenced by the work of James Scott (1978) and the *Journal of Peasant Studies,* for instance, the special 1983 issue on sharecropping.

With such exciting comparative work under way, the outlook for future studies of the postemancipation South looks promising indeed.

Albert, Bill and Adrian Graves, eds. 1984. *Crisis and Change in the International Sugar Economy, 1860–1914.* Norwich, Engl.: ISC Press.

Armstrong, Thomas F. 1980. "From Task Labor to Free Labor: The Transition Along Georgia's Rice Coast, 1820–1880." *Georgia Historical Quarterly* 64(Winter):432–47.

Aston, T. H., and C. H. E. Philpin. 1985. *The Brenner Debate: Agrarian Class Structure and Economic Development in Pre-Industrial Europe.* Cambridge: Cambridge Univ. Press.

Belz, Herman. 1973. "The New Orthodoxy in Reconstruction Historiography." *Reviews in American History* 1(Mar.):106–13.

Berlin, Ira. 1980. "Time, Space, and the Evolution of Afro-American Society on British Mainland North America." *American Historical Review* 85(Feb.):44–78.

Berlin, Ira, Joseph P. Reidy, and Leslie S. Rowland, eds. 1982. *The Black Military Experience.* Ser. II of *Freedom: A Documentary History of Emancipation, 1861–1867.* New York: Cambridge Univ. Press.

Berlin, Ira, et al., eds. 1985. *The Destruction of Slavery.* Ser. I, vol. 1 of *Freedom: A Documentary History of Emancipation, 1861–1867.* New York: Cambridge Univ. Press.

______. 1990. *The Wartime Genesis of Free Labor.* Ser. I, vols. 2 and 3 of *Freedom: A Documentary History of Emancipation, 1861–1867.* New York: Cambridge Univ. Press.

Billings, Dwight B., Jr. 1979. *Planters and the Making of a "New South": Class, Politics, and Development in North Carolina, 1865–1900.* Chapel Hill: Univ. of North Carolina Press.

Blum, Jerome. 1978. *The End of the Old Order in Rural Europe.* Princeton: Princeton Univ. Press.

Bonner, James C. 1964. *A History of Georgia Agriculture, 1732–1866.* Athens: Univ. of Georgia Press.

Brandfon, Robert L. 1967. *Cotton Kingdom of the New South: A History of the Yazoo Mississippi Delta from Reconstruction to the Twentieth Century.* Cambridge: Harvard Univ. Press.

Brenner, Robert. 1976. "Agrarian Class Structure and Economic Development in Pre-Industrial Europe." *Past and Present* 70(Feb.):30–75.

Brooks, Robert Preston. 1914. *The Agrarian Revolution in Georgia, 1865–1912.* Madison: Univ. of Wisconsin Press.

Cell, John W. 1982. *The Highest Stage of White Supremacy: The Origins of Segregation in South Africa and the American South.* New York: Cambridge Univ. Press.

Clifton, James M. 1978. "Twilight Comes to the Rice Kingdom: Postbellum Rice Culture on the South Atlantic Coast." *Georgia Historical Quarterly* 62(Summer):146–54.

Coclanis, Peter. 1988. *The Shadow of a Dream: Economic Life and Death in the South Carolina Low Country, 1670–1920.* New York: Oxford Univ. Press.

Cohen, William. 1976. "Negro Involuntary Servitude in the South, 1865–1940: A Preliminary Analysis." *Journal of Southern History* 42(Feb.):35–50.

Cooper, Fred. 1980. *From Slaves to Squatters: Plantation Labor and Agriculture in Zanzibar and Coastal Kenya, 1890–1925.* New Haven: Yale Univ. Press.

Daniel, Pete. 1972. *The Shadow of Slavery: Peonage in the South, 1901–1969.* Urbana; Univ. of Illinois Press.

______. 1979. "The Metamorphosis of Slavery, 1865–1900." *Journal of American History* 66(June):88–99.

______. 1985. *Breaking the Land: The Transformation of Cotton, Tobacco,and Rice Cultures since 1880.* Urbana: Univ. of Illinois Press.

______. 1986. *Standing at the Crossroads: Southern Life since 1900.* New York: Hill and Wang.

Davis, Ronald L. F. 1982. *Good and Faithful Labor: From Slavery to Sharecropping in the Natchez District, 1860–1890.* Westport, Conn: Greenwood Press.

Dobb, Maurice. 1947. *Studies in the Development of Capitalism.* New York: International Publishers.

Du Bois, W. E. B. 1935. *Black Reconstruction in America: An Essay toward a History of the Part Which Black Folk Played in the Attempts to Reconstruct Democracy in America, 1860–1880.* New York: Russell and Russell.

Duncan, Kenneth, and Ian Rutledge, eds. 1977. *Land and Labor in Latin America: Essays on the Development of Agrarian Capitalism in the Nineteenth and Twentieth Centuries.* Cambridge: Cambridge Univ. Press.

Eichner, Alfred E. 1969. *The Emergence of Oligopoly: Sugar Refining as a Case Study.* Baltimore: Johns Hopkins Univ. Press.

Escott, Paul. 1978. *After Secession: Jefferson Davis and the Failure of Confederate Nationalism.* Baton Rouge: Louisiana State Univ. Press.

Ferleger, Louis. 1982. "Farm Mechanization in the Southern Sugar Sector after the Civil War." *Louisiana History* 23(Winter):21–34.

______. 1984. "Cutting the Cane: Harvesting in the Louisiana Sugar Industry." *Southern Studies* 23(Spring):42–59.

Fields, Barbara Jeanne. 1983. "The Nineteenth-Century South: History and Theory," *Plantation Society in the Americas* 2(Apr.):7–17.

______. 1985a. *Slavery and Freedom on the Middle Ground: Maryland during the Nineteenth Century.* New Haven: Yale Univ. Press.

______. 1985b. "The Advent of Capitalist Agriculture: The New South in a Bourgeois World." In *Essays on the Postbellum Southern Economy,* ed. Thavolia Glymph and

John J. Kushma. College Station: Texas A&M Univ. Press for the University of Texas at Arlington.

Fite, Gilbert. 1984. *Cotton Fields No More: Southern Agriculture, 1865–1980.* Lexington: Univ. Press of Kentucky.

Flynn, Charles, Jr. 1983. *White Land, Black Labor: Caste and Class in Late Nineteenth Century Georgia.* Baton Rouge: Louisiana State Univ. Press.

Foner, Eric. 1980. "Reconstruction and the Crisis of Free Labor." In *Politics and Ideology in the Age of the Civil War.* New York: Oxford Univ. Press.

______. 1983. *Nothing But Freedom: Emancipation and Its Legacy.* Baton Rouge: Louisiana State Univ. Press.

______. 1988. *Reconstruction: America's Unfinished Revolution, 1863–1877.* New York: Harper and Row.

Ford, Lacy K. 1983. "Labor and Ideology in the South Carolina upcountry: The Transition to Free-Labor Agriculture." In *The Southern Enigma: Essays on Race, Class, and Folk Culture,* ed. Walter J. Fraser and Winfred B. Moore. Westport, Conn.: Greenwood Press.

______. 1984. "Rednecks and Merchants: Economic Development and Social Tensions in the South Carolina Upcountry, 1850–1900." *Journal of American History* 71(Sept.):294–318.

Fredrickson, George M. 1981. *White Supremacy: A Comparative Study in American and South African History.* New York: Oxford Univ. Press.

Genovese, Eugene. 1965. *The Political Economy of Slavery: Studies in the Economy and Society of the Slave South.* New York: Pantheon.

______. 1969. *The World the Slaveholders Made: Two Studies in Interpretation.* New York: Pantheon.

Gerteis, Louis. 1974. *From Contraband to Freedman: Federal Policy toward Southern Blacks, 1863–1865.* Westport, Conn.: Greenwood Press.

Glymph, Thavolia. 1985. "Freedpeople and Ex-Masters: Shaping a New Order in the Postbellum South, 1865–1868." In *Essays on the Postbellum Southern Economy,* ed. Thavolia Glymph and John J. Kushma. College Station: Texas A&M Univ. Press for the University of Texas at Arlington.

Goodwyn, Lawrence. 1971. "Populist Dreams and Negro Rights: East Texas as a Case Study." *American Historical Review* 76(Dec.):1135–71.

______. 1976. *Democratic Promise: The Populist Moment in America.* New York: Oxford Univ. Press.

Gottlieb, Peter. 1987. *Making Their Own Way: Southern Blacks, Migration to Pittsburgh, 1916–30.* Urbana: Univ. of Illinois Press.

Gould, Jeffrey. 1984. "The Strike of 1887: Louisiana Sugar War." *Southern Exposure* 12(Nov./Dec.):44–55.

Gray, Lewis Cecil. 1933. *History of Agriculture in the Southern United States to 1860.* 2 vols. Washington: Carnegie Institution.

Hahn, Steven. 1982a. "Common Rights and Commonwealth: The Stock-Law Struggle and the Roots of Southern Populism." In *Region, Race,and Reconstruction: Essays in Honor of C. Vann Woodward,* ed. J. Morgan Kousser and James M. McPherson. New York: Oxford Univ. Press.

______. 1982b. "Hunting, Fishing, and Foraging: Common Rights and Class Relations in the Postbellum South." *Radical History Review* 26:37–64.

______. 1983a. *The Roots of Southern Populism: Yeoman Farmers and the Transformation of the Georgia Upcountry, 1850–1890.* New York: Oxford Univ. Press.

______. 1983b. "Landed Elites and National Unification in Postemancipation Societies:

The Southern Planters in Comparative Perspective." John Lax Lecture, Mount Holyoke College.

Hair, William Ivy. 1969. *Bourbonism and Agrarian Protest: Louisiana Politics, 1877–1900.* Baton Rouge: Louisiana State Univ. Press.

Hammond, M. B. 1897. *The Cotton Industry: An Essay in American Economic History.* New York: Macmillan.

Heitmann, John A. 1987. *The Modernization of the Louisiana Sugar Industry 1830–1910.* Baton Rouge: Louisiana State Univ. Press.

Heyward, Duncan Clinch. 1937. *Seed from Madagascar.* Chapel Hill: Univ. of North Carolina Press.

Higgs, Robert. 1977. *Competition and Coercion: Blacks in the American Economy, 1865–1914.* New York: Cambridge Univ. Press.

Hilton, Rodney, ed. 1976. *The Transition from Feudalism to Capitalism.* London: New Left Books.

Hobsbawm, Eric. 1962. *The Age of Revolution, 1789–1848.* London: Weidenfeld and Nicolson.

______. 1975. *The Age of Capital, 1848–1875.* New York: Scribner's.

______. 1987. *The Age of Empire, 1875–1914.* New York: Pantheon.

Holmes, William F. 1975. "The Demise of the Colored Farmers' Alliance." *Journal of Southern History* 41(May):187–200.

Holt, Thomas. 1982. "An Empire over the Mind: Emancipation, Race, and Ideology in the British West Indies and the American South." In *Region, Race and Reconstruction: Essays in Honor of C. Vann Woodward,* ed. J. Morgan Kousser and James M. McPherson. New York: Oxford Univ. Press.

Jaynes. Gerald David. 1986. *Branches Without Roots: Genesis of the Black Working Class in the American South, 1862–1882.* New York: Oxford Univ. Press.

King, J. Crawford. 1982. "The Closing of the Southern Range: an Exploratory Study." *Journal of Southern History* 48(Feb.):53–70.

Klein, Herbert S., and Stanley L. Engerman. 1985. "The Transition from Slave to Free Labor: Notes on a Comparative Economic Model." In *Between Slavery and Freedom: The Spanish-speaking Caribbean in the Nineteenth Century,* ed. Manuel Moreno Fraginals, Frank Moya Pons, and Stanley L. Engerman. Baltimore: Johns Hopkins Univ. Press.

Kolchin, Peter. 1987. *Unfree Labor: American Slavery and Russian Serfdom.* Cambridge: Harvard Univ. Press.

Kousser, J. Morgan. 1974. *The Shaping of Southern Politics: Suffrage Restriction and the Establishment of the One-Party South, 1880–1910.* New Haven: Yale Univ. Press.

Lenin, V. I. 1956. *The Development of Capitalism in Russia.* Moscow: Foreign Language Publishing House.

Litwack, Leon. 1979. *Been in the Storm So Long: The Aftermath of Slavery.* New York: Alfred A. Knopf.

McDonald, Forrest, and Grady McWhiney. 1975. "The Antebellum Southern Herdsman: A Reinterpretation." *Journal of Southern History* 41(May):147–66.

McMath, Robert C. 1975. *The Populist Vanguard: A History of the Southern Farmers' Alliance.* Chapel Hill: Univ. of North Carolina Press.

Magdol, Edward. 1977. *A Right to the Land: Essays on the Freedmen's Community.* Westport, Conn.: Greenwood Press.

Mandle, Jay R. 1978. *The Roots of Black Poverty: The Southern Plantation Economy after the Civil War.* Durham: Duke Univ. Press.

Marx, Karl. 1967. *Capital.* vol. 1. New York: International Publishers.

Mintz, Sidney. 1979. "Slavery and the Rise of Peasantries." *Historical Reflections* 6(Summer):213–53.

Moore, Barrington, Jr. 1966. *Social Origins of Dictatorship and Democracy: Lord and Peasant in the Making of the Modern World.* Boston: Beacon Press.

Moore, John Hebron. 1956. *Agriculture in Ante-Bellum Mississippi.* New York: Bookman Associates.

Moreno Fraginals, Manuel. 1985. "Plantations in the Caribbean: Cuba, Puerto Rico, and the Dominican Republic in the Late Nineteenth Century." In *Between Slavery and Freedom: The Spanish-speaking Caribbean in the Nineteenth Century,* ed. Manuel Moreno Fraginals, Frank Moya Pons, and Stanley L. Engerman. Baltimore: Johns Hopkins Univ. Press.

Morgan, Philip D. 1982. "Work and Culture: The Task System and the World of Lowcountry Blacks, 1700–1880." *William and Mary Quarterly* 3d. ser. 39(Oct.):563–99.

North, Douglass C. 1966. *The Economic Growth of the United States.* New York: W. W. Norton.

Novack, Daniel A. 1978. *The Wheel of Servitude: Black Forced Labor after Slavery.* Lexington: Univ. Press of Kentucky.

Oubre, Claude F. 1978. *Forty Acres and a Mule: The Freedmen's Bureau and Black Land Ownership.* Baton Rouge: Louisiana State Univ. Press.

Palmer, Bruce 1980. *"Man over Money": The Southern Populist Critique of American Capitalism.* Chapel Hill: Univ. of North Carolina Press.

Percy, William Alexander. 1941. *Lanterns on the Levee: Recollections of a Planter's Son.* New York: Alfred A. Knopf.

Powell, Lawrence. 1980. *New Masters: Northern Planters During the Civil War and Reconstruction.* New Haven: Yale Univ. Press.

Range, Willard. 1954. *A Century of Georgia Agriculture, 1850–1950.* Athens: Univ. of Georgia Press.

Ransom, Roger, and Richard Sutch. 1977. *One Kind of Freedom: The Economic Consequences of Emancipation.* New York: Cambridge Univ. Press.

Reid, Joseph D. 1973. "Sharecropping as an Understandable Market Response: The Post-Bellum South." *Journal of Economic History* (Mar.):106–30.

Reidy, Joseph P. Forthcoming. *Masters and Slaves, Planters and Freedmen: The Transition from Slavery to Freedom in Central Georgia, 1820–1880.*

Robinson, Armstead L. 1986. "The Difference Freedom Made: The Emancipation of Afro-Americans." In *The State of Afro-American History: Past, Present, and Future,* ed. Darlene Clark Hine. Baton Rouge: Louisiana State Univ. Press.

Rodney, Walter. 1981. *A History of the Guyanese Working People, 1881–1905.* Baltimore: Johns Hopkins Univ. Press.

Rose, Willie Lee. 1964. *Rehearsal for Reconstruction: The Port Royal Experiment.* Indianapolis: Bobbs-Merrill.

Rosengarten, Theodore. 1974. *All God's Dangers: The Life of Nate Shaw.* New York: Alfred A. Knopf.

Saloutos, Theodore. 1960. *Farmer Movements in the South, 1865–1933.* Berkeley: Univ. of California Press.

Schmitz, Mark. 1979. "The Transformation of the Southern Sugar Cane Sector." *Agricultural History* 53(Jan.):270–85.

Schwartz, Michael. 1976. *Radical Protest and Social Structure: The Southern Farmers' Alliance and Cotton Tenancy, 1880–1890.* New York: Academic Press.

Scott, James C. 1978. *The Moral Economy of the Peasant.* New Haven: Yale Univ. Press.

Scott, Rebecca. 1985. *Slave Emancipation in Cuba: The Transition to Free Labor, 1860–1899.* Princeton: Princeton Univ. Press.

Shlomowitz, Ralph. 1979. "The Origins of Southern Sharecropping." *Agricultural History* 53(July):557–75.

______. 1982. "The Squad System on Postbellum Cotton Plantations." In *Toward a New South? Studies in Post Civil War Southern Communities,* ed. Orville Vernon Burton and McMath, Jr. Westport, Conn.: Greenwood Press.

______. 1984. "Bound or Free? Black Labor in Cotton and Sugarcane Farming, 1865–1880." *Journal of Southern History* 50(Nov.):569–96.

Shugg, Roger W. 1939. *Origins of Class Struggle in Louisiana: A Social History of White Farmers and Laborers during Slavery and After, 1840–1875.* Baton Rouge: Louisiana State Univ. Press.

Sitterson, J. Carlyle. 1953. *Sugar Country: The Cane Sugar Industry in the South, 1753–1950.* Lexington: Univ. of Kentucky Press.

Stone, Alfred Holt. 1905. "A Plantation Experiment." *Quarterly Journal of Economics 19(Feb.):270–87.*

Strickland, John Scott. 1983. "No More Mud Work: The Struggle for the Control of Labor and Production in Low Country South Carolina, 1863–1880." In *The Southern Enigma: Essays on Race, Class, and Folk Culture,* ed. Walter J. Fraser and Winfred B. Moore, Jr. Westport, Conn.: Greenwood Press.

______. 1985. "Traditional Culture and Moral Economy: Social and Economic Change in the South Carolina Low Country, 1865–1910." In *The Countryside in the Age of Capitalist Transformation,* ed. Steven Hahn and Jonathan Prude. Chapel Hill: Univ. of North Carolina Press.

Taylor, Arnold H. 1976. *Travail and Triumph: Black Life and Culture in the South since the Civil War.* Westport, Conn.: Greenwood Press.

Thomas, Emory M. 1971. *The Confederacy as a Revolutionary Experience.* Englewood Cliffs, N.J.: Prentice Hall.

Tilley, Nannie May. 1948. *The Bright-Tobacco Industry, 1860–1929.* Chapel Hill: Univ. of North Carolina Press.

Trotter, Joe W., Jr. 1984. *Black Milwaukee: The Making of an Industrial Proletariat, 1915–45.* Urbana: Univ. of Illinois Press.

Wallerstein, Immanuel. 1974. *The Modern World System I: Capitalist Agriculture and the Origins of the European World-Economy in the Sixteenth Century.* New York: Academic Press.

______. 1980. *The Modern World System II: Mercantilism and the Consolidation of the European World Economy, 1600–1750.* New York: Academic Press.

Wayne, Michael. 1983. *The Reshaping of Plantation Society: The Natchez District, 1860–1880.* Baton Rouge: Louisiana State Univ. Press.

Weiman, David F. 1985. "The Economic Emancipation of the Non-Slaveholding Class: Upcountry Farmers in the Georgia Cotton Economy." *Journal of Economic History* 45(Mar.):71–93.

Wiener, Jonathan M. 1975. "Planter-Merchant Conflict in Reconstruction Alabama." *Past & Present* 68(Aug.):73–94.

______. 1976. "Planter Persistence and Social Change, Alabama, 1850–1870." *Journal of Interdisciplinary History* 7(Autumn):235–60.

______. 1978. *The Social Origins of the New South: Alabama, 1860–1885.* Baton Rouge: Louisiana State Univ. Press.

______. 1979. "Class Structure and Economic Development in the American South, 1865–1955." *American Historical Review* 84(Oct.):970–1006.

Wolf, Eric. 1982. *Europe and the People without History.* Berkeley: Univ. of California Press.

Woodman, Harold D. 1968. *King Cotton and His Retainers: Financing and Marketing the*

*Cotton Crop of the South, 1800–1925.* Lexington: Univ. of Kentucky Press.

______. 1977. "Sequel to Slavery: The New History Views the Postbellum South." *Journal of Southern History* 43(Nov.):523–54.

______. 1979. "Post Civil War Southern Agriculture and the Law." *Agricultural History* 53(Jan.):319–37.

______. 1985. "The Reconstruction of the Cotton Plantation in the New South." In *Essays on the Postbellum Southern Economy,* ed. Thavolia Glymph and John J. Kushma. College Station: Texas A&M Univ. Press for the University of Texas at Arlington.

Woodward, C. Vann. 1951. *The Origins of the New South, 1877–1913.* Baton Rouge: Louisiana State Univ. Press.

______. 1978. "The Price of Freedom." In *What Was Freedom's Price?* ed. David Sansing. Jackson: Univ. Press of Mississippi.

Wright, Gavin. 1978. *The Political Economy of the Cotton South: Households, Markets, and Wealth in the Nineteenth Century.* New York: W. W. Norton & Co.

______. 1986. *Old South, New South: Revolutions in the Southern Economy Since the Civil War.* New York: Basic Books.

# PART III

# Special Topics

CHAPTER 11

# Women in Agriculture during the Nineteenth Century

ELIZABETH FOX-GENOVESE

> Woman's place in farm life is the severest test that agriculture has to face. If farm life cannot give the farm woman opportunity for real growth, for something besides drudgery, our rural civilization cannot go on.
>
> —*Kenneth L. Butterfield, 1923*

WELL INTO THE NINETEENTH CENTURY, the United States remained a rural nation. As late as 1860 the dream of independent property ownership dominated the goals of Americans and many potential immigrants (Atack and Bateman 1987). The violent struggle over Kansas in the 1850s embodied northerners' and southerners' conflicting determinations to claim that promising territory for their own system of property ownership (Robinson 1976). As early as the 1820s the growth of manufacturing and towns in Massachusetts and Rhode Island was delineating the country's industrial future, and throughout the pre–Civil War decades the industrial model steadily gained ground. But during the same period agriculture continued to expand, primarily with the opening of new lands on the successive frontiers. For most Americans agricultural life best embodied the values of independent households, which they took to be the foundations of national political, social, and moral life. Both sides in the debate over slavery—the freedom of labor—believed they were defending the authentic vision of propertied independence rooted in agriculture. Yet the outcome of the war ensured the triumph not of agriculture but of industry. Although the Homestead Act of 1862 provided the first public guarantee of the free availability of land and was followed by a massive settlement of western lands, it too confirmed the triumph and rapid acceleration of industrialization and especially the tightening hold of agricultural capitalists and speculators on the best land (Gates 1973; Pfeffer 1983).

From the beginning of the nineteenth century, the image of the independent farm household masked conflicting visions of American development (Kaufman 1982, especially Introduction). If, as widely believed, the members of most northeastern and midwestern farm households remained personally committed to an independence grounded in subsistence production, they were nonetheless demonstrably participating in the growth and expansion of a capitalist market. The members of southern farm households, in contrast, even those of large scale and high profits, were actively inhibiting the penetration of capitalist relations into their communities while paradoxically producing for the world market (Fox-Genovese and Genovese 1983; Fox-Genovese 1983, 1988, Chap. 1) The differences were not trivial, as a bloody war confirmed. For under the surface of a common commitment to "republicanism," and despite an apparently similar devotion to ideals of familial independence, northern and southern agricultural households were proceeding along very different routes with dramatically different consequences for the capitalist development of their regions (Genovese 1965; Wright 1978; Woodman 1968; Mandle 1978; Bateman and Weiss 1981; Fox-Genovese 1988).

Prevailing classifications—notably market-subsistence, urban-rural, agricultural-industrial—risk obscuring more than they clarify about the nature of nineteenth-century America. From at least the eighteenth century, rural American households had been actively engaged in production for the market, even as many also strove for a safe margin of self-sufficiency. Typically northern households sought both profit and independence, but in pursuing profit they inadvertently contributed to developments that would undermine the independence of their heirs. Many southern households sought the same conflicting ends but generated different consequences. In their case, slavery as a social system kept most households more independent of markets—both as consumers and as producers—and forestalled market penetration of the household itself. Slavery as a social system also discouraged the expansion of industry based on free labor (Genovese 1965; Bateman and Weiss 1981; Fox-Genovese 1988). The Civil War uprooted the formal obstacles to southern development, although without destroying the social and cultural obstacles (Ferleger 1989). Before the war, frontier agricultural households replicated the features and values of both the settlers' region of origin, whether North or South. But after the war, despite initial periods of independence, they succumbed to the imperatives of a vigorous, expanding national market.

The story remained empirically ambiguous, despite its underlying logic: As some farm households began to function as "agribusinesses," others persisted primarily as subsistence operations, and still others disappeared entirely (Pfeffer 1983). Secular changes, notably the irreversible triumph of industrial capitalism, carried some farm households on a rising tide of profitability, leaving others to founder on the shoals of decreasing viability, and relegating still others to an uneasy marginality. But when all complexities are taken into account, the United States entered the nineteenth century as a rural-agricultural nation and left it as an urban-industrial one.

For women, as for men, participation in agriculture turned upon a complex system of relations between households and markets. For women, as for men, the nature of those relations depended upon the larger social system to which their specific households belonged (Fox-Genovese 1983, 1988, Chap. 1). The experience of women as members of agricultural households varied as much as that of men, according to region and decade, size of farm and type of crop, wealth of household and relations to household head. And if the agricultural experience of various groups of nineteenth-century Americans affected women, women themselves were central—however "invisible" they have often been to scholars—to the history of American agriculture (Sachs 1983; Papanek 1979).

From the beginning of the nineteenth century, women have related to agricultural households in three primary ways. They have worked within households and farms to produce essential goods and services for members of farm households, notably by preserving and preparing food and making clothing, but also through direct contributions to subsistence crops and marketable surpluses. They have also worked on their own to produce goods—notably eggs, butter, textiles—directly for the market, for merchant outputters, or for local exchange. And they have worked outside the household to earn money to contribute to its support (Fink 1986, 3; Ankarloo 1979; Dublin 1985; Jensen 1986; Prude, 1983). Some women (frequently widows) ranked as the actual heads of farm households, although in most instances they probably benefited from the active assistance of men: sons, sons-in-law, overseers, male kin (Fox-

Genovese 1988; Atack and Bateman 1987, 43; Sokolow and Lamanna 1984; Hargreaves 1976, 182; Harris 1987; Snell 1971; Stewart 1914). The overwhelming majority of women participated in agriculture as members of families—of households—that shaped their roles but that could not easily have prospered without them (Atkeson 1929; Faragher 1979, 1981, 1986; Fox-Genovese 1988; Riley 1981; Jameson 1987; Frankel 1982).

In any region and in any decade throughout the nineteenth century, extraordinary circumstances could push individual women into any of a farm household's varied activities—into fields as well as barnyard, garden, and house. As a rule, however, and with the notable exception of black slave women, women's participation in agriculture followed a combination of time-honored European and American attitudes towards gender roles (Allen 1933; Fox-Genovese 1988, 1985; Segalen 1980). The traditional cleavage between men's and women's customary roles followed the distinction between the house and the fields, with some blurring in the case of dairies, gardens, and orchards. The distinction has frequently been depicted as that between reproduction and production, although such a sharp distinction woefully fails to capture the complexity of everyday relations, much less unexpected consequences of change. Originally, many such distinctions derived from the assumption that men dealt with the outside world, notably with markets, while women dealt with basic needs of the household. Even that association could break down when women's customary activities acquired a market value. Women's primary responsibilities to the household nonetheless concerned reproduction in the broad sense: childbearing and childrearing, the preparation of meals and the preservation of food, the washing and repairing of clothing, nursing the sick (Ankarloo 1979; Carrigan 1962; Flora 1981; Harris 1987; Jameson 1987; Riley 1981).

Beyond these core responsibilities, farm women participated in one or more aspects of production, even if few, including themselves, labeled them as such. Throughout the century, they were likely to make at least some clothing, frequently most of it. Intermittently, in a variety of circumstances, they produced some or all of the cloth from which the clothing and household linens were made. The association between women and textiles had an ancient history, and women frequently persisted in textile production as it came under the sway of merchant capitalists, and even followed it from the household into the factory (Dublin 1979, 1985). Women commonly tended gardens, which were viewed as an extension of the kitchen or household rather than of the fields. They commonly kept chickens, or other fowl, and disposed of the eggs, and ultimately the meat, either for their own kitchens or for sale. They sometimes engaged in dairying, especially the production of butter, which they also might sell (Riley 1981; Jensen 1986; Bateman 1978; Glasgow 1925).

Scholars have not notably attended to the roles of women in agriculture, presumably because so many of women's responsibilities have been associated with use value rather than market value and have, from one perspective, varied little according to time and place (Fox-Genovese 1981). Agriculture households have normally manifested a distinct ideological conservatism with respect to men's and women's roles, not least because for so much of the nineteenth century the respective physical attributes of women (childbearing capacity) and men (muscular strength) were of decisive importance in the success of the household. The apparent relevance of anatomy to role dramatically reinforced traditional ideas of women's and men's inevitable destinies, and of men's roles as heads of households. The importance of the farm household as a unit of production and reproduction in which all the members recognized the imperative of cooperation for the well being of all tended to reinforce the idea that individuals participated according to gender and familial roles. Overwhelmingly, allowing for the important exception of black slave women and occasional white female servants, women in agricultural households were wives, mothers, daughters whose family roles masked their roles as laborers. Scholars, in this perspective, have merely followed and perpetuated the assumptions of those they are studying.

If women's roles in nineteenth-century American agriculture tended to conform to their roles in agricultural families (although they might not), and if their labor within agricultural households tended not to produce distinct market values (although they might), those roles and that labor nonetheless varied significantly by decade, region, and wealth of household, just as women's "invisible" labor contributed to the success of individual households and changed as the conditions of the households changed. In particular, the growing participation of farms in the capitalist market tended to undercut the perceived value of women's labor (Fink 1984; Hedley 1981). Since official statistics, notably census data, did not normally record women's labor as a particular occupation nor disaggregate household production to account for women's contributions, and since many rural women had inadequate leisure to write of their experience or even remained illiterate, women's work in agricultural households remains resistant to most forms of quantitative analysis. Variations in fertility patterns can be documented and do reveal something about women's lives in addition to offering hints about their attitudes (Atack and Bateman 1987; Kulikoff 1990; Easterlin 1976a, 1976b). But, overall, quantifiable sources cannot tell us much about the quality of agricultural women's lives, about the round of tasks that made up their everyday lives, about the quality of their contributions to the success of agricultural households.[1] Even the most well-intentioned and sympathetic scholars frequently have to rely on intelligent speculation or upon the infrequent accounts of the women and other commentators. Writing the history of women in agriculture remains, in sum, a challenge to be met by the accumulation of a host of microcosmic studies. With this caveat filed, it remains possible to suggest some general outlines.

Here, I offer only a preliminary sketch of the kinds of work in which the women of agricultural households engaged and make some preliminary suggestions about the variations in their experience by class and race, by period and region. In general, I follow conventional distinctions by region (Northeast, South, Midwest, and West) and by period (pre- and post-Civil War). Despite our intuitive sense of the basic similarities of the experience of women within agricultural households across time and space, and despite the arguments of many women's historians that women's history defies conventional periodization, I do unabashedly insist upon the significance of "men's" events for the experience of agricultural women—notably the growth of industrialization and the Civil War. The growing density and sophistication of markets together with the attendant growth in industrial manufacture incontrovertibly influenced the lives of agricultural women by decreasing the value and the quantity of their household production relative to the price of marketed commodities; by transforming the economic scope and roles of the households to which they belonged; by increasing their access to literacy, to the possibility of remaining single, to new opportunities for employment or for leisure; by influencing the ages at which they married and the numbers of children they bore. The Civil War had no less of an impact, and in some instances a more dramatic one. If for northern and midwestern women its primary effect lay in increasing the tempo of industrialization and the integration of the national market, for southern women its consequences were more far-reaching. Emancipation, however bitter some of its fruits, transformed the hopes and identities of former slave women as well as the necessities and expectations of many white southern women. For western settler women the war, by cementing their households' ties to developing markets, effectively consolidated their place as full members of the nation (Gates 1965). The history of women in the West, of course, also includes the previous and poorly studied history of Spanish-American and Native American women who had long inhabited most parts of the West.

The idea of "the frontier" loomed large in the minds of the nineteenth-century women and men who confronted and settled it, but the unity of the concept belies the variety of frontiers that different Americans encountered. The issue concerns less the distinctions among mining, cattle, and agricultural frontiers, than it does the progressively shorter duration of frontier conditions with each successive move westward. Even duration of frontier conditions does not capture the whole story, for each successive frontier related to a different hinterland. During the antebellum period there were both northern and southern frontiers, with portentous consequences. Following the war, all frontiers related to exponentially expanding markets and industrial development. Thus, although some agricultural women in Wisconsin in the decades following the war apparently returned briefly to the use of the spinning wheel, as a general rule women's household textile production never enjoyed the importance that it had for earlier generations. Nor did women of the postbellum frontiers participate in cottage industry as their New England foremothers had. Even women butter makers confronted the competition of railroads, refrigerated cars, and the attendant growth of strong regional dairy industries. These examples illustrate that women's roles in agriculture are best understood in the context of the economic development of the nation as a whole.

Throughout the antebellum period agriculture as the embodiment of household independence and prosperity retained a strong hold on the imagination of the people of the northern states—those of New England and the Mid-Atlantic.[2] Yet even before the beginning of the century, the world of northern farming was experiencing the steady commercialization of all aspects of life that animated the region as a whole (Bateman 1983; Atack and Bateman 1987, 202–4; Danhof 1969; Gross 1982; Lemon 1967, 1987; Rothenberg 1981, 1985). Between 1800 and 1820 the urban population grew from 7.8 to 31.7 percent in New England and from 8 to 34.3 percent in the Mid-Atlantic (Fox-Genovese 1988, 77), testifying to the growing hold of capitalist transformation upon the rural population.[3] Although the majority of the northern population still lived on farms (Atack and Bateman 1987, 202), that transformation was altering the lives of the women of agricultural households.

Even at the beginning of the century, northern agricultural women were relinquishing many aspects of household production, notably textiles. Laurel Thatcher Ulrich (1982, 16) found that as early as 1730 the percentage of household inventories in Essex and York counties that included spinning wheels had dropped to 38 and 40 percent respectively. The percentage of households with looms was 13 for Essex and 3 for York. Ulrich also found that by the middle of the eighteenth century agricultural households no longer produced everything they needed for their own tables, much less for their own use. An important part of an agricultural housewife's skill consisted in her knowledge of the formal and informal markets on which she might barter her goods for others that she lacked (Ulrich 26–27). By the opening decades of the nineteenth century household production was in secular decline (Tryon 1917); by 1840 the per capita value of northern household production had dropped to $1.34, by 1860 to $0.72 (Atack and Bateman 1987, 205).

Aggregate statistics can admittedly cloak wide variations among households. A per capita measure that includes the rapidly growing populations of towns and cities obviously dilutes the possible persistence of production within rural households, especially those furthest from the densest markets.[4] Nonetheless, the aggregate statistics do confirm a general tendency for women to move away from many of the forms of household production in which they had typically been active, and to move instead into new kinds of work that signaled their households' growing dependence upon the capitalist market. As Thomas Dublin (1985) points out, between 1800 and 1860 industry spread its tentacles throughout the countryside from Connecticut to Maine. In some instances burgeoning manufacturers drew young women from New England farms into the mill towns of Lowell and Lawrence as laborers (Dublin 1979). In other instances, industry reached out to women who remained within farm households, putting them to work on raw materials or semifinished goods for merchant outputters (Dublin 1985). Rural women in substantial numbers worked at various aspects of textile and shoe production and, in the case studied by Dublin (1985), worked palm leaves into hats. Throughout New England, putting-out industries developed in conjunction with both the persistence of agricultural households and the growth of factories. The full scope of the process, which has so far only been studied for individual cases, remains elusive, but it was clearly considerable (Prude 1983, 1985; Nobles 1984; Clark 1979).

The ubiquitous presence of women as individuals and as members of families in the emerging New England factory systems testifies at least indirectly to growing class divisions within the rural community and to the declining importance of women's labor within agricultural households as opportunities rose to purchase inexpensive replacements for the goods traditionally produced by women. The women of agricultural households retained heavy responsibilities for the bearing and rearing of children; for the cooking, serving, and preservation of food; for the washing, finishing, sewing, and repairing of textiles; for quilting, darning, and knitting. Such tasks frequently amounted to heavy and seemingly unending labor, especially when to them are added the tending of fires, the cleaning of houses and cooking utensils, and even the carrying of water. We can, nonetheless, safely assume that however necessary to the sustenance of the farm household, the tasks increasingly declined in value as substitutes for women's labor—including the labor of servants—became available for purchase by the more successful, and as even the less successful increasingly traded or bartered for the goods and services that they could not produce for themselves. In the absence of full data on the value of women's labor in agricultural households, future historians might profitably attend to data on the prices of goods traditionally produced by women within households and especially to the records of country stores.[5]

Thanks primarily to the work of Joan Jensen (1986, 1988) and of Fred Bateman (1978), it is clear that some northern agricultural women retained and developed their roles in dairying, especially in butter making. But the transformation of this form of female labor also testifies to the growing intrusion of the market. In effect, some women—presumably from reasonably prosperous households—began to specialize in butter making for local markets. Jensen (1988), who has meticulously described the details of that heavy labor, argues that women should also be credited with substantial innovations in the technology of butter making and with the systematic transmission of skills across generations. There can be no doubt either of the value of women's labor in this area or of its having gradually been taken over by men as the possibilities for a wholesale dairy industry gradually emerged. Similarly, there can be no doubt that women's predominance in commercial butter making represented a transitional stage in the move of agricultural households from partial subsistence to full market participation and dependence (Loehr 1952).[6]

Ultimately, the transformation of northern women's labor in agricultural households must be linked to the rise of a female labor force in manufacture (Dublin 1979), to the decline in rural women's fertility rates (Atack and Bateman 1987), to the growing numbers of women who never married, and perhaps especially to the dramatic increase in women's literacy in general and their entry into the teaching profession in particular. Throughout the antebellum period northern women, rural women as well as urban, were irreversibly drawn into the industrial capitalism and urban system that was stamping northern society in general (Fox-Genovese 1988, Chap. 1). As many scholars insist, the transition was prolonged and included an abiding commitment to the maximum possible independence of households and to local networks of exchange over full participation in regional or national markets (Prude 1983). Doubtless it also included strong commitments to conventional gender roles—to the primacy of identification with family and household over individualism for women. It nonetheless gradually undermined the centrality of women's labor to household prosperity even as it also offered women new possibilities to shake off the drudgery of rural household labor (Fox-Genovese 1988; Kulikoff 1990).

Southern agricultural households differed significantly from northern because of the broad consequences of slavery as a social system. Although large slaveholding households—those of twenty or more slaves—were a small minority, slavery cast a indelible stamp upon southern society and its households. First and foremost, slavery guaranteed the persistence of the South as a predominantly agricultural society that resisted the penetration of bourgeois social relations and of their attendant values. Comparative statistics for urbanization make the point dramatically. By 1860 among the states that would join the Confederacy only Kentucky exceeded 10 percent urban, and only at 10.4. Among the others, Virginia came closest with 9.5 percent. Georgia and South Carolina followed with 7.1 and 6.9 percent respectively. The others came in at about 5 percent or less (Fox-Genovese 1988, Chap. 1). The majority of the South's urban population remained concentrated in the great port cities such as Charleston and New Orleans. In short, the South did not develop the kind of city system that was drawing northern rural households into the maelstrom of capitalist development (Fox-Genovese 1988,

Chap. 1; Pred 1973, 1980). It failed to develop a significant industrial sector (Bateman and Weiss 1981) and, of special significance, failed to develop dynamic local markets in such basic commodities as textiles, boots, or nails.

Wealthy slaveholding households participated in long-distance national and international markets in which they sold and bought in bulk, but they did little to generate local economic development (Fox-Genovese and Genovese 1983). Some smaller slaveholding and yeoman households, notably those with more than 100 improved acres, showed an increasing propensity to produce for the market towards the end of the antebellum period (Weiman 1987; Ford 1983), but even they remained primarily concerned with marketing their own (small) agricultural surpluses rather than participating in regular exchange for basic commodities. With only 7 percent of slaveholding households' owning more than ten slaves, many of them participated in general farming rather than primarily producing cotton (or sugar or rice) for the market (Otto 1981). Under these conditions all southern agricultural households, which constituted the vast majority of all southern households, remained much more dependent upon basic household production for most of their needs than their northern counterparts. The aggregate figures for per capita value of household production capture the (growing) difference between the two regions. Whereas in 1840 the per capita value for northern household production was $1.34, the southern equivalent was $2.31; in 1850, the respective figures were $0.72 and $2.04; and, in 1860, they were $0.36 and $1.60 (Atack and Bateman 1987, 205; Tryon 1917). Even those figures may not fully capture the differences for the manuscript sources suggest a much higher level of production within many southern households (Fox-Genovese 1988). Whatever the actual value of their productions, southern households remained largely impermeable to the incursions of the capitalist market, relying primarily upon the labor of their own members—which could include slaves—augmented by a local exchange of labor services and some goods.

Southern agricultural households, despite tremendous variation in size and market participation, generally followed a traditional division of labor according to gender, complicated in the case of slaveholding households by divisions by class and race. Southern households betrayed little enthusiasm for the democracy among family members that might result in northern and midwestern households in which the essential labor of women and children entitled them to some say in household decisions. With few exceptions, they fell under the command of their white male heads (Fox-Genovese 1988).

White slaveholding women carried the same responsibilities that accrued to other rural women throughout the country, with the significant difference that they performed little of the actual labor themselves. Thus, although slaveholding women continued to bear more children than their northern counterparts, and although they took their roles as mothers seriously, they did not themselves normally perform the basic tasks of child care. Slave women nursed, washed, dressed, and "minded" slaveholding children.[7] Similarly, slaveholding women rarely participated directly in the preparation of meals; on large and even middle-sized plantations they left the actual work in hot kitchens to slave women. Their "cooking" primarily consisted in putting up preserves or pickles or, less frequently, in making special pies or cakes, but even those tasks might be entirely assigned to slaves or seconded by them. In their diaries and journals white slaveholding women commonly referred to "having" something made, to "having" a task completed. Their papers abound with barrels of flour opened, gardens tended, clothes washed, candles made—all as if on their own. Without the slaves' labor the mistress could not have met her responsibilities, even if she did not "see" that labor. Her attitude resembled that of the typical planter who would note that he had "ploughed my field" (Fox-Genovese 1988, Chap. 2).

Slaveholding households required large stocks of supplies that were purchased or produced in bulk both because of the difficulty of obtaining them on a daily basis and the numbers of people to be served. If the master ordinarily saw to the purchase of the barrels of flour, whiskey, and sugar, and the sacks of coffee, the mistress oversaw their disposition. Mistresses often presided over the slaughter of dozens of chickens, the smoking of hundreds of pounds of pork and hams, the salting and cutting of hogs, and the making sausages. Many, like Mary Hamilton Campbell in Virginia in the 1820s, complained of "the unpleasant task of superintending the cutting and salting of our meat." Others, like Martha Ogle Forman in Maryland in the same period, apparently took satisfaction from the extensive processing and storing of beef, hams, and pork. One day she had seventeen hogs weighing 2,077 pounds killed and had the tallow rendered and the sausage made. No one woman could have managed supplies in such quantities without the constant assistance of slaves. (Fox-Genovese 1988, Chap. 2; Mary Hamilton Campbell to

David Campbell, 8 Jan. 1822, Perkins Library, Duke University; Wilson 1976, 6, 9, 30, 135, 321).

As with cooking and supplies, so with textiles, which southern households continued to produce in much greater quantities than northern. During the 1810s and 1820s, Martha Ogle Forman regularly reported the preparation and storing of flax. She also kept silkworms, which most slaveholding women did not. But most slaveholding women lived in households that boasted extensive textile production, even if they did not actively participate in all of its facets. By the 1850s Frederick Law Olmsted traveling in Mississippi noted that whereas in Ohio the spinning wheel and hand loom had practically disappeared and with them homespun as a material for clothing, "half the white population of Mississippi still dress in homespun, and at every second house the wheel and the loom are found in operation" (Olmsted 1970, 352; Wilson 1976, 6, 9, 271, 295, 359).

This persistence of homespun testified only to the dress of poor white and slave women. Slaveholding women, except those from the ranks of the smallest slaveholders, neither wore it nor worked at the spinning wheels or looms that produced it. Mistresses of slaveholding households might oversee the production of homespun clothing by and for the slaves, and occasionally even the preparation of wool and flax as well as cotton cloth. Mary Henderson in North Carolina, for example, had wool washed; Sara Cobb near Augusta, Georgia, noted that her overseer sometimes "gave the wool out to be spun," and also that she had had "some fine wool" spun on their place. But generally, mistresses restricted their own labors in textiles to fine sewing for the white family and to supervising the cutting and sewing of the slaves' clothing, which at least until the war persisted as a principal activity. During the war, at least for a few years, textile production appears to have substantially increased to offset the effects of the blockade (Atack and Bateman 1987; Fox-Genovese 1988, Chap. 2).

If slaveholding men observed the conventions of gender with respect to the women of their own families, no similar respect for female delicacy deterred them in the case of slave women. At most, prejudices about men's and women's roles might lead masters to avoid placing women in charge of men as "drivers;" and prejudice frequently led them, when they had a large enough force, to organize work gangs by gender.[8] If many masters seem instinctively to have assigned some tasks to women and others to men—for instance, hoeing to women and plowing to men—their choices hardly reflected the concerns that would have motivated them in the case of the women of their class and race. As "full hands" adult slave women participated in the most demanding labor of the planting cycle, including plowing (Fox-Genovese 1988, Chap. 3; Sellers 1950, 66; Olmsted 1970, 81; Olmsted 1981, 310; Easterby 1945, 290; Sydnor 1938, 113). Indeed many women plowed and apparently enjoyed it.[9]

During the slow winter months slave women's work in the fields—normally in the company of other women, although under the direction of a slave man or an overseer—included grubbing and hoeing; mending old or putting up new fences; shrubbing hedge rows and raking manure in the yard; cutting down and picking up corn stalks; burning blown-down logs and brush; picking out cotton and thrashing out oats. And always during inclement weather or when nothing else was pressing, they spun. With the beginning of the planting season, some women would be set to bedding up cotton lots and preparing beds for corn and potatoes. Others—frequently in the company of children, referred to as "small hands"—planted cotton, corn, rye, peas, Irish potatoes, and began weeding. Hoeing and weeding continued throughout the summer. During the wheat harvest in August, women spread and stacked hay and began to rake leaves. Then came the cotton harvest, which might last for more than a month. Cotton picking was interspersed with harvesting other crops, picking out the cotton, and continuing other general farm tasks such as fencing. When rain interrupted harvesting, they spun. (Fox-Genovese 1988, Chap. 3; Davis 1939, 67–68; Sydnor 1933, 11.) As the fall wore on and the crops came in, they devoted more and more time to spinning (Massenburg Farm Book 1834–39; Seaborn Hawks Plantation Book 1854–1862; Cody 1982).[10] Elsewhere, women cut stock and briars, planted pumpkin seed, shelled corn, pulled flax, cut briars, stripped clover seed out of the field, gathered clover seed, dug sweet potatoes, sprouted new ground, cut sprouts in the new ground, and hoed corn (James Hervey Greenlee Diary 1848–1853). Patterns of tasks differed by region and crop, but the general principle of grouping the women to work together and setting them to heavy work was ubiquitous (Fox-Genovese 1988, Chap. 3; Joyner 1984, 45, 74–75; Easterby 1945, 101, 271–73; David 1939; Moody 1924; Clifton 1978, 142; Northup 1970, 155–56).[11]

Although masters assigned slave women to the same kinds of work as slave men, they did not normally expect the women to complete the same quantities of work as the men. James Bertrand, a former slave, recalled that "out in the field, the man had to pick three hundred pounds of cotton and the woman had to pick two hundred pounds." But women's lighter loads normally meant that they were needed not only for other kinds of labor, notably spinning and weaving, but also for the care of slave cabins. On a noteworthy number of plantations women matched or overmatched the productivity of the men in cotton picking and other types of field work.[12]

Slave women's work in household production differed from one plantation to the next, especially with respect to specialization, and according to location and period. Many large households included spinners, weavers, carders, dyers, and sewers. Many former slaves remembered that the slave women in their households wove cotton and linen for household linens and wool for blankets. In some households slave seamstresses made clothes for the slaves and even for members of the white family.[13] In smaller households slave women were less likely to specialize but nonetheless worked extensively in textiles and other kinds of household production. Like slave women in larger households, they frequently spun, wove, sewed, and knit; they also milked and churned; preserved meat, fruits, and vegetables; sometimes made candles; prepared and served meals; washed clothes; cleaned; minded children; and worked in the fields (Fox-Genovese 1988, Chap. 3). The largest and wealthiest plantations sustained a wide and diverse complex of textile production, even if they were also likely to be the heaviest purchasers of luxury textiles from elsewhere. The smaller the plantation or farm, the more likely it was to depend almost exclusively on the skills of its members for cloth and clothing. In this respect these smaller units resembled the households of the solid yeoman with the difference that yeoman women did most of the work that slave women performed in plantation households of various sizes. Indeed, a yeoman who could afford to buy one slave more often than not chose a woman to relieve his wife of household drudgery as well as for her reproductive capacities, and even a yeoman who could not afford to purchase a slave woman might draw upon the skills of a hired slave for such specialized tasks as weaving. Laura Caldwell, a former slave, recalled her mother going "to the white folks' houses and weaving clothes for them for small pay." The white families for whom she wove did all their carding and spinning at home (Rawick 1972, *South Carolina Narratives,* vol. 2, Pt. 1, 169). By such labor yeoman women processed at least some of the cotton that their husbands produced and thus contributed to the self-sufficiency of their households.

In addition to considerable textile production, yeoman and small slaveholding women performed extensive amounts of labor for their households, even, especially during harvest in the fields.[14] On a smaller scale they performed all the tasks that slave women performed and over which slaveholding women presided. With smaller, less elaborate households—and most slaveholding households were much more primitive than the myths of moonlight and magnolias suggest—they normally processed and preserved fewer varieties of food (Herndon 1982). The staples of the yeoman diet derived from corn and pork, sometimes supplemented by the produce of a small garden and less commonly a small dairy or local exchange. Travelers' accounts invariably emphasized the simplicity of the yeoman establishment and table. Yeoman women were, nonetheless, responsible for providing the large quantities of food required by men—husbands, sons, occasionally a hired slave—who worked hard in the fields. They also frequently carried the water necessary for cooking and cleaning; made candles; kept their small houses as clean as the soil, dust, bugs, and animals permitted; did the washing; cultivated small gardens; supervised poultry; made butter; bore and reared children; nursed the sick. During the later antebellum period some yeoman daughters were drawn into work in textile mills, but never in the numbers of northern women, and frequently not in urban settings (Moragne 1951). With a persisting dearth of common schools and resistance to the establishment of normal schools throughout the period, southern women did not pursue careers in teaching as so many northern women did. And with a general lack of commercialization, they, unlike northern women, did not have much opportunity to purchase manufactured commodities.

The westward migration of agricultural households was well underway by the beginning of the century. South Carolinians who settled in Alabama, like New Englanders who settled in western New York, brought their patterns of gender relations and attendant divisions of labor with them. Successive waves of migrants who established agricultural households from Mississippi to Texas and throughout the Old Northwest and the Western Reserve did the same. Studies of women in these and subsequent waves of westward migration offer scant reason to believe that these households departed significantly from the norm of a clear division of labor by gender. Such variation as occurred primarily concerned the specific conditions of the westward journey, the rigors of initial settlement, and the crises that could strike individual households. As circumstances required, women shouldered guns against dangerous animals or threatening Indians, or simply to provide food; they helped to build sod huts, cabins, or houses; they worked in the fields. But their initiative and resourcefulness in specific situations were not taken as signs of their moving into new roles. With the passing of danger, the advent of settled routines, and especially the beginnings of prosperity, women normally resumed defined female roles. Even though the work of many women in agriculture remained burdensome and heavy by urban standards, it was women's work.

Scholars continue to differ over whether women welcomed or dreaded frontier life, and in somewhat lesser measure over whether the experience of migration and frontier offered women any greater freedom from male dominance than they had known in the more settled regions of the East (Jeffrey 1979; Faragher 1979; Schlissel 1982). At least some women went west on their own, frequently to pursue careers as teachers (Kaufman 1984). But most women migrated as members of agricultural households for which they invariably had to perform substantial amounts of domestic labor.

Just before the war western farm households of all sizes were still producing more goods domestically than their northern counterparts. Butter and cheese took pride of place, accounting for 80 to 90 percent (or more) of the total value of household production. As early as the 1850s the emergence of cheese factories in the East began to pull that production out of eastern households, but until the end of the century butter remained primarily a household production and within households disproportionately produced by women's labor (Atack and Bateman 1987, 205–7; Jensen 1986, 1988). Thus throughout the century as women's household production of textiles, candles, soap, and other goods steadily declined, their production of butter persisted and even increased. Butter remained, par excellence, the commodity that women could produce in conjunction with their other responsibilities, both for the consumption of their own households and for sale. But in some instances substantial butter production emerged as a specialized activity in which women who could afford the help to relieve them of other kinds of work engaged. It thus appears highly probable that women's ability to engage in specialized butter production testified to a growing divergence in the prosperity of households. To specialize in butter production, women would need assistance, but the butter they produced would generate income that would contribute to household prosperity and permit them to purchase rather than make other goods. In contrast to the declining production of other household goods, the persistence of butter production testifies to the growth of local markets in which it could be sold.

Midwestern women, especially in newly settled areas, also continued to produce a considerable quantity of homespun during the decades before the Civil War. At least some women in Kansas, for example, sheared the sheep, washed and picked out the wool, spun and wove the yarn, died it, cut it, and sewed clothing for their families (Stratton 1981, 66–67). In Illinois women produced homespun as well as butter, eggs, and herbs, not only for their own families, but also for a vigorous local trade. In the judgment of John Faragher (1986, 105; 1985), who has closely studied the Illinois community of Sugar Creek, "butter, cloth, and other women's goods" provided indispensable complements to the men's field crops and "measurably added to the ability of families not only to practice self-sufficiency but to enter the market." Beyond indispensable contributions to household survival and comfort, women's household labor also contributed to the payment of taxes and the possibilities for further growth through investment in machinery or purchase of additional land. In this respect, as in so many others, women played central roles in the success of the individual farm owner-operator (Loehr 1952).

In the single most comprehensive study of the lives of rural midwestern women at midcentury, Glenda Riley (1981, 58, 71, 87) chronicles the wide variety of their normal responsibilities. Typically, the Iowa women she studied bore and cared for babies; attended to the cleaning of houses that could range from sod huts to eastern-style frame houses, and for which they might also help to build furniture; tended gardens; dressed turkeys; cured deer flesh and made candles from the fat; gathered and dried or preserved wild fruits; oversaw and frequently did the spinning, knitting, and making of clothes. Food required the most constant attention and providing it for the household (family and possibly hired hand[s] or boarder[s]) might extend beyond cooking and serving to providing vegetables, fruits, eggs, milk, and herbs. Yet even as some women struggled to provide their families with clothing through their own spinning and sewing (although apparently not weaving), other women purchased lengths of cloth, ribbons, gloves, collars, combs, shoes, bonnets, mitts, and even ready-made dresses. Increasingly, even in frontier Iowa, manufactured commodities were available to those who had the money to buy them. In this rural society women might contribute to their households by selling the products of their industry, especially butter, eggs, herbs, or by taking in boarders. In Iowa in the 1850s agricultural women lived in a society that was divided by class, race, and ethnicity and that increasingly depended upon the market.

The Civil War signaled the beginnings of the demise of that independent household in which the dream of American rural life had been anchored. The dream, to be sure, persisted, even acquiring renewed force with the increase in immigrants from Europe and Asia, migrants from the ever-more urbanized East, and former slaves who sought to establish independent households. The contours of the transformation of American agriculture during the final three decades of the nineteenth century have been remarkably traced, notably by Paul Gates, Allan Bogue, and their students. From the investment required for equipment that would permit the initial breaking of the prairies, to the rise of the railroads and the attendant large agricultural markets, to the interventions of speculators and the attendant divides between landlords and tenants, and beyond, the story is familiar (Gates 1973). No less familiar are the general contours of the story of the emergence of the tenant and sharecropping systems in the postbellum South, although much remains to be done (Ferleger 1984, 1990). The emergence of an agricultural wage-labor force, which had begun even before the Civil War with Chinese laborers in California, accelerated rapidly with the capitalization of agriculture during the postbellum period (Pfeffer 1983). But the excellent historical work on these topics has, to date, paid little or no attention to women. And women's historians, notwithstanding promising beginnings, have yet to fill in the gaps, especially the variations in women's experience by class, race, and ethnicity.

Women did not participate in the same numbers as men in the initial postbellum settling of the West. By 1870 the ratio of women to men in Kansas, Nebraska, and the Dakotas had significantly exceeded one to two, but it did not reach that proportion in the Territory of Colorado until nearly 1880, and not until nearly 1890 in Montana and Wyoming (Hargreaves 1976). Women, Mary Hargreaves (1976) has pointed out, arrived on the plains in tandem with the development of the settled agriculture to which their labor was indispensable. Typically, the women of the agricultural households of the plains, like their counterparts elsewhere, kept chickens, milked cows, gardened, churned butter, baked bread, salted meat, brined kraut and corn, canned and preserved fruits and vegetables, cooked meals, sewed clothing for women and much of that for men, washed, ironed, and cleaned the house. When necessary they also helped in the fields. Living in the isolation of the plains, they also provided essential medical services, especially nursing and midwifery; worked to get schools and churches established; and provided lodging and meals for passers-through and new arrivals.

Even on the plains, women's household productions (with the possible exception of butter) and their productions for informal local exchange and barter declined steadily with the increase in towns and markets. Laura Ingalls Wilder (1941, 109), recalling her youth in the Dakotas in the 1890s as the daughter of a man who spent his life struggling to be an independent farmer, emphasized the difficulties of a life in sod huts and claim shanties that required the efforts of both parents and four daughters merely to survive. She, her mother, and her sisters all worked according to their capacities, even in the fields. Washing alone demanded great effort. The first time that Laura tried to do it with her sisters to surprise their mother, the effort astounded her. "She had not realized how heavy a quilt is, to lift soaked and dripping from a tub, and to wring out, and to hang on a line." A thorough housecleaning further required washing and ironing all the curtains, scrubbing the floors, blacking the stove, washing the straw ticks and filling them with fresh hay (1941, 110–13). Never, throughout her girlhood, was there any money to spare: The bountiful crops that would take them over the edge to solid prosperity never materialized. Her mother made all their clothes, in addition to quilting and knitting. But the clothes were made from store-bought fabrics, and cut according to store-bought patterns (1941, 80–88). The yarn for knitting was purchased, as was the four-burner cook stove. Eventually they even got a sewing machine. Laura's blind sister, Mary, was sent to a special college. And Laura herself attended school to earn the teaching certificate that would permit her to contribute to the family income (Wilder 1941, 1943).

The world of the Ingalls women, like that of so many other agricultural women of more or less prosperous households, was embedded in the market. By the turn of the century even butter production was being taken over by commercial producers. The greatest variation among households may have concerned textile production. In Wisconsin, for example, the proliferation of fairly inexpensive yard goods led to a significant decline in spinning by the 1860s. But the difficulties of procuring goods during the war led to a marked return to spinning. And even after the war, immigrant women continued to spin their own yarn. During the 1870s and beyond, Norwegian immigrant women commonly spun all year for the socks and stockings for the members of their households, and frequently for caps as well. Towards the end of the century, German immigrant women in Dodge County, Wisconsin, were still spinning in sufficient numbers to support several makers of German-Style spinning wheels (Hilts and Hilts 1982).

The growing density and sophistication of national local markets, in conjunction with the proliferation of towns by the end of the century had decisively undermined agricultural women's household production but had not decisively altered the quantity and character of their labor. The great improvements—notably the automobile, plumbing, and electricity—all arrived in the twentieth century, and for many farm women, several decades into it (Hargreaves 1973). By the late nineteenth century the more prosperous farm women were enjoying some of the advantages of store-bought clothing and furniture; various kinds of processed foods, even canned goods; improved schools and medical services; and possibly the assistance of a hired girl in their basic chores. Without the money to pay for such "luxuries," less prosperous farm women continued to perform most of the backbreaking labor that their foremothers had.

During the 1910s in Wisconsin, Nellie Sawyer Kedzie Jones published a series of letters to an imaginary niece for "The Country Gentlewoman," a department of the nationally circulated farm magazine *The Country Gentleman* (Delgado 1973, 6, 7). Her advice about how a young bride should organize and manage her responsibilities—including personal relations—in a farm household captures both the persistence of women's labor and the intrusion of "urban" attitudes and commodities. Her advice begins with the disposition of the kitchen since, for "the next twenty years more of your waking hours will be spent in it than out of it." The old, large kitchen the young bride had inherited must be rationalized and subdivided: Two-thirds for the kitchen itself, the other third to be divided between a laundry room, which would really serve as a washroom for her husband and his hired hands, and a storeroom. The central feature of the kitchen must be an excellent six-hole range with a large reservoir for abundant supplies of hot water. In the cold weather it is impossible to have too much hot water, what "with chicken mashes to make, calf feed to prepare, and dairy utensils to wash and scald." The storeroom also requires serious attention, for, unlike the city women, the farm women cannot just telephone for whatever they need. The storeroom must be big enough to house hundred-pound sacks of sugar; sacks of flour; a ninety-pound bag of oatmeal; a sack of calf meal and possibly food for baby chicks; a vinegar jug; and some canned fruits and vegetables.

After planning the layout, the farm bride must attend to the chores themselves. Never plan two big jobs for the same day: Do not bake on the day that a crew of hired men must be fed. To forestall being taken by surprise, it could be useful to bury a batch of unbaked bread in the garden. Ordinarily, the farm wife must count on two full baking days a week, normally Wednesday and Saturday. On Wednesday Aunt Nellie advised baking, in addition to the bread, a batch of drop cookies, and in winter a crock of doughnuts. The extra bread dough should also yield a batch of cinnamon rolls. Saturday's baking should regularly include bread and layers for cakes, which can be filled when needed, and at least every two weeks, pies. Normally it will also include good batches of baked beans and Indian pudding. The other big jobs, notably the laundry, must be worked in around the regular responsibilities of getting the meals, keeping the house clean, tending the chickens, and eventually caring for the children. Aunt Nellie had special advice on the rearing of girls. They must be allowed to play with the boys at least until they are nine or ten and must be encouraged to play outside. The girl's body needs just the same kind of exercise and development as the boy's, and she should be able to climb a tree as fast as any boy of her size. The prudish notions of "our New England mothers and grandmothers" will not do for today's farm girls. Dolls are all very well for a rainy day, but then they are good for boys too. Otherwise, weather permitting, girls should engage in real athletics. "Housework is not very good for the health of a growing girl" (Delgado 1973, 25).

Nellie Jones belonged to the dynamic and growing group of rural women who were becoming teachers, writers, editors, political spokeswomen both for women and for the farming population as a whole (Hargreaves 1973; Marti 1982; Blumberg 1978; Springer and Springer 1986). These women disproportionately wrote the memoirs, letters, and accounts that shape our knowledge of farm women's experience. In important ways they obviously were not typical of farm women as a group.

Yet in equally important ways they did identify with the experience and values of ordinary farm women, who also frequently combined attachment for farm life with a deep commitment to the full development of women's potential. To be sure, not all women who were born or married into farm households shared that commitment. At the end of the century, as indeed at its beginning, poorer farm households could offer women an unending round of drudgery, meager access to literacy, and exclusion from the alluring goods and comforts of a dynamic, advanced industrial economy.

Throughout the country, but perhaps especially in the South, young women often preferred the difficulties and uncertainties of town employment to the burdensome routine and the loneliness of farm life (Allen 1931). Other southern farm women belonged to households that could not hold their own against the perils of sharecropping and the intrusion of capitalism into agriculture and were drawn into the mills (Wright 1986; Hall, et al. 1987). Still others belonged to households that struggled to hold their own, often with indifferent success and mounting bitterness, against the alien urban world that threatened to undermine the communities and values they cherished (Hahn 1983, 1985). Whatever the variations, these women knew lives of unending toil and frequently of dire poverty.

Ruth Allen's splendid, pioneering study of the labor of Texas women in the production of cotton (1933) remains the best and most comprehensive discussion of women's agricultural labor but concerns the 1920s. Nonetheless, it and Ellen Glasgow's novel, *Barren Ground* (1925), together provide a picture of the lives of southern farm women that probably pertains to the late nineteenth century as well. Allen (45–49), for example, emphasizes the persisting burden of washing for rural southern women. Although two-fifths of the women she studied had piped water in their houses, forty-five of them hauled their own water for washing, some took their clothes to the river or creek, and others drew water from deep wells. Allen (78, 79) also found that by the early twentieth century there was a marked division between white women who worked in the fields (46.2 percent) and white women who did not (53.8 percent), with the nonfield-workers more likely to be members of households of men who owned their own land, and more likely to be married. Allen and Glasgow implicitly concur that even prosperous southern farm women were still performing substantial amounts of labor.

*Barren Ground* traces the career of Dorinda, the daughter of a white farmer in the Valley of Virginia, who takes over her parents' farm, conquering the invading brush and turning it into a prosperous concern. Part of her success derives from her move into dairying, especially the successful marketing of butter. But even as an owner-manager who could afford to employ a number of black hands, she continued to perform much of the heavy work herself. In the mornings she rose before dawn and, after eating a hurried breakfast, milked the cows. "Even the scalding of the churns, the straining and skimming of the milk in the old-fashioned way without a separator—all these simple tasks came under her watchful eyes." She found her reward in successfully demanding a high price for her butter, and, during the succeeding years, slowly working the farm out of debt. The milestones of her prosperity included turning an 18-acre field into pasture, building a new dairy and cowbarns, gradually increasing her seven cows into a herd. Her father, she recalled, "had worked harder than she could ever do, toiling summer and winter, day and night, over the crops, which always failed because they were expected to thrive on so little." But for his effort she substituted her own entrepreneurial sense. Borrowing money and hiring milkers, she gradually transformed the failing farm into a successful dairy (Glasgow 1925, 240, 268–69).

Even in the face of accelerating industrial capitalism and urbanization, the dream of rural independence retained a strong hold on the imagination of many Americans, perhaps especially the former slaves of the rural South. In 1878 Frances Ellen Harper voiced the sentiments of many when she wrote that with intelligence and land a man—a black man—"holds in his hand the basis of power and elements of truth" (Brown 1976, 206). Harper knew full well that the white community, which fully understood the relation between land and power, was doing its best to deprive blacks of both. Immediately following the war, white plantation owners had vigorously attempted to get their former slaves to engage in gang labor for wages. The massive resistance of the freedmen and women eventually led to the development of tenancy and sharecropping, and to the desperate poverty of rural southern blacks. At the core of the resistance lay the blacks' determination to establish their own rural households, including their determination that black women work within their own household for their own families—that they never again work in the fields and households of white men (Frankel 1982; Litwack 1979).

Although we know little about the experience of individual black farm women during the late nineteenth century, we do know that the experience of the vast majority was a grinding poverty, exacerbated by growing dependence upon white credit. Under these conditions the majority of black farm women experienced lives that made those of the Ingalls women look comfortable. Twentieth-century writers like Maya Angelou (1969) and Jean Wheeler Smith (1975) convey a sense of their people's desperate struggles, the relentless struggle merely to survive. Like white farm women throughout the country, rural southern black women performed the full range of domestic labor from bearing and rearing children to preparing food to washing and cleaning. Like so many others, they also sewed, darned, mended, and quilted. Like some others, they gardened and tended poultry, if they had any. But, much more than others, they were also likely to work either in their own fields or in those of whites cultivating cotton (Allen 1931). In the postbellum South, as throughout the country, cotton yard goods were readily available for purchase, but for the southern black woman such purchases meant increased indebtedness to white merchants. For her, only the prolonged labor of spinning and weaving—provided her family could afford to retain some of the cotton they grew—could free her from the indebtedness; only the freedom from indebtedness could free her from the labor (Jones 1985; Frankel 1982).

For blacks, as much as or more than for whites, the agricultural household embodied the essence of the American ideal of independence. But in the postbellum South the attempt to realize that dream led black people back toward servitude—not personal servitude to individual masters, but servitude to ruthless landlords who represented in their own persons the implacable intrusion of capitalism into southern agriculture. By the close of the nineteenth century that capitalist market had irreversibly penetrated American agriculture as a whole. From the 1890s more Americans resided in urban than in rural areas; from the 1920s at the latest, more Americans worked for others than for themselves. The class cleavages among those who persisted in agriculture was relentlessly widening. And even the successful were being drawn into the world of markets and credit that would decisively color the experience of American farmers in our own time.

Throughout the nineteenth century women were everywhere in American agriculture. If for brief periods on particular frontiers they were outnumbered by men, that disproportion did not last. Throughout the century the core of American agriculture lay in farm households to which the labors and reproductive capacities of women were essential. Yet the gradual penetration and transformation of American agriculture by capitalism did inexorably transform women's position. Women persisted as the wives of agricultural capitalists and owner-operators. They have increased—although still in small numbers—as owner-operators themselves. They have persisted—indeed may have increased—as wage laborers and servants. And in growing numbers they have opted to work off the farm even as they supported their husbands' preference for farming (Barlett 1988). But unless they worked in agriculture for wages, the value of their labor gradually eroded. By and large, women's labor in agricultural households had always been invisible—had always been viewed through the prism of use value. Early in the century, however, as in the early stages of the successive frontiers, that use value had enjoyed some measure of appreciation and earned the women who contributed it a say in farm affairs. As the produce of farms increasingly came to be reckoned by the measures of the market, the labor of women increasingly lost in status. If at the end of the century the use value of women's labor in agricultural households retained a distinct importance, it was, nonetheless, increasingly coming to resemble the labor of women in urban households. In agriculture, as in all other spheres, capitalism relegated use value—unpaid labor—to the margins, and with it the women who embodied it. Nonetheless, a small but significant number of American women have continued to value agriculture as a way of life.

## APPENDIX

### *Excerpts From Nellie Kedzie Jones's Advice*

With hired men at your table, make up your mind to three hearty meals every day in the year. I know the prospect is appalling to one who is doing her own work, as you are; but face the fact, prepare for the worst, and any respite you may get later count as pure gain. Most of the writing of recipes, the estimates of quantities for families of various sizes, and marketing directions, have been done by Home-Economics experts who have gotten their data from col-

Source: Delgado (1973, 17, 18, 19, 21).

leges, public institutions, city homes and city eating places, but not from farms. They are too small for farm conditions. Add about fifty per cent to the usual standard recipe, or even double it and you will not be far off. From the hired man's point of view a short ration is the unpardonable sin. He believes that a pie ought not be cut into more than four pieces. I get round that difficulty, though, by often serving easy puddings.

Meat, meat and more meat is what you will have to provide, and it will cost you a lot of money. This is wrong; we are, in town and country, eating altogether too much meat, but I am trying not to substitute in these letters the ideal for the real. You want to know actual country conditions, and I reply by saying that Ben [her niece's husband] will not be able to keep his hired men unless you provide them all the meat they want. And what a lot that will look like to you! However, by raising your own meat and learning how to cook appetizingly the poorer cuts, you can usually work up the whole animal and so help keep down your meat bill. . . .

Chickens are expensive in time and money, but you will eat some for a change of meat and, of course, when your town friends come they will expect some of your home-grown fried chickens as a characteristic farm treat. Train your town friends to let you know a day ahead when they are coming. Many do not appreciate how much work it is to kill and dress and cook chickens. . . .

In the old days it took three women three days to get ready for threshers. "I am almost as tired as though I had cooked for threshers," was a household proverb in our mothers' day.

But all that has changed, and while on the subject I might as well devote this letter to the easiest way to cook for a crew on the farm. An old thresherman said to me last fall; 'We used to get so much chicken that we grew tired of it, but now we scarcely ever get it.' The modern farm wife is finding out that there is an easier way to feed a big crew of hungry men than to pick, pinfeather and draw a lot of chickens.

In your part of the country, where threshers stay for but three or four meals at most, the farm wife will have to feed them for some time still [whereas, in the West, chuck wagons follow threshers]. And then there are silo fillers, hay balers, harvest hands, wood-sawing crews, and so on, so that you must make up your mind to cook for a crew of from five to fifteen men every now and then.

I would begin my baking about three days beforehand with a big batch of drop cookies. At the same time the pie crust, if you want to use pie, can be mixed up ready for the wetting and set away in the ice box. Do the bread baking the day before. At the same time boil and bake a whole ham. This will make your cold sliced meat. A big ham out to boil at least four hours, and bake in a slow oven two more. The night before peel 'a heap' of potatoes for each man per meal and a few more for the dish. Set them away, covered with cold water. . . .

Just as soon as you can find out when the threshers will be getting into your neighborhood, even a week or ten days beforehand, I would make up some pork cake or some other moderately rich fruit cake. If you should not need it there will be no loss, for it will keep for some time. Better have too much than too little. Ice cream is one of the easiest desserts for a crowd. The mixing is practically nothing; one of the men will freeze it and pack it away hours before it is needed. It is easily served. . . .

Having the help live right in the home with us is where we farmer-folks have it hard. In the city a man appears when the whistle blows and disappears when it blows again, and that is the end of it. But not so on the farm. The hired man is at the table three meals a day, unless the place is big enough to have a tenant house and a family man can be contracted for by the year. And then, oh Janet, a year is an awfully long time, when waiting to get rid of an undesirable citizen without a lawsuit.

Most of your helpers, both indoors and out, will be the children of neighboring farmers, for fortunately you live in a part of the United States where rural life is not broken into classes, landlord and tenant. The one is prosperous, the other poor; but the social distinction is even greater than the financial. You are in a democratic region of small- or middle-sized farms, for the most part worked by the owners themselves, and that is the real ideal of farm life.

## NOTES

1. Literacy of rural women lagged behind that of men in all regions. Literacy of northeastern rural women improved dramatically during the early nineteenth century. Joan Jensen (1988) attributes some of the blindness to the malevolence of men.

2. Notwithstanding significant intraregional variations in detail, the New England and Mid-Atlantic states underwent structurally comparable developments during the antebellum period.

3. The appropriate definition of urban remains subject to dispute. My figures are based on the census classification of places of 2,500 inhabitants. For a discussion of the problems with that measure, see Williamson (1971), and, for a calculation based on "highly urbanized counties," Gibson (1987).

4. In the absence of specific studies, we cannot be sure that considerable household production did not persist in, say, Vermont. Similarly, we still lack adequate studies of southern yeomen women.

5. Dublin (1985), for example, has used this data, although primarily to attempt to calculate the value of the labor performed by women in hatmaking.

6. Jensen's work should be read in the context of Lindstrom's (1978) fine study of the transformation of the economy of the Philadelphia region during the antebellum period.

7. I discuss the vigorous debate that persists about the presence and role of mammies in slaveholding households in *Within the Plantation Household* (1988). Whatever the variations in terminology, slave women and girls did nurse and care for the children of slaveholders. Even if many slaveholding women did not have an established Mammy, few of them cared for their own children without considerable assistance from slave women.

8. Above all, they differed according to size, for small slaveholders could not divide their few hands by gender. A slaveholder would have to have upwards of fourteen working hands before he could organize gangs at all. And even on larger holdings planters would behave differently. Those of the South Carolina Rice Coast relied on the task system, which provided greater flexibility in labor assignments than the gang labor system did. In contrast, the hemp growers of Kentucky regarded women as physically unfit for the essential

tasks. Farmers with only a few slaves had to send men and women into the fields to work side by side. Not everyone with a large slave force organized the gangs by gender, but most seem to have done so.

9. The evidence for slave women's plowing is scattered through the antebellum plantation diaries and daybooks. For references, see Fox-Genovese 1988, Chap. 3. The WPA narratives with former slaves (Rawick 1972, 1977) confirm that women plowed and that many enjoyed and took pride in their work (Rawick 1972, vol. 15, Pt. 2, 57, 73, 149, 364; Rawick 1972 vol. 8, Pt. 1, 19, 317; Fisk University 1968, 103).

10. Extensive citations to the primary sources can be found in Fox-Genovese 1988. For confirmation from the narratives of former slaves (Rawick 1972, 1977), see *Georgia Narratives,* suppl., ser. 1, vol. 3, 97; *Oklahoma Narratives,* vol. 7, Pt. 1, 314; *North Carolina Narratives,* vol. 15, Pt. 2, 74, 149; *Arkansas Narratives,* vol. 8, Pt. 1, 157.

11. The tendency to organize gangs by gender also prevailed on the sugar plantations of Louisiana and the cotton plantations of Mississippi (Fox-Genovese 1988, Chap. 3).

12. Plantation account books usually recorded with special care the amount of cotton picked, and women regularly figured among the top pickers. Some planters made pointed reference to a woman as their best picker or best general field hand (Fox-Genovese 1988, Chap. 3; Sydnor 1965, 96–97; bonner 1956). Evidence also abounds in the narratives (Rawick, 1972, 1977, for examples, *Arkansas Narratives,* vol. 8, Pt. 1, 19, 157, 317; *North Carolina Narratives,* vol. 15, Pt. 2, 149, 364; *Georgia Narratives,* vol. 13, Pt. 3, 210; *Texas Narratives,* vol. 4, Pt. 2, 16).

13. In the absence of detailed studies, it remains difficult to determine the actual extent of household textile production. During the late antebellum period cheap cloth was easily available and many planters purchased it for slave clothes. Nonetheless, planters' farm and plantation books from the antebellum period confirm that weaving was a regular activity on at least some plantations, and spinning and sewing on most if not all. And there is some evidence that at least on some plantations slave men and boys wore homespun (McDaniel 1982, 114). Certainly slave women were much more likely than their mistresses to comment upon the household's self-sufficiency in textiles. The narratives of former slaves include numerous references to spinning wheels, cards, and looms, as well as the hours spent at them to produce homespun for the slaves' clothes, and the additional hours spent in dyeing, sewing, and knitting. Former slaves also recalled the production of calicos and the production of shoes, from cleaning and tanning to sewing the leather (Rawick 1972, 1977, for example, *Texas Narratives,* vol. 4, Pt. 2, 1, 61, 67, 136, 160, 292; vol. 5, Pt. 3, 259; *North Carolina Narratives,* vol. 14, Pt. 1, 168–70; vol. 15, Pt. 2, 114–15; *South Carolina Narratives,* vol. 2, Pt. 1, 58, 100, Pt. 2, 114, 172; *Georgia Narratives,* supp., ser. 1, vol. 3, Pt. 1, 126–27, 240; *Mississippi Narratives,* supp., ser. 1, vol. 6, Pt. 1, 36, vol. 10, Pt. 5, 1989, 2006; *Arkansas Narratives,* vol. 8., Pt. 1, 46, vol. 9, Pt. 4, 151, Pt. 8, 6).

14. Even during the war, when their men were off fighting, they found the responsibilities of getting in the crop daunting and their difficulties in this regard accounted for massive desertions on the part of yeoman men, who could not accept that their wives and children starve for lack of male farm labor (Robinson 1976).

## REFERENCES AND FURTHER READINGS

Allen, Ruth Alice. 1931/1975. *The Labor of Women in the Production of Cotton.* Chicago: University of Chicago Libraries. Reprint.

Angelou, Maya. 1969. *I Know Why the Caged Bird Sings.* New York: Random House.

Ankarloo, Bengt. 1979. "Agriculture and Women's Work: Directions of Change in the West, 1700–1900." *Journal of Family History* 4:111–20.

Atack, Jeremy, and Fred Bateman. 1987. *To Their Own Soil: Agriculture in the Antebellum North.* Ames: Iowa State Univ. Press.

Atkeson, Mary Meek. 1929. "Women in Farm Life and Rural Economy." *Annals of the American Academy of Political and Social Science* 143:188–94.

Barlett, Peggy F. 1988. "Motivations of Part-Time Farmers." Paper Presented at the Symposium of Multiple Job-Holding Among Farm Families in North America. Rosslyn, Va., May 16–17.

Barron, Hal. S. 1984. *Those Who Stayed Behind: Rural Society in Nineteenth-Century New England.* Cambridge: Harvard Univ. Press.

Bateman, Fred. 1978. "The 'Marketable Surplus' in Northern Dairy Farming: New Evidence by Size of Farm in 1860." *Agricultural History* 52(July):145–63.

______. 1983. "Research Developments in American Agriculture since 1960: The Northern Farm Economy." *Agricultural History Review* 31:132–48.

Bateman, Fred, and Thomas Weiss. 1981. *A Deplorable Scarcity: The Failure of Industrialization in the Slave Economy.* Chapel Hill: Univ. of North Carolina Press.

Billington, Ray Allen. 1964. *Westward Expansion: A History of the American Frontier.* New York: Macmillan.

Blumberg, Dorothy Rose. 1978. "Mary Elizabeth Lease, Populist Orator: A Profile." *Kansas History* 1, no. 1(Spring):3–15.

Bogue, Allan G. 1963. *From Prairie to Corn Belt: Farming on the Illinois and Iowa Prairies in the Nineteenth Century.* Chicago: Univ. of Chicago Press.

Bonner, James C. 1956. "Plantation Experiences of a New York Woman." *North Carolina Historical Review* 33, nos. 2, 3(July, Oct.):384–412, 520–46.

______. 1964. *A History of Georgia Agriculture, 1732–1860.* Athens: Univ. of Georgia Press.

Brown, Minnie Miller 1976. "Black Women in Agriculture." *Agricultural History* 50, no. 1(Jan.):202–12.

Calkins, Earnest Elmo. 1939. *They Broke the Prairie: Being Some Account of the Settlement of the Upper Mississippi Valley by Religious and Educational Pioneers, Told in Terms of One City, Galesburg, and of One College, Knox.* New York: Charles Scribner's.

Carrigan, Jo Ann. 1962. "Nineteenth-Century Rural Self-Sufficiency: A Planter's and Housewife's 'Do-it-Yourself' Encyclopedia." *Arkansas Historical Quarterly* 21, no. 2(Summer):132–45.

Chirot, Daniel. 1975. "The Growth of the Market and Service Labor Systems in Agriculture." *Journal of Social History* 8(Winter):76–80.

Clark, Christopher. 1979. "Household Economy, Market Exchange and the Rise of Capitalism in the Connecticut Valley, 1800–1860." *Journal of Social History* 13, no. 2(Winter):169–89.

Clifton, James M. ed. 1978. *Life and Labor on Argyle Island: Letters and Documents of a Savannah River Plantation, 1833–1867.* Savannah, Ga.: Beehive Press.

Cody, Cheryll Ann. 1982. "Slave Demography and Family Formation: A Community Study of the Ball Family Plantations, 1720–1896." Ph.D. diss., University of Minnesota.

Danhof, Clarence H. 1969. *Change in Agriculture: The Northern United States, 1820–1870.* Cambridge: Harvard Univ. Press.

Davis, Charles S. 1939. *The Cotton Kingdom in Alabama.* Montgomery: Alabama Department of Archives and History.

Delgado, Jeanne Hunnicutt, ed. 1973. "Nellie Kedzie Jones's Advice to Farm Women." *Wisconsin Magazine of History* 57, no. 1(Autumn):3–27.

Drake, Daniel. 1840/1948. *Pioneer Life in Kentucky, 1785–1800,* ed. Emmet Field Horine. New York: Henry Schuman. Reprint.

Dublin, Thomas 1979. *Women and Work: The Transformation of Work and Community in Lowell, Massachusetts, 1826–1860.* New York: Columbia Univ. Press.

______. 1985. "Women and Outwork in a Nineteenth-Century New England Town: Fitzwilliam, New Hampshire 1830–1850." In *The Countryside in the Age of Capitalist Transformation: Essays in the Social History of Rural America,* ed. Steven Hahn and Jonathan Prude, 51–70. Chapel Hill: Univ. of North Carolina Press.

Easterby, J. H. ed. 1945. *The South Carolina Rice Plantation as Revealed in the Papers of R. W. Allston.* Chicago: Univ. of Chicago Press.

Easterlin, Richard. 1976a. "Population Change and Farm Settlement in the Northern United States." *Journal of Economic History* 36(Mar.):47–75.

______. 1976b. "Factors in the Decline of Farm Fertility in the American North." *Journal of American History* 63(Dec.):600–14.

Faragher, John Mack. 1979. *Women and Men on the Overland Trail.* New Haven: Yale Univ. Press.

______. 1981. "History From the Inside-Out: Writing the History of Women in Rural America." *American Quarterly* 33:537–57.

______. 1985. "Open-Country Community: Sugar Creek, Illinois, 1820–1850." In *The Countryside in the Age of Capitalist Transformation: Essays in the Social History of Rural America,* ed. Steven Hahn and John Prude, 233–58. Chapel Hill: Univ. of North Carolina Press.

______. 1986. *Sugar Creek: Life on the Illinois Prairie.* New Haven: Yale Univ. Press.

Ferleger, Louis. 1984. "Self Sufficiency and Rural Life on Southern Farms." *Agricultural History* 58, no. 3(July):314–29.

______. 1990. *Tools and Time: Southern Farmers after Reconstruction.* (Forthcoming).

Fink, Deborah. 1984. "Rural Women and Family in Iowa." *International Journal of Women's Studies* 7:57–69.

______. 1986. *Open Country, Iowa: Rural Women, Tradition and Change.* Albany: State Univ. of New York Press.

Fisk University. 1945/1968. *Unwritten History of Slavery: Autobiographical Accounts of Negro Ex-Slaves.* Washington, D.C.: Howard Univ. Press. Reprint.

Fite, Gilbert C. 1966. *The Farmer's Frontier, 1865–1900.* New York: Holt, Rinehart and Winston.

Flora, Cornelia Butler. 1981. "Farm Women, Farming Systems and Agricultural Structure: Suggestions for Scholarship." *The Rural Sociologist* 1:383–86.

Ford, Lacy K., Jr. 1983. "Social Origins of a New South Carolina: The Upcountry in the Nineteenth Century." Ph.D. diss., University of South Carolina.

Fox-Genovese, Elizabeth. 1981. "Women and Work: A Preliminary Contribution to the Terms of the Discussion." Fernand Braudel Center, *Working Papers,* Seminar I: 'Group Formation and Group Conflict in the Historical Development of the Modern World System.' Mar.

______. 1983. "Women and Work." In *French Women in the Age of Enlightenment,* ed. Samia Spencer, 111–27. Bloomington: Indiana Univ. Press.

______. 1988. *Within the Plantation Household: Black and White Women of the Old South.* Chapel Hill: Univ. of North Carolina Press.

Fox-Genovese, Elizabeth, and Eugene D. Genovese. 1983. *Fruits of Merchant Capital: Slavery and Bourgeois Property in the Rise and Expansion of Capitalism.* New York: Oxford Univ. Press.

Frankel, Noralee. 1982. "Workers, Wives, and Mothers: Black Women in Mississippi, 1860–1879. Ph.D. diss., George Washington University.

Friedmann, Harriet. 1978. "World Market, State and Family Farm: Social Bases of Household Production in the Era of Wage Labor." *Comparative Studies in Society and History* 20:545–86.

______. 1980. "Household Production and the National Economy: Concepts for the Analysis of Agrarian Formations." *Journal of Peasant Studies* 7, no. 2(Jan.)158–84.

Gallman, Robert. 1970. "Sufficiency of the Cotton Economy of the Antebellum South." *Agricultural History* 44(Jan.):5–24.

Gates, Paul W. 1960. *The Farmer's Age: Agriculture, 1815–1860.* New York: Holt, Rinehard and Winston.

______. 1965. *Agriculture and the Civil War.* New York: Alfred A. Knopf.

______. 1972. "Problems of Agricultural History, 1790–1840." *Agricultural History* 46, no. 1(Jan.):33–58.

______. 1973. *Landlords and Tenants on the Prairie Frontier: Studies in American Land Policy.* Ithaca: Cornell Univ. Press.

Genovese, Eugene D. 1965. *The Political Economy of Slavery: Studies in the Economy and Society of the Slave South.* New York: Pantheon Books.

Gibson, Campbell. 1987. "The Population in Large Urban Concentrations in the United States, 1790–1980: A Delineation Using Highly Urbanized Counties." *Demography* 24, no. 4(Nov.):601–14.

Glasgow, Ellen. 1925/1957. *Barren Ground.* New York: Harcourt, Brace. Reprint.

Gray, Lewis Cecil. 1933/1958. *History of Agriculture in the Southern United States to 1860.* 2 vols. Gloucester, Mass.: Peter Smith.

Greenlee, James Hervey. Diary. Southern Historical Collection, Chapel Hill, N.C.

Gross, Robert A. 1982. "Culture and Cultivation: Agriculture and Society in Thoreau's Concord." *Journal of American History* 69, no. 1(June):42–55.

Hahn, Steven. 1983. *The Roots of Southern Populism: Yeoman Farmers and the Transformation of the Georgia Upcountry, 1850–1890.* New York: Oxford Univ. Press

______. 1985. "The 'Unmaking' of the Southern Yeomanry: The Transformation of the Georgia Upcountry, 1860–1890." In *The Countryside in the Age of Capitalist Transformation: Essays in the Social History of Rural America,* ed. Steven Hahn and Jonathan Prude, 179–203. Chapel Hill: Univ. of North Carolina Press.

Hahn, Steven, and Jonathan Prude, eds. 1985. *The Countryside in the Age of Capitalist Transformation: Essays in the Social History of Rural America.* Chapel Hill: Univ. of North Carolina Press.

Hall, Jacquelyn Dowd et al. 1988. *Like A Family.* Chapel Hill: Univ. of North Carolina Press.

Hampsten, Elizabeth. 1982. *Read This Only to Yourself: The Private Writings of Midwestern Women, 1880–1910.* Bloomington: Indiana Univ. Press.

Hargreaves, Mary W. M. 1973. "Homesteading and Homemaking on the Plains: A Review," *Agricultural History* 47, no. 2(Apr.):156–63.

______. 1976. "Women in the Agricultural Settlement of the Northern Plains." *Agricultural History* 50, no. 1(Jan.):179–89.

Harris, Katherine. 1987. "Homesteading in Northeastern Colorado, 1873–1920." In *The Women's West,* ed. Susan Armitage and Elizabeth Jameson, 165–78. Norman: Univ. of Oklahoma Press.

Hedley, Max. 1981. "Relations of Production of the 'Family Farm': Canadian Prairies." *Journal of Peasant Studies* 9:71–85.

Herndon, G. Melvin. 1982. "Elliott L. Story: A Small Farmer's Struggle for Economic Survival in Antebellum Virginia." *Agricultural History* 56, no. 3(July):516–27.

Hilts, Victor L., and Patricia A. Hilts. 1982. "Not for Pioneers Only: The Story of Wisconsin's Spinning Wheels." *Wisconsin Magazine of History* 66, no. 1(Autumn):3–24.

Jameson, Elizabeth. 1987. "Women as Workers, Women as Civilizers: True Womanhood in the American West." In *The Women's West,* ed. Susan Armitage and Elizabeth Jameson, 145–64. Norman: Univ. of Oklahoma Press.

Jeffrey, Julie Roy. 1979. *Frontier Women: The Trans-Mississippi West, 1840–1880.* New York: Hill and Wang.

Jensen, Joan. 1981. *With These Hands: Women Working on the Land.* Old Westbury: Feminist Press.

______. 1985. "The Role of Farm Women in American History: Areas for Additional Research." *Agriculture and Human Values* 2, no. 1:13–17.

______. 1986. *Loosening the Bonds: Mid-Atlantic Farm Women, 1750–1850.* New Haven: Yale Univ. Press.

______. 1988. "Butter-Making and Economic Development in Mid-Atlantic American from 1750–1850." *Signs* 13, no. 3(Summer):813–29.

Jones, Jacqueline. 1985. *Labor of Love, Labor of Sorrow: Black Women, Work, and the Family from Slavery to the Present.* New York: Basic Books.

Joyner, Charles. 1984. *Down by the Riverside: A South Carolina Slave Community.* Urbana: Univ. of Illinois press.

Kaufman, Allen. 1982. *Capitalism, Slavery, and Republican Values: Antebellum Political Economists, 1819–1848.* Austin: Univ. of Texas Press.

Kaufman, Polly Welts. 1984. *Women Teachers on the Frontier.* New Haven: Yale Univ. Press.

Kulikoff, Allan. 1990. *The Rise and Destruction of the American Yeoman Classes.* Forthcoming. Working manuscript.

Lemon, James T., 1967. "Household Consumption in Eighteenth-Century American and Its Relationship to Production and Trade: The Situation Among Farmers in Southeastern Pennsylvania." *Agricultural History* 41, no. 1(Jan.):59–70.

______. 1987. "Agriculture and Society in Early America. *Agricultural History Review* 35, Pt. 1:76–94.

Lindstrom, Diane. 1978. *Economic Change in the Philadelphia Region, 1810–1850.* New York: Columbia Univ. Press.

Litwack, Leon. 1979. *Been in the Storm So Long: The Aftermath of Slavery.* New York: Alfred A. Knopf.

Loehr, Rodney C. 1952. "Self-Sufficiency on the Farm." *Agricultural History* 26, no. 2(Apr.):37–41.

McDaniel, George W. 1982. *Hearth and Home: Preserving a People's Culture.* Philadelphia: Temple Univ. Press.

Mandle, Jay R. 1978. *The Roots of Black Poverty: The Southern Plantation Economy after the Civil War.* Durham, N.C.: Duke Univ. Press.

Marti, Donald B. 1982. "Woman's Work in the Grange: Mary Ann Mayo of Michigan, 1882–1903." *Agricultural History* 56, no. 2(Apr.):439–52.

Massenburg, Nicolas Bryor. Farm Book. Southern Historical Collection, Chapel Hill, N. C.

Moody, V. Alton. 1924. "Slavery on Louisiana Sugar Plantations." *Louisiana Historical Quarterly* 7, no. 2(Apr.):190–301.

Moragne, Mary E. 1951 *The Neglected Thread: A Journal of the Calhoun Country, 1836–42,* ed. Delle Mullen Craven Columbia: Univ. of South Carolina Press.

Myres, Sandra L. 1982. *Westering Women and the Frontier Experience, 1800–1925.* Albuquerque: Univ. New Mexico Press.

Nobles, Gregory. 1984. "Commerce and Community: A Case Study of the Rural Broom-making Business in Antebellum Massachusetts. *Journal of the Early Republic* 4, no. 3:287–308.

Northrup, Solomon. 1854/1970. *Twelve Years a Slave,* ed. Philip S. Foner. New York: Peter Smith. Reprint.

Olmsted, Frederick Law. 1970. *A Journey in the Back Country, 1853–1854,* ed. Clement Eaton. New York: B. Franklin. Reprint.

———. 1981. *The Papers of Frederick Law Olmsted.* vol. 2, *Slavery and the South, 1852–1857,* ed. Charles E. Beveridge and Charles Capen McLaughlin. Baltimore, Md.: Johns Hopkins Univ. Press.

Otto, John Solomon. 1981. "Slaveholding General Farmers in a 'Cotton County'." *Agricultural History* 55, no. 2(Apr.):167–78.

Papanek, Hanna. 1979. "Family Status Production: The 'Work' and 'Non-Work' of Women." *Signs* 4:775–81.

Pfeffer, Max J. 1983. "Social Origins of Three Systems of Farm Productions in the United States." *Rural Sociology* 48, no. 4:540–62.

Pred, Allan. 1973. *Urban Growth and the Circulation of Information: The United States System of Cities, 1790–1840.* Cambridge: Harvard Univ. Press.

———. 1980. *Urban Growth and City-Systems in the United States, 1840–1860.* Cambridge: Harvard Univ. Press.

———. 1985. "Town-Factory Conflict in Antebellum Rural Massachusetts." In *The Countryside in the Age of Capitalist Transformation: Essays in the Social History of Rural America,* ed. Steven Hahn and Jonathan Prude, 179–203. Chapel Hill: Univ. of North Carolina Press.

Prude, Jonathan. 1983. *The Coming of the Industrial Order: Town and Factory Life in Rural Massachusetts, 1810–1860.* Cambridge: Cambridge Univ. Press.

Rawick, George P., ed. 1972. *The American Slave: A Composite Autobiography.* 19 vols. Westport, Conn.: Greenwood Press.

———. 1977. *The American Slave: A Composite Autobiography. Supplement.* 12 vols. Westport, Conn.: Greenwood Press.

Riley, Glenda 1981. *Frontierswoman: The Iowa Experience.* Ames: Iowa State Univ. Press.

Robinson, Armstead Louis. 1976. "Day of Jubilo: Civil War and the Demise of Slavery in the Mississippi Valley, 1861–1865." Ph.D. diss., University of Rochester.

Robinson, Sara T. L. 1857. *Kansas; Its Interior and Exterior Life. A Full View of Its Settlement, Political History, Social Life, Climate, Soil, Productions, Scenery, Etc.* 7th ed. Boston: Crosby, Nichols.

Rothenberg, Winifred B. 1981. "The Market and Massachusetts Farmers, 1750–1855." *Journal of Economic History* 41, no. 2(June):283–314.

———. 1985. "The Emergence of a Capital Market in Rural Massachusetts, 1730–1838." *Journal of Economic History* 45, no. 4(Dec.):781–808.

Sachs, Carolyn E. 1983. *The Invisible Farmers: Women in Agricultural Production.* Totowa, N.J.: Rowman and Allanheld.

Schelbecker, John T. 1975. *Whereby We Thrive: A History of American Farming, 1607–1977.* Ames: Iowa State Univ. Press.

Schlissel, Lillina. 1982. *Women's Diaries of the Westward Journey.* New York: Schocken Books.

Seaborn Hawks Plantation Book. Georgia Department of Archives and History, Atlanta, Georgia.

Segalen, Martine. 1980. *Mari et Femme dans la Societe Paysanne.* Paris: Flammarion.

Sellers, James B. 1950. *Slavery in Alabama.* 2d ed. University: Univ. of Alabama Press.

Shannon, Fred A. 1945. *The Farmer's Last Frontier: Agriculture, 1860–1897.* New York: Farrar and Rinehart.

Smith, Jean Wheeler. 1975. "Frankie Mae." In *Black-Eyed Susans: Classic Stories By and About Black Women,* ed. Mary Helen Washington, 3–22. Garden City, N.Y.: Anchor Books.

Snell, Joseph W., ed. 1971. "Roughing It on Her Kansas Claim: The Diary of Abbie Bright, 1870–71." *Kansas Historical Quarterly* 37, no. 3(Autumn): 233–68, and no. 4(Winter):394–28.

Sokolow, Jayme A., and Mary Ann Lamanna. 1984. "Women and Utopia: The Woman's Commonwealth of Belton, Texas." *Southwestern Historical Quarterly,* 87, no. 4 (April):371–92.

Springer, Marlene, and Haskell Springer, eds. 1986. *Plains Woman: The Diary of Martha Farnsworth, 1882–1922.* Bloomington: Indiana Univ. Press.

Stewart, (Mrs.) Elinore Pruitt. 1914. *Letters of a Woman Homesteader.* Boston: Houghton Mifflin.

Stratton, Joanna L. 1981. *Pioneer Women: Voices from the Kansas Frontier.* New York: Simon and Schuster.

Sydnor, Charles S. 1933/1965. *Slavery in Mississippi.* Gloucester, Mass.: Peter Smith. Reprint.

Tryon, Rolla. 1917/1966. *Household Manufactures in the United States, 1640–1860.* New York: Augustus Kelley.

Ulrich, Laurel Thatcher. 1982. *Good Wives: Image and Reality in the Lives of Women in Northern New England, 1650–1750.* New York: Oxford Univ. Press.

Weiman, David. N.d. "Families, Farms and Rural Society in Pre-Industrial America: An Interpretation." Unpublished paper.

______. 1985. "Slavery, Plantation Settlement, and Regional Development in the Antebellum Cotton South." Paper delivered at the annual meeting of the Economic History Association, New York, September.

______. 1987. "Farmers and the Market in Antebellum America: A View from the Georgia Upcountry." *Journal of Economic History* 47, no. 3(Sept.):627–47.

Wilder, Laura Ingalls. 1941. *Little Town on the Prairie.* New York: Harper and Row.

______. 1943. *These Happy Golden Years.* New York: Harper and Row.

Williamson, Jeffrey G. 1971. "Urbanization in the American Northeast, 1820–1870." In *The Reinterpretation of American History,* ed. Robert W. Fogel and Stanley L. Engerman, 426–36. New York: Harper and Row.

Wilson, W. Emerson, ed. 1976. *Plantation Life at Rose Hill: The Diaries of Martha Ogle Forman, 1814–1845.* Wilmington, Del.: Scholarly Resources.

Woodman, Harold D. 1968. *King Cotton and His Retainers: Financing and Marketing the Cotton Crop of the South, 1800–1925.* Lexington: Univ. of Kentucky Press.

Wright, Gavin. 1978. *The Political Economy of the Cotton South: Households, Markets, and Wealth in the Nineteenth Century.* New York: Oxford Univ. Press.

______. 1986. *Old South, New South: Revolutions in the Southern Economy since the Civil War.* New York: Oxford Univ. Press.

CHAPTER 12

# Immigrants in Nineteenth-Century Agricultural History

KATHLEEN NEILS CONZEN

THROUGHOUT AMERICAN HISTORY the yeoman farmer has often spoken with a non-English accent. The great rural transformations of the Old World shook loose a vast migration to help cultivate the New. In the eighteenth century German-speaking peasants plowed the rich lands of southeastern Pennsylvania and joined the Scotch-Irish to penetrate the western valleys from New York down through the Carolinas. The lure of American land helped attract the mass immigration of the nineteenth century, and everywhere outside the slaveholding South the farmers' frontier became also an immigrant frontier. By century's end, Asian farmers were tilling the soil of California, and Italians and Poles were moving onto the older farmlands of the East.

It is fair to estimate that something like 10 to 15 percent of all American farms were operated by immigrants in the decades between 1870 and 1920.[1] The immigrant presence was weakest in the South and New England, heaviest in those northern and western states opened for settlement in the era of mass immigration. By 1880 more than half of all farmers in Wisconsin, Minnesota, Dakota Territory, and Utah were foreign born, as were about a quarter to a third of the farmers in eleven other midwestern and western states (Table 12.1). By one calculation, some 363 largely rural counties in twelve midwestern states reached at least 25 percent foreign born at some point between 1850 and 1930 (calculated from Hart 1986, 207). Over time, as the children of immigrant farmers replaced their Anglo-American neighbors, these concentrations could intensify: in North Dakota in 1965, to take an extreme example, only 13.5 percent of all farm operators identified themselves as

TABLE 12.1. Foreign-born Farmers and Planters by State, 1880

| State | Foreign-born Farmers | Foreign-born as % of all Farmers |
|---|---|---|
| United States | 610,180 | 14.4 |
| New England | | |
| Maine | 3,414 | 5.7 |
| New Hampshire | 1,116 | 3.7 |
| Vermont | 3,678 | 10.3 |
| Massachusetts | 4,571 | 11.6 |
| Rhode Island | 443 | 7.0 |
| Connecticut | 3,259 | 12.1 |
| Middle Atlantic | | |
| New York | 42,897 | 17.8 |
| New Jersey | 3,724 | 11.1 |
| Pennsylvania | 16,973 | 8.7 |
| East North Central | | |
| Ohio | 32,822 | 12.6 |
| Indiana | 20,692 | 9.8 |
| Illinois | 68,217 | 24.3 |
| Michigan | 52,206 | 31.2 |
| Wisconsin | 83,719 | 60.5 |
| West North Central | | |
| Minnesota | 63,322 | 65.5 |
| Iowa | 60,839 | 28.7 |
| Missouri | 23,287 | 9.8 |
| Dakota | 13,274 | 58.4 |
| Nebraska | 24,775 | 36.1 |
| Kansas | 28,349 | 19.2 |
| South Atlantic | | |
| Delaware | 334 | 3.8 |
| Maryland | 1,876 | 5.0 |
| District of Columbia | 42 | 12.6 |
| Virgina | 1,115 | 0.9 |
| West Virginia | 1,536 | 2.3 |
| North Carolina | 433 | 0.3 |
| South Carolina | 380 | 0.4 |
| Georgia | 352 | 0.2 |
| Florida | 548 | 2.1 |
| East South Central | | |
| Kentucky | 3,427 | 2.0 |
| Tennessee | 1,222 | 0.8 |
| Alabama | 546 | 0.4 |
| Mississippi | 772 | 0.6 |
| West South Central | | |
| Arkansas | 841 | 0.8 |
| Louisiana | 1,018 | 1.8 |
| Texas | 14,843 | 7.4 |
| Mountain | | |
| Montana | 773 | 32.8 |
| Idaho | 766 | 30.8 |
| Wyoming | 253 | 27.8 |
| Colorado | 1,315 | 20.2 |
| New Mexico | 505 | 7.3 |
| Arizona | 582 | 33.7 |
| Utah | 4,707 | 51.6 |
| Nevada | 845 | 41.6 |
| Pacific | | |
| Washington | 2,345 | 26.0 |
| Oregon | 2,871 | 15.6 |
| California | 14,356 | 33.0 |

Source: Compiled from U.S. Census (1880, 730).

of Anglo-American stock (Sherman 1983, 129–30). Those areas where family farming remains most dominant today are generally those that received the heaviest foreign imprint (Hart 1987, 200; 1986, 207).

Immigrants, to be sure, generally remained more urban in their employment than the national average throughout the period (Table 12.2). Nevertheless, a fifth to a quarter of all immigrant workers were probably employed in agriculture in the middle decades of the nineteenth century, perhaps an eighth by the early twentieth century (Hutchinson 1956, 79, 98, 121, 159). The relative appeal of farming varied by group: close to half of all Scandinavians found their way into agriculture, compared with something like a quarter of the Germans, British, or Canadians. Even among groups with smaller proportions in agriculture, such as the Irish among the earlier immigrants or the Italians or Poles among the later comers, a tenth to a sixth of all immigrants found their way to the land, a proportion that sometimes increased among the second generation. In absolute terms the Germans and (by the early twentieth century) Scandinavians were the most numerous among the foreign-born farmers. Varying regional settlement patterns gave distinctive ethnic character to the different regions; Canadian dominance in New England contrasted with the German tone of the Midwest and the Scandinavian imprint on the Great Plains (Table 12.3).

So dramatic an immigrant presence poses important issues for historians newly anxious to thread the rural back into the tapestry of American history after decades of neglect (Hahn and Prude 1985, 7). The rise of a "new rural history" situated at the center of current historiographical efforts to understand the contours of the transition from the traditional to the modern (Swierenga 1981) is raising anew old questions about the pace and pervasiveness of the rural immigrant assimilation process. The lively debate that has erupted during the last decade over agrarian complicity in capitalist transformation, for example, cannot be fully resolved until the immigrant farmer is taken into account. The central points at issue concern the extent to which farmers resisted or embraced the market, and the degree to which market involvement compelled change in rural society willy-nilly (Hahn and Prude 1985; Rothenberg 1987). If immigrant values could shape a distinctive rural way of life, the same freedom must have existed for their native-born counterparts. Conversely, if distinctive ethnic values survived, they too are part of the inheritance that any assessment of rural patterns and their implications for American historical development must survey.

TABLE 12.2. Relative Immigrant Group Concentration in Agriculture

| Birthplace | Agricultural Employment as Percent of all Employment of Group Members | | |
|---|---|---|---|
| | 1870 1st Generation | 1900 1st Generation | 1900 2nd Generation |
| All U.S. Gainfully Employed | 47.3 | 32.7 | |
| All Immigrants/2nd Generation | 22.9 | 13.1 | 13.5 |
| Germans | 26.8 | 23.9 | 23.4 |
| Irish | 14.6 | 10.9 | 12.5 |
| British | 25.6[a] | 16.5 | 22.1 |
| Scandinavians | 46.0 | 33.8 | 11.8 |
| British Canadians | 25.5 | 17.5 | 24.9 |
| French Canadians | | 10.8 | 17.0 |
| Italians | | 14.4 | 7.7 |
| Poles | | 9.7 | 17.3 |

Source: Calculated from Hutchinson (1956, 79: U.S. Immigration Commission Reports 1:799–800, 830–37).
[a]English and Welsh only.

TABLE 12.3. Nativity of Foreign-born Farm Owners within Each Region, 1920
Farm Owners within Each Group as a Percent of All Foreign-born Farm Owners in Each Region

| Foreign-born Farm Owners | Region | | | | | | | | | Total |
|---|---|---|---|---|---|---|---|---|---|---|
| | NE | MA | SA | ENC | WNC | ESC | WSC | M | P | |
| All Foreign-born | 25,138 | 38,308 | 6,193 | 122,469 | 161,022 | 2,821 | 22,274 | 34,905 | 51,223 | 464,333 |
| Germany | 1,658 | 8,884 | 1,640 | 39,433 | 42,930 | 982 | 8,368 | 4,842 | 8,219 | 116,956 |
| | (6.6%) | (23.2%) | (26.5%) | (32.2%) | (26.7%) | (34.8%) | (37.6%) | (13.9%) | (16.0%) | (25.2%) |
| Ireland | 1,653 | 2,857 | 274 | 2,517 | 3,232 | 163 | 328 | 1,177 | 1,723 | 13,924 |
| | (6.6%) | (7.5%) | (4.4%) | (2.1%) | (2.0%) | (5.8%) | (1.5%) | (3.4%) | (3.4%) | (3.0%) |
| Great Britain | 2,049 | 4,041 | 991 | 6,065 | 5,703 | 281 | 1,049 | 4,958 | 4,774 | 29,911 |
| | (8.2%) | (10.5%) | (16.0%) | (5.0%) | (3.5%) | (10.0%) | (4.7%) | (14.2%) | (9.3%) | (6.4%) |
| Scandinavia | 3,006 | 2,953 | 500 | 27,303 | 67,362 | 327 | 1,517 | 10,470 | 113,744 | 127,182 |
| | (12.0%) | (7.7%) | (8.1%) | (22.3%) | (41.8%) | (11.6%) | (6.8%) | (30.0%) | (26.8%) | (27.4%) |
| Canada | 8,866 | 2,589 | 554 | 14,236 | 7,037 | 104 | 433 | 3,125 | 4,920 | 41,864 |
| | (35.3%) | (6.8%) | (8.9%) | (11.6%) | (4.4%) | (3.7%) | (1.9%) | (9.0%) | (9.6%) | (9.0%) |
| Italy | 1,489 | 3,712 | 272 | 1,018 | 422 | 163 | 1,555 | 1,576 | 3,195 | 13,402 |
| | (5.9%) | (9.7%) | (4.4%) | (0.8%) | (0.3%) | (5.8%) | (7.0%) | (4.5%) | (6.2%) | (2.9%) |
| Poland | 1,340 | 2,342 | 201 | 7,090 | 2,429 | 37 | 672 | 298 | 465 | 14,874 |
| | (5.3%) | (6.1%) | (3.2%) | (5.8%) | (1.5%) | (1.3%) | (3.0%) | (0.9%) | (0.9%) | (3.2%) |
| Austria-Hungary | 1,068 | 3,645 | 531 | 6,618 | 6,431 | 210 | 3,066 | 2,033 | 1,708 | 25,310 |
| | (4.2%) | (9.5%) | (14.1%) | (5.4%) | (4.0%) | (7.4%) | (13.8%) | (5.8%) | (3.3%) | (5.5%) |
| Russia | 1,638 | 2,206 | 188 | 2,926 | 11,809 | 48 | 1,141 | 2,301 | 1,987 | 24,244 |
| | (6.5%) | (5.8%) | (3.0%) | (2.4%) | (7.3%) | (1.7%) | (5.1%) | (6.6%) | (3.9%) | (5.2%) |
| Other Europe | 1,800 | 4,817 | 876 | 14,703 | 12,649 | 472 | 2,789 | 3,320 | 8,072 | 49,498 |
| | (7.2%) | (12.6%) | (14.1%) | (12.0%) | (7.8%) | (16.7%) | (12.5%) | (9.5%) | (15.8%) | (10.7%) |
| Mexico | 2 | 5 | 1 | 34 | 31 | 1 | 1,054 | 394 | 103 | 1,625 |
| | ... | ... | ... | ... | ... | ... | (4.7%) | (1.1%) | (0.2%) | (0.3%) |
| Other Countries | 569 | 257 | 165 | 526 | 967 | 33 | 302 | 411 | 2,313 | 5,543 |
| | (2.3%) | (0.7%) | (2.7%) | (0.4%) | (0.6%) | (1.2%) | (1.4%) | (1.1%) | (4.5%) | (1.2%) |

Source: Calculated from Brunner (1929, 253–59).

Key to Regional Designations: NE = New England; MA = Middle Atlantic; ENC = East North Central; WNC = West North Central; SA = South Atlantic; ESC = East South Central; WSC = West South Central; M = Mountain; P = Pacific.

Percentages do not always add up to 100, owing to rounding errors.

How much difference then did the involvement of immigrants make to the development of American agriculture and the character of rural society? Viewed from the perspective of the immigrants, this means asking whether and under what conditions American agriculture encouraged either adaptation to prevailing norms or the creation and evolution of ethnic variants. Were immigrants significant for their numbers alone, or did Old World customs and values survive to influence the way they lived and farmed in rural America? If so, how and why? What accounts for group variations in farm settlement, and for regional variations in rural immigrant concentration, and were such variations related to variations in cultural persistence? How different was the process of rural immigrant adaptation from the urban immigrant experience? Did ethnic culture condition the involvement of farmers in national political and social movements in any way? Has ethnic difference left any permanent mark on rural America still visible today?

As recently as a decade ago the answers to such questions seemed clear. Most American scholars thought they could discount the rural immigrant presence almost entirely.[2] Although social scientists and historians early called for greater attention to rural immigrants, leading texts in both rural sociology and agricultural history largely ignored the ethnic dimension.[3] Conditioned by Turnerian assumptions of a frontier melting pot, immigration historians posited a rural assimilation process so smooth that it could be safely ignored, and agricultural historians were able to focus on crops, practices, and policies secure in the assumption that all American farmers shared a common social order and set of values. Historians of ethnic groups and local communities continued to publish significant work relating to the rural immigrant experience, to be sure, but most were content simply to document settlement patterns and organizational life without much interpretive focus. At worst filiopietistic, at best containing an implicit argument for an immigrant "contribution" defined by the simple presence of the foreign born in agriculture and the success that they achieved, these historians' work did little to challenge prevailing interpretive conventions.[4]

That challenge would come, however, when the trajectory of a revitalized immigration history finally led back to the countryside. The urban immigrant experience had always been a much more visible and problematic phenomenon, and by the 1970s its study was flourishing under the impact of new social history methods and new national interest in the persistence of ethnicity (Vecoli 1979). As immigration historians extended new questions and methods to an ever-wider range of groups and communities, it was perhaps inevitable that they would reencounter the rural immigrant, particularly among groups like the Scandinavians and Dutch for whom farming was especially significant. There was also a methodological lure. When the computer finally offered historians the possibility of following Frank Thistlethwaite's 1960 injunction to penetrate the "salt-water curtain" and treat emigration as a unified process from point of origin to destination, among the groups for whom this proved most feasible were those who settled in cleanly delineated rural clusters (Thistlethwaite 1960; Swierenga and Stout 1976; Runblom and Norman 1976). Add to these trends a growing interest in the cultural landscape among historical geographers, the methodological and interpretive appeal of new European scholarship on peasant communities, the very belated acceptance of social history approaches among historians of the American West, and renewed public focus on a contemporary "farm problem," and the stage was set for the current reexamination of the role of immigrants in rural America (for example see Swierenga 1981).

To survey this recent work is to find little support for the old simplistic models of rapid assimilation and cultural uniformity. It is clear that the new rural history will have to come to terms with a complex patchwork of ethnic communities whose cultures—emerging, transmuting, and evolving at varying paces under varying circumstances—constituted a distinctive and determining element within the heritage of rural America. This essay makes no pretense of being a complete review of all recent work or all topics relating to the rural immigrant. Rather, it focuses on a selected set of issues in which the new historiographical confluence has significantly extended our range of knowledge and calls into question established assumptions. It concentrates heavily on the Midwest and on Germans and Scandinavians because these have received the most scholarly attention. Synthesis remains a distant goal. Studies separated by academic discipline and dedicated to particular immigrant groups often have little sense of common or comparative questions. Important kinds of evidence remain underused, a basis for comparison in studies of nonimmigrant farming communities is only beginning to appear, and finding have yet to be integrated into the historiographical mainstream. The study of rural immigrant communities has a long way to go before it can match in completeness or sophistication the scholarship on urban ethnicity. Theoretical self-consciousness is virtually nonexistent, and there is as yet no consensus on the implications of the rural immigrant experience for understanding the changes wrought by the expanding market economy.

It is, however, possible to discern the outlines of a new set of generalizations concerning measures and causes of the survival of ethnic differences among rural communities. We can see these outlines emerging in new scholarship dealing with interrelationships among rural immigrant family structure and goals, settlement patterns, agricultural practices, persistence of ethnic communities on the same land, and cultural change. A review of that scholarship makes it possible also to suggest areas that would reward further exploration.

Historians reached their original conclusions regarding the rapidity of rural assimilation in the face of significant evidence to the contrary generated by social scientists whose work they largely ignored. That work retains considerable value for historical scholarship today. Colonial and nineteenth-century Americans, native and immigrant alike, were convinced that national traits determined farming behavior. They drew the contrast most readily, perhaps, between Yankees and Germans: the former speculative, mobile, exploitative of the soil, comfort-loving, profit-oriented; the latter cautious, penurious, sedentary, conserving of the soil but exploitative of family labor. But the repertoire of stock figures could extend as well to the Englishman who farmed by the book, the careless Irish farmer, the short-sighted Scandinavian, and the like (Ross 1914, 52–53, 86, 167; see Conzen 1984, 357–61; Erickson 1964).

The first modern scholars to address the issue posed by these stereotypes were not historians but social scientists—geographers testing theories of "racial" impress on the landscape, sociologists assessing the "success" of the assimilation process, economists searching for models of farm ownership achievement, practical agronomists seeking to explain and influence observed behavior (see Speek 1921; Brunner 1929; Parsons 1948; Marshall 1948). Emil Meynen, for example, a German geographer writing in 1939, analyzed both descriptive historical testimony and surviving artifacts and patterns to give lie to the "mirage" of the melting pot by charting a distinctive German presence in Pennsylvania agriculture (Meynen 1939, 254). The distinctiveness did not lie in crops, tools, or settlement patterns. Meynen pointed out that Germans quickly abandoned accustomed open-field villages for individual farms, influenced in this choice by the prevailing land-sale system, the ready availability of land, and the greater farming efficiency it offered. (Even in Germany, newly cleared lands were usually cultivated in individual farms rather than village fields.) Rather, it was the Pennsylvania German tendency to pass on farms to succeeding generations, their commitment to mixed farming and intensive investment, their concern for maintaining soil quality and self-sufficiency in foodstuffs, their preference for assured income rather than quick profit, their large barns and choice of durable brick or stone rather than wooden houses, and their reliance upon the labor force of a large family that defined a distinctive if gradually lessening ethnic presence. Meynen was led to the study of such phenomena by the German racial ideology of the period, but his analysis was relatively straightforward and found confirmation from methodologically similar studies of German immigrant farmers in other settings and time periods (Gehrke 1935; Kollmorgan 1941, 1942; Cozzens 1943; Gagliardo 1959).

Sociologists in the interwar years were more interested in the process of assimilation than in the ethnic imprint on the landscape in and of itself, but like the geographers they recognized the significance of cultural persistence in areas of concentrated settlement. Rural life, William Carlson Smith noted, "usually tends to conserve and continue immigrant heritages in America." In comparison with urban life, "there are fewer disorganizing contacts . . . and the assimilation process goes on more slowly but nevertheless surely" (1939, 214, 186). This was not the result of rurality per se, but of concentrated settlement patterns. Theirs was essentially a rural variant of Robert Parks's "Americanization via the neighborhood" model: immigrant farmers who settled with others of their own kind assimilated, to be sure, but more gradually than did those who lived outside ethnic clusters; the larger the cluster, the slower the process. "There are foreign provinces in the agricultural sections of the country . . .," Peter Speek noted in a 1921 Carnegie Commission-sponsored study. "Such provinces have become self-sufficient: they have their own towns, their own schools, churches, industries, stores, select local public officials of their own nationality, speak their own tongue, and live according to the traditions and spirit of their home country. These traditions and this spirit are kept alive by their schools, churches, and libraries, and by the absence of any direct contact with American customs and traditions" (130–31). Konrad Bercovici (1925) stressed the physical distinctiveness of such colonies; Edmund deS. Brunner found in them clear differences in such matters as farm ownership rates, intensity of farming, willingness to incur farm debt, and school attendance. In general, he argued, factors like the degree of isolation of a community; its relative homogeneity, size, and age; religious organization; and the attitude of the native population all affected the pace of assimilation. But more than two-thirds of the seventy sample areas he studied seemed on a "well-charted course" to complete assimilation, "making haste slowly" in a process of "self-Americanization" eased by ethnic churches and other organizations (1929, 41–74, 115, 124).[5]

More detailed local studies contributed additional variables to the list of those influencing the pace of assimilation, variables that began to include culture as well as structure. John P. Johansen (1937), for example, explored the link between group settlement goals and the choice of specific locations. Given the role of group colonization and chain migration in South Dakota settlement, he argued, earliest arriving groups were most likely to find a settlement context within which they could achieve the levels of size and concentration necessary to "project" an enduring influence; but equally important was the extent to which a group achieved a unified or divided social organization. Also, some researchers noted that rural assimilation was far more a group process than its individualistic urban counterpart, and studies of assimilation as measured by farming practices, family patterns, and organizational involvement found different rates of change within similar structural contexts that seemed explicable only in terms of differing group values. These studies tended to place German and Czech communities on the conservative end of a spectrum that led through the more rapidly assimilating Scandinavians to generally individualistic Irish and British (Whetten and Green 1942; Marshall 1948, 1950; Munch 1949; Deininger 1958; Skrabanek 1949, 1951; Hill and Christensen 1942). Students of land-tenure processes were also beginning to realize the extent to which ethnic family values influenced the achievement of farm ownership (Wehrwein 1932; Hollingshead 1938; Parsons and Waples 1945; Hoffman 1949; Pedersen 1950; Deininger and Marshall 1955).

The volume of social science work on rural immigrants was not great, nor was it very central to the parent disciplines that generated it, and there was little effort to test regional variations or to ground it in knowledge of group cultures or historical processes. But its verdict by the 1950s was clear. A range of studies suggested that concentrated ethnic settlements had been created and persisted in the countryside, that they fostered distinctive values evident in areas of life ranging from ways of farming to family behavior and social cooperation, and that the content of these values and the pace of their acculturation varied not only with the structural context of a given community but with the inherited culture of the group in ways that bore some relationship to earlier stereotypes. Few social scientists would retain much interest in the topic in the ensuing decades, but a problem had been defined and a historical research agenda had been established.

Among historians, however, not much attention was paid. It is true that as early as 1939 one historian (Shryock) was willing to derive from this work a significant challenge to environmental interpretations of American agricultural history. But environmentalism was too pervasive among the few historians with any interest in rural immigrants, the need to assert the dominance of assimilative forces too strong. The conventional case for the rapid assimilation of rural immigrants originated among historians as a matter of Turnerian faith. When Frederick Jackson Turner first called historians' attention to the social order of rural America, it was not ethnic differences that he stressed, but "the free lands that made the process of absorption easy." If the frontier stripped native-born Americans of their cultural baggage to leave them in democratic equality, could that all-powerful environment do any less for the immigrant? "In the crucible of the frontier the immigrants were Americanized, liberated, and fused into a mixed race. . . ." (Turner 1901; Turner, 1920, 51).

Though Turner's students pioneered techniques for mapping and analyzing ethnic settlement patterns on the frontier, and like the master frequently had personal acquaintance with rural ethnic life, they remained faithful to his interpretive dicta (see Everest 1892; [Everest] Levi 1898; Luebke 1983, 388–93). Thus Marcus Lee Hansen preserved Turner's Americanizing frontier by effectively denying the immigrant presence on it, compressing the frontier period to argue that ethnic colonies were the product not of frontiersmen but of later fillers-in (1940, 60–76). Even more palpable are Joseph Schafer's (1922/23, 1927, 1937) mental contortions in his pioneering census-based studies of frontier Wisconsin. He documents the establishment of large areas of German settlement in the state; he shows that German communities persisted and even expanded over time in contrast to higher levels of mobility among Yankees and Irish; he argues as did the social scientists that the influence of ethnicity could be seen in everything from land selection, pace of clearance, and crop mix to housing choice and investment strategies. Yet at the same time his general survey of American agriculture asserts that there was little resistance to Americanization except in "massed centers"; that competition for land prevented much "massing"; and that distinctiveness lapsed after a single generation (Schafer 1936, 216, 215, 30; Schafer 1927, 84).

Schafer's innovative use of census data provided the next generation of historians, now equipped with increasingly sophisticated data-processing technology, with a more rigorous way to test Turner's assertion of the assimilative role of the frontier. Most readily quantified were the solid measures of farm inputs, output, and persistence that could be derived from the population and agriculture schedules of the manuscript federal censuses. Thus when Merle Curti and his students restated Turner's arguments in testable form, it was to these measures that they turned, looking for and finding evidence of Americanization in improving economic status and converging patterns of farming. Even the evidence they found of social segregation and minimal intermarriage could be interpreted as indicating "conditions favorable to the intellectual as well as the general acculturation of the foreign-born. Here they faced the same challenge and the same general range of problems as the native born. The came in, not as rivals and competitors, but as people needed to help build up the country. The very fact that the country was thinly settled . . . tended to minimize friction. They were able to have their own churches and to get together socially: They had time to get their bearings before being drawn into the thick of American life" (Curti et al. 1959, 138).

Curti still interpreted his findings as confirming not only the assimilative but also the democratic character of the frontier. Subsequent quantitative studies grew increasingly less comfortable with the whole concept of frontier democracy, but remained no less environmental and assimilative in emphasis. Regardless of setting, students of frontier farming could find little substantial difference attributable to ethnicity alone in crop production, farming practices, types of tenure, or rates of persistence and success. Whether in clustered enclaves or scattered among settlers of different ethnic backgrounds, immigrants responded equally with their native-born counterparts to the imperatives of climate, land survey, market, and technology (Bogue 1968; M. Conzen 1971; Cogswell 1975; Winters 1978; Swierenga 1980, 332–38). "If there ever was an American melting pot, a concept modern urban-based sociological studies increasingly question," the most recent social history of rural

America asserts, "the place to look might be the American countryside. . . . a beyond the melting pot hypothesis is more difficult to apply to rural America: assimilation and acculturation, given the commonalities of the groups involved, proceeded more swiftly and surely than in the nation's cities" (Shover 1976, 48).

In effect, the Turnerian trail had taken historians of rural society to the assimilative, consensual view of American life that dominated contemporaneous immigration historiography. The implicit environmental explanatory factor was now not so much Turner's free land as Oscar Handlin's modernizing industrial economy, the outcome more readily to be viewed as painful deracination than optimistic progress. Handlin's was a joyless model of peasant farmers uprooted from a close village community and forced to subsist as individuals in a harsh physical environment and market-dominated economy in which old habits and skills were useless. "Wrapped up in the unfamiliar landscapes of prairie distance or forest solitude, the peasants found nowhere an equivalent of the village, nowhere the basis for re-establishing the solidarity of the old communal life." Except in the "occasional" closely-knit township, sparseness of settlement insured intermarriage and a separation of children from the ethnic world of their parents even more rapid and "decisive" than in the city with its rich associational life; success could be achieved only at the expense of grinding self-exploitation and complete cultural transformation (Handlin 1951, 83–105, 257).

Historians and social scientists seemed to be looking at two different rural worlds. To some extent, theirs was a difference of opinion on whether the glass was half empty or half full. Social scientists were struck by the extent of remaining ethnic difference; historians, more aware of Old World backgrounds, were impressed by the cultural distance immigrants and their children had traversed. But both were generalizing with too little in the way of real data or carefully articulated theory, and both had interpretive frameworks that to some extent predetermined the answers they found. Nor were data and settings fully comparable. Most of the historical studies were set in areas of some ethnic mix and concentrated on measures of farming behavior readily derived from the manuscript census, while most of the social science studies that found greater ethnic difference examined a wider range of behavior in more ethnically homogeneous settings.

Recent historical scholarship has begun to find its way out of this impasse by shifting its focus from measures of outcome to analyses of process. Rather than searching for differences from American models it traces evolution from homeland ones. Like immigration historians more generally (Vecoli 1979; Higham 1982; Bodnar 1985), students of immigrants in agriculture are no longer willing to see immigrants as passive victims or beneficiaries of external forces nor content to disregard evidence for enduring ethnic pluralism. Building on the established historiography, scholars have continued their efforts to delineate with greater sophistication areas of rural ethnic difference or assimilation; setting the problem of rural immigrant adaptation within the context of the more general transformation of traditional rural societies, they have extended their research into areas like familial behavior and religion; and they have successfully devised methods for tracing immigrants from origin to destination. Such issues were first joined directly in an important series of detailed settlement area studies by historical geographers that provided new hard evidence and explanatory models for the persistence of ethnic difference, as well as systematic explorations of intergroup differences. As historians have moved in similar directions, belatedly joined by rural sociologists, a body of new work sufficient to support generalization is becoming available.[6]

First of all, this work confirms the older findings that most groups quickly accepted the logic of American settlement systems, crops, and farming methods. Only in isolated instances where religious communitarianism was exceptionally strong did immigrants run against the logic of the land survey system and the efficiencies of the isolated farmstead to form common-field village settlements (see Bennett 1967; Barthel 1984; Johnson 1984).[7] Seldom, whether in Texas or Missouri, Nebraska, Kansas, Minnesota or Wisconsin, did they fail to conform quickly to local cropping and livestock patterns; immigrants who migrated to achieve economic security for their families found little reason to challenge the imperatives of environment and market (Jordan 1966; Kamphoefner 1987; Baltensperger 1980; Flora and Stitz 1985; McQuillan 1975; Gjerde 1985; Ostergren 1979). There is also evidence, however, for a pattern that Terry Jordan in his pathbreaking study of German Texas termed "cultural rebound": the partial revival of some old world trait (such as half-timbered buildings, favorite food crops) after a period of dormancy in the initial frontier stage. Immigrants learned new patterns and assessed the value of old from guidebooks and local advice, from working for farmers already in the area, from direct observation and market experience; after surviving the hardships of the first years, they would then have the luxury of adding Old World crops or techniques to their repertoire should they so choose (Jordan 1966, 195–201; Baltensperger 1980, 187; Gjerde 1985, 175–92).

There were instances, of course, where groups did carve out ethnically distinctive niches for themselves within the agricultural sector. These niches, however, often had less to do with transplanted Old World agricultural habits than with New World dilemmas or opportunities. A group sufficiently distinctive in its demographic structure or restricted in its range of options by the force of prejudice or its own goals, might find a place for itself only within a limited range of local agricultural options. Sucheng Chan's meticulous reconstruction (1986) of Chinese agricultural patterns in California, for example, documents Chinese dominance in truck gardening and clarifies the logic of the symbiotic relationship that they developed with white landowners as the rich Sacramento Delta was drained and brought into production. Well known also is the Basque specialization in sheepherding (Douglas and Bilbao 1975). The emergence over time of other local ethnic specializations, such as those of the Michigan Dutch in vegetable farming or of Wisconsin Norwegians in tobacco growing, can be similarly explained by circumstances that made a given type of agriculture particularly suitable to the needs of a particular group; thus many German Catholics found in dairy farming by the late nineteenth century a stable, labor-intensive enterprise well-suited to their conservative values and large family labor pools (Bjorkland 1964; Strickon 1979; Salamon 1985; Conzen 1985b).

But if, unlike such exceptional cases, most immigrant farmers quickly adopted farming practices little different from local norms, it is also clear that the social scientists were correct in noting that ethnic differences could remain evident far longer in other areas of farm life. Thus Jordan (1966, 1980) documented subtle differences from non-Germans in the greater intensity, productivity, and locational stability of German farmers, in their higher rates of land ownership, and in the layout and architecture of their communities. Russell Gerlach's survey of present-day German settlement areas in the Missouri Ozarks (1977) arrived at similar conclusions. Missouri Germans formed settlements that expanded over time and were characterized by distinctive landscape features like street villages, regular farmstead distribution, and preference for certain house types. Both farmsteads and farming patterns indicate a greater willingness to engage in substantial long-term investment, more intensive farming, and greater stability on the land. Also, German communities are more church centered and exhibit patterns of sociability different from those of non-Germans. Nebraska evidence likewise shows German and particularly Russian-German farmers distinguishing themselves by tenure patterns and scale of operations, while in frontier Kansas, German and Russian-German ethnicity was found to be associated with farm persistence through family size and reliance on family labor as well as crop choices and subsistence activities (Flora and Stitz 1985; McQuillan 1978). In Illinois today, outside of as well as within ethnically defined communities, ethnic differences in family farm persistence, landownership trends, and farm diversification remain evident (Salamon 1984, 1985; Foster et al. 1987).

In effect, scholars have explored the association of ethnic background with measurable differences in farm operation, long-term investment strategies, and persistence of the family on the land. Ethnicity in this context is indexed by a cluster of traits; assimilation becomes the reduction of observable differences in such traits. By these measures, it would seem, the most notable areas of differences have been found to lie not in the quotidian details of farm operation but in the strategies for familial security and persistence or for individual profit maximization: not in farm patterns, but in the reasons why families farm. To achieve their goals, immigrants would readily adopt some aspects of American agriculture, like cropping patterns and machinery. Other adaptations, like conformity to the existing settlement system, would be virtually forced upon them. But they had greater degrees of freedom in other aspects of life, and so choice of neighbors, investment strategies, family patterns, or religious behavior could both reflect and reinforce accustomed patterns of living—an ethnic culture—for those who so chose.

In his important study of several Norwegian communities in Wisconsin and Minnesota, however, Jon Gjerde (1985) has pointed out that these areas of life were themselves altered as a consequence of the dialectic between imported values and American circumstance. Thus the availability of land in America enabled Norwegian youth to retain the custom of marrying only when a livelihood was available, yet it also changed marriages. New World Norwegians married earlier, produced more children, and often migrated away from their own parents. Land availability also kept farm labor scarce and expensive, forcing immigrants to rely far more on the labor of their now larger families than they had in Norway. At the same time American cropping patterns, by expanding traditionally male jobs but creating little demand for the types of farm work normally carried out by women, supported new patterns of female domesticity focused on family rearing and the bourgeois life-style made possible by new prosperity. Those bourgeois values were ones that Norwegian ministers had been urging even in Norway, but their adoption stemmed in large part from the new divisions of labor and new ease of livelihoods that removed the occasions for sexual license that had been part of traditional culture. Thus in Gjerde's schema, the encounter with American opportunity by immigrants who had migrated to retain an accustomed way of life resulted in self-reinforcing changes in farming patterns, mobility, gender division of labor, family size, and sexual mores. The trajectory remained essentially Norwegian, but the culture it promoted increasingly resembled that of their American neighbors.

The long-term persistence of such ethnic trajectories was in the most immediate sense dependent upon the intergenerational persistence of farm families in a given area. Persistence is not only an indication of ethnic difference but its precondition: unless families remain on the land or are replaced by others of similar background, by definition the ethnic community will disappear. A recent Wisconsin study, finding that 47 to 72 percent of farm transfers in sample areas were among family members, concluded that "contemporary rural ethnic groups persist largely through the forces of inertia" (Vogeler 1975, 12). Farm families are "at risk" of nonpersistence when they fail to achieve ownership, when they carry too much debt, when members are responsive to the lure of better livings elsewhere, or when they simply fail to have enough children to guarantee an heir (McQuillan 1978). The patterns of behavior associated with German farmers, to take one example, seemed designed to avoid such risks. Another moment of risk is the point when the farm must be transferred from one generation to the next. Studies of German farming communities are beginning to suggest that transplanted habits of cooperative father-son farming, intervivos farm transfer, and compensation to other heirs were important factors in assuring intergenerational ethnic stability (Conzen 1985b; Salamon and O'Reilly 1979; Salamon 1980; see Barton 1980). Similar strategies used in Minnesota by a settlement of farmers from an isolated parish in Sweden likewise supported high levels of persistence and ethnic distinctiveness. But less homogeneous Swedish communities seem to have exhibited more individualistic values, postponing intergenerational land transfer until after death and requiring sale of the estate to insure equal inheritances, thereby endangering both familial and community persistence on the land (Ostergren 1981c; Salamon 1984).[8]

These inheritance strategies themselves require explanation. Why do some groups import and retain strategies to insure intergenerational continuity in farming while other do not? Part of the answer must lie in the immigrants' homeland culture and the content of the value systems they brought with them. Several studies suggest, for example, that Scandinavian immigrants laid relatively great weight on education for their sons, helping to insure their departure from farming and the shrinkage of rural settlements as the national opportunity structure altered in the twentieth century. German Catholic farmers, by contrast, often seemed determined to minimize the educational opportunity available to their children (Salamon 1984; Brown 1967; Munch 1949; Marshall and Peterson 1948; Conzen 1985b). German Catholics, it could be argued, were doubly insulated from secular, individualistic values by the traditional posture of a religion that saw itself at odds with American society and by their defensive stance within religion vis-à-vis the dominant Irish. Conversely, the bourgeois thrust within Norwegian churches noted by Gjerde meant that Norwegian immigrants had far less quarrel with modern society from the start. The contrast between the village-based communal agriculture that many (though by no means all) German peasants had practiced and the more individualistic system in Norway might also be relevant.

But another part of the answer must be related to differential patterns of clustered settlement. As the early social scientists observed and Jordan (1966, 196–97) and numerous other have since confirmed, it was when rural immigrants clustered in settlements of their own kind that they were most able to transplant accustomed ways of life and maintain group identity. Why and how did clustered settlement patterns emerge? Gjerde (1985, 138–67) notes the utility of community-based settlement: the ethnic community could provide vital assistance to the family in satisfying its own needs for labor exchange, land transfer, welfare, and socializing. Such practical benefits, along with patterns of migrant selectivity and the sheer ease of life among others speaking one's language and understanding one's values, go far towards explaining the highly localized migration chains that recent scholarship has documented. Long-standing concern for the role of general economic, demographic, social, and ideological push and pull factors in determining the pace and direction of emigration has now been given much greater specificity by studies tracing individual migrants from origin to destination and beyond. These studies have clarified the ways in which ties of kinship and local acquaintance channeled migration both to and within America and created the clustered settlements necessary to nurture rural ethnic cultures.

Building on earlier research by Scandinavian scholars, for example, John G. Rice and Robert C. Ostergren (1978; see also Runblom and Norman 1976) were able to show in a sample area of Sweden that the emigration impulse spread from parish to parish along normal routes of trade and communication, taking root if conditions were favorable at the time it arrived, and then within villages spreading from family to family along ties of marriage and kinship and down the social ladder. Data generated in Robert P. Swierenga's massive linkage of Dutch emigration records to shiplists and American censuses demonstrate how differences in structures of opportunity at the very local level influenced receptivity to the emigration impulse, and Gjerde's Norwegian work underlines the critical role of family structures in translating local economic differences into differential migration decisions (Swierenga and Saueressig-Schreuder 1983; Saueressig-Schreuder 1985a; Gjerde 1985, 116–36). Historians refer to such patterns of immigration by extended families or neighborhoods or villages (rather than by isolated member families) as "chain migration." These and other studies make clear the extent to which chains so formed led families from the same village to the same destinations in rural America, and to the subsequent formation of daughter colonies in other American locations as the quest for land continued (Scheben 1939; Rice 1973; Ostergren 1980; Maire 1980, 146–49; Kamphoefner 1987, 70–105). Indeed, Robert Ostergren shows that even subvillage level neighborhood concentrations could be replicated in America (1979; but see also Saueressig-Schreuder 1985b). Walter Kamphoefner has argued from German evidence that the more intense the emigration rate from an area, the greater the probability of chain migration; that emigration intensity was highest in the areas of protoindustrialization; and that the tightly homogeneous settlements created by chain migration tended to impede both assimilation and social mobility. Kamphoefner (1987, 70–105, 170–200) has used cross-national comparisons to hypothesize a consistent linkage between emigration intensity, chain migration, clustered settlement, and impeded assimilation (defined by socioeconomic status) that remains to be specifically tested.

The mechanisms by which clustered communities influenced persistence and cultural change have not been as carefully studied. Presumably the survival of ethnic patterns much beyond the second generation would depend on a settlement concentration sufficiently large and isolated to support the churches and schools required to socialize succeeding generations in group values, create a large enough pool of potential marriage partners within the community, permit the ethnic group to dominate local governments and the local land market and discourage much casual contact with outside ways. Where strong religious belief promoted self-segregation, a settlement could be quite small and still retain its ethnic self-definition; where the community was internally more heterogeneous or more accepting of secular values, it would require greater size or geographic isolation to preserve much cultural distinctiveness.[9]

One can, in fact, begin to derive from the literature a typology of rural settlement patterns more or less favorable to ethnic distinctiveness. Categories of classification would include geographic isolation and internal size and homogeneity, the latter measured in terms of religious affiliation as well as closeness of homeland origins. Settlement "generation" would also play a role. The initial foundation of a given migration chain might miscarry because of imperfect information that located a cluster on poor soil or so close to other groups that little autonomy or expansion was possible (see Palmer 1983; Gonner 1889; Jordan 1966). The subsequent major settlement could expect to bear the strongest imprint of traditional culture, receiving as it would heavy family migration directly from the sending region. As it sent out its own daughter colonies, though, their populations would be increasingly composed of persons already somewhat accustomed to American ways, or of the young immigrants directly from Europe who comprised larger portions of maturing migration chains; the odds of such a later colony attracting a mix of settlers from different migration chains previously clustered in different parts of the United States would also increase as information flows became more generalized. Such factors would explain a lighter initial impress of the traditional culture in succeeding daughter colonies than in the major settlement (see Gjerde 1985; Ostergren 1980; Conzen 1984).

Such differences among settlements, defined not in terms of regional location or date of founding but of state in the maturation of the migration may help clarify intra- and intergroup differences in ethnic distinctiveness that scholars have identified: why, for example, German communities generally seem more distinctive in their original settlement areas in the Midwest than on the Great Plains, where contemporaneous Russian-Germans, establishing their first settlements, were able to form sharply defined communities (Johansen 1937; Sherman 1983). It should also be possible to clarify the experience of British and Irish immigrant farmers in similar terms (Erickson 1972; Akenson 1982): have they been largely "invisible" to historians through failure to create migration chains (unlikely), because chains were usually distended by the greater breadth of their information fields (quite possible), or because initial clusters failed to endure or send out daughter colonies (very probably)?

However useful the characteristics of migration chains and settlement clusters might be, they cannot fully explain group and community differences. Insofar as it is possible to compare group patterns as they emerge from the existing literature, such comparison seems to confirm the older notion of a spectrum of ethnic intensity ranging from Germanic groups on one end, through Slavs and Scandinavians to Irish and British on the other, a spectrum defined in terms of both initial transfer and long-term persistence of ethnic differences. Settlement characteristics were critical constraining or enabling factors, but a range of other factors could also influence where along that spectrum a given group or community found itself. The degree of difference in fundamental values and desired life-styles between those imported by immigrants and found among their American neighbors obviously played an important role. The more consonant a group's values with those of modernizing American society, the more likely the group's own trajectory to take its members away from the farm. Also, the greater the wealth gained from farming, the more group members could afford new life-styles. Alternatively, differences in language or preferred life-style could create needs that could be satisfied only within an ethnic community. Equally important, a given group's heritage of values and institutions might be more or less well-suited to maintaining familial and community persistence even in the absence of a conscious sense of difference. Inheritance customs could play such a role; so too could differences in the nature of primary kinship linkages (that is, lineal or lateral) or a heritage of mutual benefit organizations, as Josef Barton demonstrates in his comparison of Czechs and Mexicans in south Texas (Barton 1980). If some elements only of the traditional local social structure (age groups, classes, occupations, religious orientations) were transplanted, some aspects of traditional culture would be unavoidably lost, others overemphasized. The circumstances of American farming might force a greater change in the familial relations at the heart of traditional culture for one group, as Gjerde has shown, than might be the case for another whose old patterns were more functional in the new environment.

Finally, the defining contours of ethnic difference could themselves change. Most historical studies of rural immigrant communities still rely implicitly on a "straight-line" model of individual assimilation: ethnic culture is defined as the sum total of differences in the characteristics of community members from those of the native born and similarities to those of homeland compatriots, assimilation as either the recession of such characteristics from the standard set by homeland culture or the diminution of differences from the native born. Such a model implies a gradual but inevitable approach to the standard set by the host society. But as some recent studies have recognized, part at least of the cultural change within rural immigrant communities can be seen as an independent response to the same processes of embourgeoisement that were also transforming homeland and host societies. Moreover, all aspects of group culture did not necessarily change at the same pace or in the same direction. A community could adopt the forms of an American middle–class life-style while retaining distinctive familial and religious values; or its values could undergo significant change but its sense of social isolation and communal identity could endure (Useem 1945). Persistent values could find new forms appropriate for their expression, as when German Catholics in the late nineteenth century found in dairying an ideal fit for their familial culture. Equally, old forms could take on new meaning, as the shifting basis of the Wisconsin Norwegian identification with tobacco growing demonstrates (Salamon 1985; Strickon 1979).

Thus metaphors of grafting and pruning, cultivating and winnowing, crossbreeding and selection may be as useful to understanding the rural immigrant community as the more familiar ones of uprooting and transplanting.[10] The culture and identity nurtured in rural immigrant colonies was the result of a selective process of ethnicization that created a new set of values, behavior, and sense of community born out of American circumstances but tapping elements of Old World heritage. Over time, greater or lesser numbers of community members would leave, to make their way into the future without community support. The community itself might disintegrate in the process, in a kind of straight-line assimilation. But it could also survive and evolve, responding to its own internal trajectory as well as to forces from the wider world, in a process that might be better interpreted as one of parallel integration than assimilation. The community may retain a distinctive sense of self-identity and distinctive values and customs at the same time as it modernizes and loses conscious identification with its immigrant past, so that what was once ethnic becomes local, one of the innumerable local cultures with their own ways of doing things that survive even in mass society, gaining strength for that survival from their unvoiced common heritage (see Ostergren 1983; Wheeler 1986). Or it may even consciously reach back to its ethnic past for a source of identity and meaning, though the cultural referents of that identity may be long since transformed and unrecognizable.[11] It is in those local cultures today that the long-term influence of ethnicity on agriculture can be sought; their persistence, and the emphasis on family farming as a way of life that persisted with them, may in the long run be the most significant "contribution" that immigrants have made to American agriculture.

Any new rural history, therefore, will have to deal with an economic and social landscape often contoured by ethnic difference. The decisive impulses that transformed rural America emanated from beyond the ethnic communities, to be sure. In most areas native-born Americans set the initial farming pattern and dominated local economic and political institutions, and their values molded the broader context within which the dynamic commercial transformation of farming unfolded. Not even the most isolated of rural ethnic communities remained long immune to the insistent influences of markets, transportation improvement, urban opportunity, national institutions, or popular culture (see Thompson and Warber 1913; Moline 1971; Barron 1984). Yet, as this review has suggested, responses to those influences were not predetermined, unidirectional, or without effect on the future course of local development. Thus the story of Western settlement must be a story of European as well as American migration chains. Standard generalizations on such old issues as farm-making costs, tenure systems, mortgaging habits, or the determinants of persistence deserve systematic reexamination in the light of ethnically varying values that could skew economic behavior. The same is true for newer questions regarding inheritance patterns or gender division of labor.

Recent excursions into rural history, despite the fineness of data and subtleties of interpretation that have emerged in the detailed investigation of local communities, have too often been hampered by an overly schematic insistence on the overwhelming force of the market in transforming futilely resistant traditional farming cultures. The more critical question, as scholars are coming to realize, has to do with the logic of American family farming itself, and the historical forces that have shaped its character and evolution as one among a number of possible forms of agricultural production (for example Friedmann 1978; Buttel 1980; Pfeffer 1983; Gilbert and Akor 1988). As central to that logic as economic, political, demographic, or environmental factors are the varying strategies adopted by farmers themselves, and the world views that informed them (Long and Van der Ploeg 1988; Vandergeest 1988). To deny the influence of ethnicity in the name of an all-obliterating hegemonic market is to adopt a new kind of environmental determinism. To integrate the immigrant experience into rural history is to underscore the cultural degrees of freedom that farming communities could enjoy in making their own paths to the present; it is also, however, to appreciate not only the gains of familial and communal rootedness, but also some of the sacrifices of self-exploitation and narrowed horizons that such paths could entail.

The immigrant experience cannot be fully integrated into the story of the "complex and often conflicted development of commercial and industrial capitalism in the countryside" (Hahn and Prude 1985, 15) without additional research. Certainly, generalizations concerning the relationships between migration, settlement concentration, ethnic cultural evolution, and American agricultural development will remain largely speculative in the absence of much in the way of comparable—let alone explicitly comparative—studies of different groups in different settings over more than just the initial generation of settlement. It is extremely difficult, for example, to estimate group variations in levels of chain migration and settlement concentration at any scale below that of the state, yet it is clear that in such variations lies much of the explanation for differential ethnic impact. Studies of individual groups have helped to clarify the factors that can account for overall distribution patterns—German settlement locations, to take one example, evidence the influence of land availability, transportation routes, market access, and the often accidental presence of countrymen in a given area—but even they have seldom addressed relative degrees of concentration or scatter (Hannemann 1936; Johnson 1941, 1951; Kloss 1974; Kamphoefner 1987).[12] State-level mapping of the ethnic mosaic, both historical and contemporary, is a vital tool in understanding the complexity of the settlement context, but in the absence of indices of relative concentration or dispersion of settlements, it is difficult to compare such mosaics from place to place (see Carman 1962; Jordan 1970; Sherman 1983; Holmquist 1981). Moreover, as Ostergren (1981a, 1983) has shown, such studies must be placed in the broader context of the uneven advance of the settlement frontier and changes in transportation and information flows, if variations in the mosaic from place to place are to be understood and their influence upon the persistence of ethnic communities is to be assessed. Chan's detailed examination (1986) of the varying structures of opportunity offered to Chinese in different sectors of California agriculture raises an important issue for other groups as well.

Because the same kinds of immigrants who settled in rural America made their way also to expanding settlement areas elsewhere in the world, international comparison offers another valuable strategy for clarifying the factors influencing settlement patterns, farming choices, and cultural continuity. Donald H. Akenson's demonstration (1982) of the frequency and success with which the Canadian Irish entered agriculture, for example, casts doubt on standard explanations for the Irish avoidance of farming in the United States, suggesting that the attraction of American urban opportunity may have been more critical than any supposed Irish distaste for or lack of skill in farming. Contrasts in village settlement patterns between Germans in South Australia and their Wisconsin counterparts highlight the consequences of differing systems of land alienation; the differing values placed on parochial schools by German Catholics in Minnesota and Saskatchewan emphasize the role that state policy could play in ethnic preservation (Erdmann 1984; White 1984). Different climates in Kansas and Brazil help account for varying degrees of architectural transplantation among Volga Germans (Kloberdanz 1980). Historians have taken too little advantage of the opportunity for controlled comparison that nineteenth century international wanderlust created.

There are other aspects of the rural immigrant world that also merit more attention than they have received in recent scholarship. For one thing, in moving from its older preoccupation with farming practice to its newer concern for family patterns, rural immigration history has continued to neglect the local community itself. There is a certain tendency to view it as little more than an organic extension of the family relationships and kin networks that channeled immigration and molded individual aspirations—as a supplement of the family in its quest for self-determination and self-perpetuation. But the community could also have an autonomous expression in institutions, organizations, and even in legally defined functions. The role of the church as the center of the rural ethnic community—as the fulcrum of settlement, the locus of voluntary organization, the distributor of social statuses, the focus of group consciousness—has received considerable attention, but it was not necessarily the only possible expression of community, nor was it the only one whose boundaries were voluntarily created.[13]

Road districts and school districts, for example, could shift boundaries in response to changing perceptions of common interest and they controlled resources so vital to the perpetuation of the community and its goals that "tipping points" in such jurisdictions may have played a role in the achievement of settlement homogeneity (Conzen 1984). Where the settlement cluster was large enough to dominate local government, the ethnic community could achieve some measure of control over its tax assessment, its stock laws, its care for the indigent, its public celebrations, its local law enforcement and morality (see Krier 1987). The political system also held out the possibility of formal community representation in higher levels of government, and thus a means to define and defend group culture before a broader forum. Public archives are rich with the sources for studying such activities, yet Curti's book still remains the most careful study of rural immigrant political integration, and even it is relatively unconcerned with implications for the ethnic community itself. Too often political activity is still seen as the final sign of Americanization, the acceptance of new habits of political self-definition that are uniquely American (McGregor 1988). Yet contrary to stereotype, European peasants often had extensive experience with communal self-government, expectations of local government intervention in significant areas of daily life, and assumptions about how such governments should function.[14] Where they could dominate local governments in rural America, political participation could thus become a significant instrument in the definition and preservation of ethnic culture, and local political cultures could be expected to bear an enduring ethnic mark.[15]

Preoccupation with quantifiable measures of difference has also obscured the potential of other kinds of sources for tracing the meaning and evolution of rural ethnic cultures, particularly sources offering more direct evidence of attitudes and beliefs as opposed to behavior. Religion, again, is a major exception. Church language usage, doctrinal debate, and liturgical practice have long provided historians with indices for rural cultural change, though only recently have there been efforts to link such factors systematically to other measures of community structure, and there has been little effort to explore how pious belief was integrated into the worldviews of immigrant farmers and influenced their daily lives (Legried and Ward 1982; Pedeliski 1986). Gjerde (1985) has shown how changes in gender division of labor and in sexual morality can be used to chart the process of cultural integration. When letters, diaries, and memoirs are lacking, as they so often are for rural immigrant communities, researchers can turn instead to the kinds of church documents Gjerde relied upon, as well as to wills, court case files, and other public documents. Newspapers too can still yield much to the imaginative researcher. These same sorts of sources can be subjected to folkloristic analysis, with its emphasis on the entire range of expressive behavior, interaction, and performance. The stories people told, the ways they spoke and interacted with one another, their beliefs, crafts, and foodways, all served to define the boundaries of the ethnic community and can be used to chart selective maintenance and adaptation processes (Dorson 1971, 15–48; Abrahams 1980; Limon and Young 1986). Unfortunately, the occasional folklorist who ventures into rural ethnic communities often has little sense for either Old World cultural roots or New World historical change (see Coggeshall 1986; but see also Borie 1986), while the methods and theory of the folklorist remain inaccessible to most historians.

The time is certainly ripe for a closer integration of newly popular studies of material culture into interpretations of rural immigrant history. Housing and artifacts can serve as eloquent witnesses to the values of those who created them. They can also testify to the constraints upon autonomous ethnic evolution. Assumptions similar to those underlying Wilbur Zelinsky's (1973, 13–14) "doctrine of first effective settlement"—that an area's initial settlers determined its vernacular culture, and that the foreign born were seldom the initial settlers—long precluded much effort to look for evidence of ethnic architectural traditions other than among the Pennsylvania Germans. Likewise, the adoption of new settlement forms, the availability of different materials and tools, and the demands of different climates and farming regimes necessitated significant architectural innovation. But recent surveys have uncovered in many parts of the nation an unsuspected wealth of surviving house, barn, and other building forms—defined in terms of construction method and materials, elevations, floor plans, room usage, interior furnishing and decoration—associated with different immigrant groups and raise significant interpretive questions regarding both initial transplantation of architectural vocabulary and its subsequent evolution (Noble 1984; Upton 1986).

Too many of these studies, content as they are simply to confirm the existence and present-day survival of a distinctive ethnic "house type" and with it an ethnic presence in an area, are burdened with significant conceptual and methodological limitations.[16] European vernacular building traditions could vary greatly within a given country, by region and class, and were themselves caught up in a complex evolutionary process that has to be taken into account before appropriate baselines for cultural transfer can be determined (Weaver 1986; Carter 1986). Presumably, the more direct the migration chain, the more isolated the community, the more similar the locally available building materials and agricultural patterns, the greater the likelihood there would be that a given community would draw upon its European architectural heritage to a significant extent (Roberts 1986). Certainly, the same group that could support a distinctive architecture in one community might fail to do so in another, a pattern that might also be related to generation of settlement but in any case serves to index attitudes toward assimilation.[17] This suggests also that groups varied from one another in their propensity to retain traditional architectural elements, a question that remains virtually unexplored. Attempts to determine the extent of traditional architecture within a community and the precise elements of the traditional vocabulary that were transplanted have been hampered by reliance upon the highly selective survival of physical evidence; as Scott T. Swank has shown for German Pennsylvania (1983), only the documentary reconstruction of the complete housing stock of a settlement area can resolve these issues.

Such evidence would be equally important in determining the pace and direction of subsequent architectural change. Charles Van Ravenswaay, in his beautifully detailed study of Missouri German material culture (1977), has pointed to the role of improved transportation and the penetration of mass market goods in the dissolution of ethnic architectural and craft traditions, while A:son-Palmqvist (1983) shows how prosperity combined with newly available prepared plans and materials to encourage the abandonment of traditional styles when it came time to replace the original dwellings of the pioneer period. New materials, tools, and environments might cause some initial architectural adaptation, but as Edward A. Chappell's important study (1986) of a German area in the Shenandoah Valley demonstrates, subsequent evolution could be linked to the gradual acceptance of regional norms, evident first in external appearance, then in room functions and arrangement (a critical stage that coincide with changes in living style and language), and lastly in the most intimate area of interior decoration.[18] Even after that evolution was complete, ethnic tradition could still influence the embellishment of American housing styles, as when wealthy Norwegian-Americans added porches to their farmhouses reminiscent of styles favored by the Norwegian gentry, or Dutch and Belgians retained a preference for polychromed brick decoration (Peterson 1984; Nobel 1984, 2:72–80). An ultimate step of sorts is reached when a group, in a gesture of ethnic assertiveness aimed more outwards than inwards, self-consciously adopts revival styles or elements of what it has learned through popular culture to consider the architecture of the homeland (Kennedy 1967, 61–77). Housing could also combine with other distinctive landscape elements—whether settlement forms like the German hamlets of Wisconsin and street villages of Missouri, field plats like the long lots of the Iowa Czechs and Illinois Welsh, characteristic farmstead organization, or building types like Finnish saunas or Belgian wayside chapels—to create visible mirrors of local cultural evolution whose variety and contribution to American landscape sensibilities have yet to be explored (Bercovici 1925; Bertrand 1942; Gerlach 1977; Klimesh 1983; Alanen and Tishler 1980; Long 1972; Mather and Kaups 1963; Laatsch and Calkins 1986; Glass 1986; Vogeler 1976).

Finally, the relationship between the rural ethnic community and the broader evolution of the meaning of ethnicity in American society also deserves consideration. For all the immigrant farmers' children who remained in the community or established themselves in daughter farming colonies, many more found their way to the city. As Kamphoefner (1987) has shown, many an immigrant farmer had himself used urban labor to earn his farm. The roots of urban ethnic cultures have been almost exclusively sought in Europe, yet by the late nineteenth century a sizeable number of people in many cities must have been nurtured in the ethnic cultures of rural America. How did these cultures mediate their urban experience? What difference did they make in the evolution of urban ethnicity? Second-generation Finns in Detroit, for example, continued to structure their lives at work and at home in terms of the migration chains that linked them to their rural communities of origin and often drew them back at retirement (Loukinen 1981). The contribution of rural immigrant communities to the emergence of ethnicity as a conscious form of self-identity within American society remains equally obscure. Speedier assimilation and greater frequency of nongroup contract seems to have fostered a stronger and more formal sense of ethnic identity among urban than among more isolated rural immigrants (Useem 1945). Lack of conscious ethnic self-definition would have encouraged the gradual metamorphosis of ethnic into local identities. It also raises questions about the relative roles of rural and urban elements in national ethnic organization, and may help explain regional differences in the weight and meaning accorded self-consciously pluralist ideologies. With scholars beginning to pay more attention to ethnicity as a consciously invented form of identity, rather than just as a quantifiable collection of traits, urban and rural differences in the formal meaning accorded ethnic identity will have to be taken into account (Berthoff 1982; Conzen 1985a).

There are numerous other areas that merit additional research. This essay has not begun to explore the major set of issues posed by the immigrant character of much of the agricultural wage labor force, for example, whether as hired hands or migrant workers. It has not directly focused upon such significant topics as the evolution of localized ethnic specialization in particular areas of agriculture, the implications of ethnicity for farm financing, or the role of ethnic difference in exacerbating tensions between town and countryside. Only in passing has it touched upon the ways in which ethnicity could affect the lives and consciousness of rural women, the creation of rural educational systems, or the relationships among different groups in rural America. The essay's regional and group biases are obvious. But at least it should be clear that students of rural America are starting to listen to the foreign accents with which so many farmers spoke, to the ethnic shadings that still color rural speech patterns in so many parts of the country. The story of the "great transformation" of rural America is the familiar story of yeomen and planters, mountaineers and bondsmen, to be sure, but it is also a story of the transformation of European peasants into American farmers and the transformation of American farming by them, a story that is now finally beginning to be told.

## NOTES

1. Census data on this point are inconsistently presented, making general summary difficult (see U.S. Census 1870, 699; U.S. Census 1880, 730; Brunner 1929, 253–54). Statistics on immigrants in agriculture are notoriously difficult to compile from published census documents because of frequent changes in definitions and lack of comparability in tabulations from census to census. Thus, present figures are cited for illustrative purposes only, and make no claims to provide a complete analysis of all available census data.

2. For essays calling attention to this state of affairs, see Saloutos (1976), Conzen (1980), Swierenga (1980), Luebke (1983).

3. Early commentators noting this neglect include Brunner (1929, xvi), and Hansen (1940, 200), (see Conzen 1980, 1–2 for examples). Even Atack and Bateman (1987, 12) in their recent census-based analysis of antebellum farming in the North fail to take advantage of the possibilities offered by the linkage of population and agriculture schedules to explore ethnic variation in farming patterns; despite their stress on agriculture as "an engine of familial and social organization with strong noneconomic motivations" they confine their attention to the influence of ethnicity to questions of fertility and migration only.

4. For a sophisticated version of this approach, see Saloutos (1975, 182–201; Saloutos 1976, 45–67). Swierenga (1980) and Luebke (1977) also provide excellent summaries of the information and insights that can be derived from such work. See also such significant state-level compilations as DeJong (1985, 66–114), Hendrickson (1977), and Holmquist (1981).

5. Caroline Ware (1937, 589) shows the acceptance by a sociologically aware historian of this perception: "Rural adjustments involve principally the bare struggle for existence. But with the transfer of a peasant people to foreign industrial centers the usual strains set up by industrialization are increased by the pressure of alien speech and alien mores which bears more immediately upon them. In rural regions the traditional language can easily be maintained and the inherited customs can remain largely undisturbed. . . ."

6. The following discussion represents an expansion and updating of suggestions initially developed in Conzen (1980); for related interpretive suggestions, see Luebke (1983).

7. There may, however, have been more experimentation with field replatting and central village residence than scholars have yet identified; see for example observations in Bercovici (1925) and Kelly (1939).

8. This is a pattern also frequently characteristic of less-persistent, native born farmers; see the discussion in Conzen (1985b, 283–85).

9. Ostergren (1973) and Rice (1978), for example, have shown that communities drawing the bulk of their members from a very circumscribed area of the homeland had higher levels of persistence and endogamy than those with wider catchment areas; since the homeland culture of the more homogeneous communities was also the most traditional, however, the role of local homogeneity per se remains unclear. Rice (1973) has also shown the greater fragility of ethnic communities located near towns and presumably more subject to modernizing urban influences.

10. The uprooting metaphor is, of course, Handlin's (1950); the transplanting correction has been made by Bodnar (1985) and Gjerde (1987).

11. For an evocative perception of the complexities of conscious and unconscious ethnicity in one such local community, see Hampl and Sorman (1987); note also Useem (1945), Hoffmann (1949), Kloberdanz (1988).

12. For a general discussion and bibliography of ethnic distribution maps, see Raitz (1978).

13. In contrast to Rice (1973, 33–39) and the references cited there, see also Ostergren (1981b), Legreid and Ward (1982). Johnson (1941) has pointed out, for example, that German communities that did not support their own church were the most likely to disappear over time.

14. Contrast, for example, the view of the peasant world that emerges from Handlin (1951) with that presented in Kaschuba and Lipp (1982).

15. The influence of ethnicity on voting behavior was a central concern of the ethnocultural school of political history, and remains an issue worth studying in the local rural context despite conceptual and methodological problems that have left this approach in some disarray; see Luebke (1983, 398–89) for an introduction to scholarship on the influence of ethnicity on Populism, for example.

16. See Henry Glassie's comment, "It seems necessary to emphasize that artifacts are not studied to run a check on demography" (1975, 17). Representative examples of recent careful studies surveying surviving examples of local ethnic architecture and defining their basic features include A:son-Palmqvist (1983), Koop and Ludwig (1984), Breisch and Moore (1986), Tischler and Witmer (1986), Gammerdinger and Rikoon (1986).

17. Contrast, for example, the rapid architectural assimilation in a Wisconsin Norwegian community described in Selkurt (1985) with what appears to be the more enduring distinctiveness of a Texas counterpart, as described in Breisch and Moore (1986); a similar contrast seems to emerge between Swedes in Kansas (Jaderborg 1981), and Minnesota (A:son-Palmqvist 1983); the issue will only be resolved through explicitly comparative research.

18. Note also Touart's (1986) demonstration that the separation of decision-making processes between homeowner and artisan could lead to "assimilated" plans executed with ethnic construction techniques.

## REFERENCES AND FURTHER READINGS

Abrahams, Roger D. 1980. "Folklore." In *Harvard Encyclopedia of American Ethnic Groups,* ed. Stephan Thernstrom, 370–79. Cambridge: Harvard Univ. Press.

Akenson, Donald H. 1982. "Ontario: Whatever Happened to the Irish?" *Canadian Papers in Rural History* 3:204–56.

Alanen, Arnold R., and William H. Tishler. 1980. "Finnish Farmstead Organization in Old and New World Settings." *Journal of Cultural Geography* 1:66–81.

A:son-Palmqvist, Lena. 1983. *Building Traditions among Swedish Settlers in Rural Minnesota: Material Culture—Reflecting Persistence or Decline of Traditions.* Växjö: Nordiska Museet.

Atack, Jeremy, and Fred Bateman 1987. *To Their Own Soil: Agriculture in the Antebellum North.* Ames: Iowa State Univ. Press.

Baltensperger, Bradley H. 1980. "Agricultural Change among Nebraska Immigrants, 1880–1900." In *Ethnicity on the Great Plains,* ed. Frederick C. Luebke, 170–89. Lincoln: Univ. of Nebraska Press.

Barron, Hal S. 1984. *Those Who Stayed Behind: Rural Society in Nineteenth-Century New England.* Cambridge: Cambridge Univ. Press.

Barthel, Diane L. 1984. *Amana: From Pietist Sect to American Community.* Lincoln: Univ. of Nebraska Press.

Barton, Josef J. 1980. "Land, Labor, and Community in Nueces: Czech Farmers and Mexican Laborers in South Texas, 1880–1930." In *Ethnicity on the Great Plains,* ed. Frederick C. Luebke, 190–209. Lincoln: Univ. of Nebraska Press.

Bennett, John W. 1967. *Hutterian Brethren: The Agricultural Economy and Social Organization of a Communal People.* Stanford: Stanford Univ. Press.

Bercovici, Konrad. 1925. *On New Shores.* New York: Century.

Berthoff, Rowland. 1982. "Under the Kilt: Variations on the Scottish-American Ground." *Journal of American Ethnic History* 1:5–34.

Bertrand, Kenneth. 1942. "Rural Agglomerated Settlements in the Eastern Lake Shore Red Clay Dairy Region of Wisconsin." *Transactions of the Wisconsin Academy of Sciences, Arts and Letters* 34:47–62.

Bjorklund, Elaine M. 1964. "Ideology and Culture Exemplified in Southwestern Michigan." *Annals of the Association of American Geographers* 54:227–41.

Bodnar, John. 1985. *The Transplanted: A History of Immigrants in Urban America.* Bloomington: Indiana Univ. Press.

Bogue, Allan G. 1963. *From Prairie to Cornbelt: Farming on the Illinois and Iowa Prairies in the Nineteenth Century.* Chicago: Univ. of Chicago Press.

Borie, Beauveau IV. 1986. *Farming and Folk Society: Threshing among the Pennsylvania Germans.* Ann Arbor: UMI Research Press.

Breisch, Kenneth A., and David Moore. 1986. "The Norwegian Rock Houses of Bosque County, Texas: Some Observations on a Nineteenth-Century Vernacular Building Type." In *Perspectives in Vernacular Architecture, II,* ed. Camille Wells, 64–71. Columbia: Univ. of Missouri Press.

Brown, Robert H. 1967. "The Upsala, Minnesota Community: A Case Study in Rural Dynamics." *Annals of the Association of American Geographers* 57:267–300.

Brunner, Edmund deS. 1929. *Immigrant Farmers and Their Children.* Garden City, N.Y.: Doubleday, Doran.

Buttel, Frederick H. 1980. "W(h)ither the Family Farm? Toward a Sociological Perspective on Independent Commodity Production in U.S. Agriculture." *Cornell Journal of Social Relations* 15:10–37.

Carman, J. Neale. 1962. *Foreign-Language Units of Kansas, Vol. 1. Historical Atlas and Statistics.* Lawrence: Univ. of Kansas Press.

Carter, Thomas. 1986. "Danes." In *America's Architectural Roots,* ed. Dell Upton, 118–23. Washington, D.C.: Preservation Press.

Chan, Sucheng. 1986. *This Bittersweet Soil: The Chinese in California Agriculture, 1860–1910.* Berkeley: Univ. of California Press.

Chappell, Edward. 1980. "Acculturation in the Shenandoah Valley: Rhenish Houses of the Massanutten Settlement." *Proceedings of the American Philosophical Society* 124:55–89.

Coggeshall, John M. 1986. "'One of those Intangibles': The Manifestation of Ethnic Identity in Southwestern Illinois." *Journal of American Folklore* 99:177–207.

Cogswell, Seddie, Jr. 1975. *Tenure, Nativity, and Age as Factors in Iowa Agriculture, 1850–1880.* Ames: Iowa State Univ. Press.

Conzen, Kathleen Neils. 1980. "Historical Approaches to the Study of Rural Ethnic Communities." In *Ethnicity on the Great Plains,* ed. Frederick C. Luebke, 1–18. Lincoln: Univ. of Nebraska Press.

______. 1984. "Deutsche Einwanderer im ländlichen Amerika: Problemfelder und Forschungsergebnisse." In *Auswanderer—Wanderarbeiter—Gastarbeiter: Bevölkerung, Arbeitsmarkt und Wanderung in Deutschland seit der Mitte des 19 Jahrhunderts,* ed. Klaus J. Bade, 1:350–77. Ostfildern, West Germany: Scripta Mercaturae Verlag.

______. 1985a. "German-Americans and the Invention of Ethnicity." In *America and the Germans: An Assessment of a Three-Hundred-Year History,* ed. Frank Trommler and Joseph McVeigh, 1:131–47. Philadelphia: Univ. of Pennsylvania Press.

______. 1985b. "Peasant Pioneers: Generational Succession among German Farmers in Frontier Minnesota." In *The Countryside in the Age of Capitalist Transformation,* ed. Steven Hahn and Jonathan Prude, 259–92. Chapel Hill: Univ. of North Carolina Press.

Conzen, Michael P. 1971. *Frontier Farming in an Urban Shadow: The Influence of Madison's Proximity on the Agricultural Development of Blooming Grove, Wisconsin.* Madison: State Historical Society of Wisconsin.

Cozzens, A.B. 1943. "Conservation in German Settlements of the Missouri Ozarks." *Geographical Review* 33:286–98.

Curti, Merle et al. 1959. *The Making of an American Community: A Case Study of Democracy in a Frontier County.* Stanford: Stanford Univ. Press.

Deininger, Marian McNeil. 1958. "Some Differential Characteristics of Minnesota's Major Ethnic Groups in Selected Rural Townships." Ph.D. diss., University of Minnesota.

Deininger, Marian McNeil, and Douglas Marshall. 1955. "A Study of Land Ownership by Ethnic Groups from Frontier Times to the Present in a Marginal Farming Area of Minnesota." *Land Economics* 31:351–60.

DeJong, Gerald F. 1985. "Dakota Resources: A Preliminary Guide for Studying European Ethnic Groups in South Dakota: A Bibliographic Essay." *South Dakota History* 15:66–114.

Dorson, Richard M. 1971. *American Folklore and the Historian.* Chicago: Univ. of Chicago Press.

Douglas, William A., and Jon Bilbao. 1975. *Amerikanuak: Basques in the New World.* Reno: Univ. of Nevada Press.

Erdmann, Claudia. 1984. "Deutsche Siedlungen in Südaustralien: Entstehung und Kulturlandschaftliche Entwicklung eines Kolonisationsraumes." *Erdkunde* 38:302–14.

Erickson, Charlotte. 1964. "Agrarian Myths of English Immigrants." In *In the Trek of the Immigrants: Essays Presented to Carl Wittke,* ed. O. Fritiof Ander, 59–80. Rock Island, Ill.: Augustana College Library.

______. 1972. *Invisible Immigrants: The Adaptation of English and Scottish Immigrants in Nineteenth-Century America.* Coral Gables: Univ. of Miami Press.

Everest, Kate Asaphine. 1892. "How Wisconsin Came by Its Large German Element." *Collections of the State Historical Society of Wisconsin* 12:299–334.

Flora, Jan L., and John M. Stitz. 1985. "Ethnicity, Persistence, and Capitalization of Agriculture in the Great Plains during the Settlement Period: Wheat Production and Risk Avoidance." *Rural Sociology* 50:341–60.

Foster, Gary, Richard Hummel, and Robert Whittenbarger. 1987. "Ethnic Echoes Through 100 Years of Midwestern Agriculture." *Rural Sociology* 52:365–78.

Friedmann, Harriet. 1978. "World Market, State, and Family Farm: Social Bases of Household Production in the Era of Wage Labor." *Comparative Studies in Society and History* 20:545–86.

Gagliardo, John. 1959. "German Agriculture in Colonial Pennsylvania." *Pennsylvania Magazine of History and Biography* 83:192–218.

Gammerdinger, Harry, and J. Sanford Rikoon. 1986. "The Log Barns of German Settlers in Jefferson Township, Dubois County, Indiana." *Indiana Folklore and Oral History* 15:70–92.

Gehrke, William H. 1935. "The Ante-Bellum Agriculture of the Germans in North Carolina." *Agricultural History 9:143–60.*

Georges, Robert A., and Stephen Stern. 1982. *American and Canadian Immigrant and Ethnic Folklore: An Annotated Bibliography.* New York: Garland.

Gerlach, Russell L. 1976. *Immigrants in the Ozarks.* Columbia: Univ. of Missouri Press.

Gilbert, Jess, and Raymond Akor. 1988. "Increasing Structural Divergence in U.S. Dairying: California and Wisconsin since 1950." *Rural Sociology* 53:56–72.

Gjerde, Jon. 1985. *From Peasants to Farmers: The Migration from Balestrand, Norway, to the Upper Middle West.* Cambridge: Cambridge Univ. Press.

Glass, Joseph W. 1986. *The Pennsylvania Culture Region: A View from the Barn.* Ann Arbor, Mich.: UMI Research Press.

Glassie, Henry. 1975. *Folk Housing in Middle Virginia.* Knoxville: Univ. of Tennessee Press.

Gonner, Nicholas. 1889. *Die Luxemburger in der neuen Welt.* Dubuque: Luxemburger Gazette.

Hahn, Steven, and Jonathan Prude. 1985. *The Countryside in the Age of Capitalist Transformation: Essays in the Social History of Rural America.* Chapel Hill: Univ. of North Carolina Press.

Hampl, Patricia, and Steven Sorman. 1987. *Spillville.* Minneapolis: Milkweed Editions.

Handlin, Oscar. 1951. *The Uprooted.* New York: Grosset and Dunlap.

Hannemann, M. 1936. "Das Deutschtum in den Vereinigten Staaten: Seine Verbreitung und Entwicklung seit der Mitte des 19. Jahrhunderts." *Petermanns Mitteilungen,* Ergänzungsheft nr. 225.

Hansen, Marcus Lee. 1940. *The Immigrant in American History.* Cambridge: Harvard Univ. Press.

Hart, John Fraser. 1986. "Facets of the Geography of Population in the Midwest." *Journal of Geography* 85:201–11.

______. 1987. "The Persistence of Family Farming Areas." *Journal of Geography* 86:198–203.

Hendrickson, Gordon O., ed. 1977. *Peopling the High Plains: Wyoming's European Heritage.* Laramie: Wyoming State Archives and Historical Department.

Higham, John. 1982. "Current Trends in the Study of Ethnicity in the United States." *Journal of American Ethnic History* 2:5–15.

Hill, George W., and Harold T. Christensen. 1942. "Some Factors in Family Fertility Among Selected Wisconsin Farmers." *American Sociological Review* 7:498–504.

Hoffman, Oscar F. 1949. "Cultural Changes in a Rural Wisconsin Ethnic Island." *Rural Sociology* 14:39–50.

Hollingshead, A. B. 1938. "Changes in Land Ownership as an Index of Succession in Rural Communities." *American Journal of Sociology* 43:764–77.

Holmquist, June Drenning, ed. 1981. *They Chose Minnesota: A Survey of the State's Ethnic Groups.* St. Paul: Minnesota Historical Society Press.

Holzhueter, John O. 1986. "Aslak Lie and the Challenge of the Artifact." *Wisconsin Magazine of History* 70:3–20.

Hutchinson, Edward P. 1956. *Immigrants and their Children, 1850–1950.* New York: John Wiley.

Jaderborg, Elizabeth. 1981. "Swedish Architectural Influence in the Kansas Smoky Valley Community." *Pioneer Historical Quarterly* 1:65–79.

Johansen, John P. 1937. *Immigrant Settlements and Social Organization in South Dakota.* South Dakota Agricultural Experiment Station Bulletin, no. 313. Bookings: South Dakota State College.

Johnson, Hildegard Binder. 1941. "Distribution of German Pioneer Population in Minnesota." *Rural Sociology* 6:16–34.

______. 1951. "The Location of German Immigrants in the Middle West." *Annals of the Association of American Geographers* 41:1–41.

Johnson, Paul R. 1984. "Communal Imprints on the Wisconsin Landscape: St. Nazianz, 1854–1873." *Bulletin of the Wisconsin Council for Geographic Education* (Spring):34–43.

Jordan, Terry G. 1966. *German Seed in Texas Soil.* Austin: Univ. of Texas Press.

______. 1970. "Population Origin Groups in Rural Texas." *Annals of the Association of American Geographers* 60:404–5 and map suppl. no. 13.

______. 1980. "A Religious Geography of the Hill Country Germans of Texas." In *Ethnicity on the Great Plains,* ed. Frederick C. Luebke, 109–28. Lincoln: Univ. of Nebraska Press.

Kamphoefner, Walter D. 1987. *The Westfalians: From German to Missouri.* Princeton: Princeton Univ. Press.

Kaschuba, Wolfgang, and Carola Lipp. 1982. *Dörfliches Überleben: Zur Geschichte materieller und sozialer Reproduktion ländlicher Gesellschaft im 19. und frühen 20. Jahrhundert.* Tübingen, West Germany: Tübinger Vereinigung für Volkskunde E.V.

Kelly, Sister Mary Gilbert, O.P. 1939. *Catholic Immigrant Colonization Projects in the United States, 1815–1860.* New York: United States Catholic Historical Society.

Kennedy, Roger. 1967. *Minnesota Houses: An Architectural and Historical View.* Minneapolis: Dillon Press.

Klimesh, Cyril M. 1983. *They Came to This Place: A History of Spillville, Iowa and its Czech Settlers.* Sebastopol, Calif.: Methodius Press.

Kloberdanz, Timothy J. 1980. "Plainsmen of Three Continents: Volga German Adaptation to Steppe, Prairie, and Pampa." In *Ethnicity on the Great Plains,* ed. Frederick C. Luebke, 54–72. Lincoln: Univ. of Nebraska Press.

______. 1988. "Symbols of German-Russian Ethnic Identity on the Northern Plains." *Great Plains Quarterly* 8:3–15.

Kloss, Heinz ed. 1974. *Atlas der im 19. und frühen 20. Jh. entstandenen deutschen Siedlungen in U.S.A.* Marburg, West Germany: N.G. Elwert.

Kollmorgen, Walter M. 1941. *The German Settlement in Cullman County, Alabama: An Agricultural Island in the Cotton Belt.* Washington, D.C.: U.S. Department of Agriculture, Bureau of Agricultural Economics.

______. 1942. "The Pennsylvania German Farmer." In *The Pennsylvania Germans,* ed. Rudolph Wood, 27–55. Princeton: Princeton Univ. Press.

Koop, Michael, and Stephen Ludwig. 1984. *German-Russian Folk Architecture in Southeastern South Dakota.* Vermillion, S.D.: State Historical Preservation Center.

Krier, Beatrice Wester, ed. 1987. *Tapestry of Luxembourgers: The Making of Belgium.* Belgium, Wis.: by the author.

Laatsch, William G., and Charles F. Calkins. 1986. "The Belgian Roadside Chapels of Wisconsin's Door Peninsula." *Journal of Cultural Geography* 7:116–28.

Legried, Ann M., and David Ward. 1982. "Religious Schism and the Development of Rural Immigrant Communities: Norwegian Lutherans in Western Wisconsin, 1880–1905." *Upper Midwest History* 2:13–29.

Lehr, John C. 1976. *Ukrainian Vernacular Architecture in Alberta.* Edmonton: Alberta Culture, Historical Resources Division.

Levi, Kate Everest. 1898. "Geographical Origin of German Immigration to Wisconsin." *Collections of the State Historical Society of Wisconsin* 14:341–93.

Limon, J. E. and M. J. Young. 1986. "Frontiers, Settlements, and Development in Folklore Studies, 1972–1985." *Annual Reviews in Anthropology* 15:437–60.

Long, Amos. 1972. *The Pennsylvania German Family Farm.* Breinigsville, Penn.: Pennsylvania German Society.

Long, Norman. 1988. "New Challenges in the Sociology of Rural Development: A Rejoinder to Peter Vandergeest." *Sociologia Ruralis* 28:30–41.

Loukinen, Michael M. 1981. "Second Generation Finnish-American Migration from the Northwoods to Detroit, 1920–1950." In *Finnish Diaspora II: United States,* ed. Michael G. Karni, 107–26. Toronto: Multicultural Society of Ontario.

Luebke, Frederick C. 1977. "Ethnic Group Settlement on the Great Plains." *Western Historical Quarterly* 8:405–30.

______. 1983. "Ethnic Minority Groups in the American West." In *Historians and the American West,* ed. Michael P. Malone, 387–413. Lincoln: Univ. of Nebraska Press.

Luebke, Frederick C., ed. 1980. *Ethnicity on the Great Plains.* Lincoln: Univ. of Nebraska Press.

McGregor, Robert Kuhn. 1988. "Cultural Adaptation in Colonial New York: The Palatine Germans of the Mohawk Valley." *New York History* 69:5–34.

McQuillan, D. Aidan. 1975. "Adaptation of Three Immigrant Groups to Farming in Central Kansas, 1875–1925." Ph.D. Diss., Univ. of Wisconsin-Madison.

______.1978. "Territory and Ethnic Identity: Some New Measures of an Old Theme in the Cultural Geography of the United States." In *European Settlement and Development in North America: Essays on Geographical Change in Honour and Memory of Andrew Hill Clark,* ed. James R. Gibson, 136–69. Toronto: Univ. of Toronto Press.

Maire, Camille. 1980. *L'émigration des Lorrains en Amérique 1815–1870.* Metz, France: Centre de Recherches Relations Internationales de l'Université de Metz.

Marshall, Douglas G. 1948. "Nationality and the Emerging Culture." *Rural Sociology* 13:40–47.

______. 1950. "The Decline in Farm Family Fertility and Its Relationship to Nationality and Religious Background." *Rural Sociology* 15:42–49.

Marshall, Douglas G., and Milo Peterson. 1948. *Factors Associated with Variations in School Attendance of Minnesota Farm Boys.* Minnesota Agricultural Experiment Station Paper No. 635, Miscellaneous Journal Series.

Mather, Cotton, and Matti Kaups. 1963. "The Finnish Sauna: A Cultural Index to Settlement." *Annals of the Association of American Geographers* 53:494–504.

Meynen, Emil. 1939. "Das Pennsylvaniendeutsche Bauernland." *Deutsches Archiv für Landes- und Volksforschung* 3:253–92.

Moline, Norman T. 1971. *Mobility and the Small Town 1900–1930: Transportation Change in Oregon, Illinois.* Chicago: Univ. of Chicago, Department of Geography, research paper no. 132.

Munch, Peter A. 1949. "Social Adjustment among Wisconsin Norwegians." *American Sociological Review* 14:780–87.

Noble, Allen G. 1984. *Wood, Brick, and Stone: The North American Settlement Landscape,* 2 vols. Amherst: Univ. of Massachusetts Press.

Ostergren, Robert 1973. "Cultural Homogeneity and Population Stability among Swedish Immigrants in Chisago County." *Minnesota History* 43:255–69.

______. 1979. "A Community Transplanted: The Formative Experience of a Swedish Immigrant Community in the Upper Middle West." *Journal of Historical Geography* 5:189–212.

______. 1980. "Prairie Bound: Migration Patterns to a Swedish Settlement on the Dakota Frontier." In *Ethnicity on the Great Plains,* ed. Frederick C. Luebke, 73–91. Lincoln: Univ. of Nebraska Press.

______. 1981a. "Geographic Perspectives on the History of Settlement in the Upper Middle West." *Upper Midwest History* 1:27–39.

______. 1981b. "The Immigrant Church as a Symbol of Community and Place in the Upper Midwest." *Great Plains Quarterly* 1:225–38.

______. 1981c. "Land and Family in Rural Immigrant Communities." *Annals of the Association of American Geographers* 71:400–11.

______. 1983. "European Settlement and Ethnicity Patterns on the Agricultural Frontiers of South Dakota." *South Dakota History* 13:49–82.

Palmer, Howard. 1983. "Escape from the Great Plains: The Icelanders in North Dakota and Alberta." *Great Plains Quarterly* 3:219–33.

Parsons, Kenneth H. 1948. "Research in the Succession of Farms: A Comment on Methodology." *Land Economics* 24:293–302.

Parsons, Kenneth H., and Eliot O. Waples. 1945. *Keeping the Farm in the Family: A Study of Ownership Processes in a Low Tenancy Area of Eastern Wisconsin.* Wisconsin Agricultural Experiment Station Research Bulletin 157.

Pedersen, Harold A. 1950. "A Cultural Evaluation of the Family Farm Concept." *Land Economics* 26:52–64.

Pedeliski, Theodore B. 1986. "Ukrainians on the Prairies: Old World Cultural Values and the Demands of a New Land." *North Dakota History* 53:17–25.

Peterson, Fred W. 1984. "Norwegian Farm Homes in Steele and Traill Counties, North Dakota: The American Dream and Retention of Roots, 1890–1914." *North Dakota History* 51:4–13.

Pfeffer, Max J. 1983. "Social Origins of Three Systems of Farm Production in the United States." *Rural Sociology* 48:540–62.

Raitz, Karl B. 1978. "Ethnic Maps of North America." *Geographical Review* 68:335–50.

Rice, John G. 1973. *Patterns of Ethnicity in a Minnesota County, 1880–1905.* Umeå Sweden: Department of Geography, University of Umeå, Geographical Reports 4.

______. 1978. "Marriage Behavior and the Persistence of Swedish Communities in Rural Minnesota." In *The Swedish Heritage in the Upper Midwest,* ed. Nils Hasselmo. Proceedings of an International Conference, University of Minnesota-Duluth, April 1–3, 1976.

Rice, John G. and Robert C. Ostergren. 1978. "The Decision to Emigrate: A Study in Diffusion." *Geografiska Annaler* 60, ser. B:1–15.

Roberts, Warren E. 1986. "German American Log Buildings of Dubois County, Indiana." *Winterthur Portfolio* 21:265–74.

Ross, Edward A. 1914. *The Old World in the New.* New York: Century.

Rothenberg, Winifred B. 1987. "The Bound Prometheus." *Reviews in American History* 15:628–37.

Runblom, Harald, and Hans Norman, eds. 1976. *From Sweden to America: A History of the Migration.* Minneapolis: Univ. of Minnesota Press.

Salamon, Sonya. 1980. "Ethnic Differences in Farm Family Land Transfers." *Rural Sociology* 45:290–308.

______. 1984. "Swedish-Americans and the Persistence of the Family Farm and Ethnic Farming Community." Paper presented at the University of Wisconsin-Madison Conference on Scandinavian Immigration, Madison, August 27–30.

______. 1985. "Ethnic Communities and the Structure of Agriculture." *Rural Sociology* 50:323–40.

Salamon, Sonya and Shirley M. O'Reilly. 1979. "Family Land and Developmental Cycles among Illinois Farmers." *Rural Sociology* 44:525–42.

Saloutos, Theodore. 1975. "The Immigrant in Pacific Coast Agriculture, 1880–1940." *Agricultural History* 49:182–201.

______. 1976. "The Immigrant Contribution to American Agriculture." *Agricultural History* 50:45–67.

Saueressig-Schreuder, Yda. 1985a. "Dutch Catholic Emigration in the Mid-Nineteenth Century: Noord-Brabant, 1847–1871." *Journal of Historical Geography* 11:48–69.

______. 1985b. "Dutch Catholic Immigrant Settlement in Wisconsin." In *The Dutch in America: Immigration, Settlement and Cultural Change,* ed. Robert P. Swierenga, 105–24. New Brunswick, N.J.: Rutgers Univ. Press

Schafer, Joseph. 1922/23. "The Yankee and the Teuton in Wisconsin." *Wisconsin Magazine of History* 6:125–45, 261–79, 386–402, 7:3–19.

______. 1927. *Four Wisconsin Counties: Prairie and Forest.* Madison: State Historical Society of Wisconsin.

______. 1936. *The Social History of American Agriculture.* New York: Macmillan.

______. 1937. *The Winnebago-Horicon Basin: A Study in Western History.* Madison: State Historical Society of Wisconsin.

Scheben, Joseph. 1939. *Untersuchungen zur Methode und Technik der deutschamerikanischen Wanderungsforchung: an Hand eines Vergleichs der Volkszählungslisten der Township Westphalia, Clinton County, Michigan, vom Jahre 1860 mit Auswanderungsakten des Kreises Adenau (Rheinland).* Bonn, West Germany: Ludwig Röhrscheid Verlag.

Selkurt, Claire. 1985. "The Domestic Architecture and Cabinetry of Luther Valley." *Norwegian-American Studies* 30:247–72.

Sherman, William C. 1983. *Prairie Mosaic: An Ethnic Atlas of Rural North Dakota.* Fargo: North Dakota Institute for Regional Studies.

Shover, John L. 1976. *First Majority—Last Minority: The Transforming of Rural Life in America.* Dekalb: Northern Illinois Univ. Press.

Shryock, Richard H. 1939. "British Versus German Traditions in Colonial Agriculture." *Mississippi Valley Historical Review* 26:39–54.

Skrabanek, Robert A. 1949. "Forms of Cooperation and Mutual Aid in a Czech-American Rural Community." *Southwestern Social Science Quarterly* 30:183–87.

______. 1951. "The Influence of Cultural Backgrounds on Farming Practices in a Czech-American Rural Community." *Southwestern Social Science Quarterly* 31:258–66.

Slocum, Walter L. 1940. "Ethnic Stocks As Cultural Types in Rural Wisconsin." Ph.D. diss., University of Wisconsin.

Smith, William Carlson. 1939. *Americans in the Making: The Natural History of the Assimilation of Immigrants.* New York: D. Appleton-Century.

Speek, Peter A. 1921. *A Stake in the Land.* New York: Harper.

Strickon, Arnold. 1979. "Ethnicity and Entrepreneurship in Rural Wisconsin." In *Entrepreneurs in Cultural Context,* ed. Sidney M. Greenfield, Arnold Strickon, and Robert T. Aubey, 159–89. Albuquerque: Univ. of New Mexico Press.

Swank, Scott T. et al. 1983. *Arts of the Pennsylvania Germans.* New York: W. W. Norton.

Swierenga, Robert P. 1980. "Ethnicity and American Agriculture." *Ohio History* 89:323–44.

______. 1981. "The Rural History: Defining the Parameters." *Great Plains Quarterly* 1:211–23.

Swierenga, Robert P., and Yda Saueressig-Schreuder. 1983. "Catholic and Protestant Emigration from the Netherlands in the 19th Century: A Comparative Social Structural Analysis." *Tijdschrift voor Economische en Sociale Geografie* 74:25–40.

Swierenga, Robert P., and Harry S. Stout. 1976. "Socio-Economic Patterns of Migration from the Netherlands in the Nineteenth Century." *Research in Economic History* 1:298–333.

Thistlethwaite, Frank. 1960. "Migration from Europe Overseas in the Nineteenth and Twentieth Centuries." *Rapports du XIe Congrès Internationale des Sciences historiques* (Stockholm), *Histoire Contemporaine* 5:32–60.

Thompson, Carl W., and G. P. Warber. 1913. *Social and Economic Survey of a Rural Township in Southern Minnesota.* Minneapolis: University of Minnesota, Bureau of

Research in Agricultural Economics, Department of Agriculture, Studies in Economics no 1.

Tishler, William, and Christopher S. Witmer. 1986. "The Housebarns of East-Central Wisconsin." In *Perspectives in Vernacular Architecture, II,* ed. Camille Wells, 102–10. Columbia: Univ. of Missouri Press.

Touart, Paul B. 1986. "The Acculturation of German-American Building Practices of Davidson County, North Carolina." In *Perspectives in Vernacular Architecture, II,* ed. Camille Wells, 72–80. Columbia: Univ. of Missouri Press.

Turner, Frederick Jackson. 1901. "The Stream of Immigration into the United States." Chicago *Record-Herald,* Sept. 25. Quoted in Edward Saveth. 1965. *American Historians and European Immigrants 1875–1925,* 132. New York: Russell and Russell.

______. 1920/1961. *Frontier and Section: Selected Essays.* Englewood Cliffs, N.J.: Prentice-Hall.

United States Census Office. 1870. Ninth Census. Vol. I, *Population.* Washington, D.C.: Government Printing Office.

______. 1880. Tenth Census. Vol. I, *Population.* Washington, D.C.: Government Printing Office.

Upton, Dell, ed. 1986. *American's Architectural Roots: Ethnic Groups that Built America.* Washington, D.C.: Preservation Press.

Useem, John. 1945. "Minority-Group Pattern in Prairie Society." *American Journal of Sociology* 50:377–85.

Vandergeest, Peter. 1988. "Commercialization and Commoditization: A Dialogue between Perspectives." *Sociologia Ruralis* 28:7–29.

Van Ravenswaay, Charles. 1977. *The Art and Architecture of German Settlements in Missouri.* Columbia: Univ. of Missouri Press.

Vecoli, Rudolph J. 1979. "The Resurgence of American Immigration History." *American Studies International* 17:46–66.

Vogeler, Ingolf. 1975. "Ethnicity, Religion, and Farm Land Transfers in Western Wisconsin." *Ecumene* 7:6–13.

______. 1976. "The Roman Catholic Culture Region of Central Minnesota." *Pioneer America* 8:71–83.

Ware, Caroline F. 1937. "Immigration." *Encyclopedia of the Social Sciences,* 7:587–95. New York: Macmillan.

Weaver, William Woys. 1986. "The Pennsylvania German House: European Antecedents and New World Forms." *Winterthur Portfolio* 21:243–64.

Wehrwein, Carl F. 1932. "Bonds of Maintenance as Aids in Acquiring Farm Ownership." *Journal of Land and Public Utility Economics* 8:396–403.

Wells, Camille ed. 1986. *Perspectives in Vernacular Architecture, II.* Columbia: Univ. of Missouri Press.

Wheeler, Wayne, 1986. *An Analysis of Social Change in a Swedish Immigrant Community.* New York: AMS Press.

Whetten, Nathan L., and Arnold W. Green. 1942. "Field Research and the Concept of Assimilation." *Rural Sociology* 7:252–60.

White, Clinton O. 1984. "Education among German Catholic Settlers in Saskatchewan, 1903–1918: A Reinterpretation." *Canadian Ethnic Studies* 16:78–95.

Winters, Donald L. 1978. *Farmers Without Farms: Agricultural Tenancy in Nineteenth-Century Iowa.* Westport, Conn.: Greenwood Press.

Zelinsky, Wilbur. 1973. *The Cultural Geography of the United States.* Englewood Cliffs, N.J.: Prentice-Hall.

APPENDIX

# Statistics of Agriculture in the United States

ALL TABLES in this appendix are reproduced from U.S. Department of Commerce, Bureau of the Census, *Historical Statistics of the United States: Colonial Times to 1970,* Part I (Washington, D.C.: U.S. Government Printing Office, 1975), Series K, pp. 458–525. Specific references to Series K tables accompany each of the tables in this appendix.

TABLE A.1. Farm population, farms, land in farms, and value of farm property and real estate, 1850–1910

| Year | Farm population | | | Number of farms (1,000) | Land in farms [1] | | | | | Value of all farm property (mil. dol.) | | | | Average value per farm of land and buildings (dol.) | Average value per acre of land and buildings (dol.) | Index of average value of farm real estate per acre (1967=100) |
|---|---|---|---|---|---|---|---|---|---|---|---|---|---|---|---|---|
| | Total (1,000) | Percent of total population | Net change through migration (1,000) | | Total (1,000 acres) | Percent of total land area | Average acreage per farm (acres) | Cropland (1,000 acres) | Pastureland (1,000 acres) | Total | Land and buildings [2] | Implements and machinery | Livestock [3] | | | |
| 1910 | 32,077 | 34.9 | -------- | [4] 6,366 | [4] 881,431 | [4] 38.8 | [4] 139 | [5] 311,293 | -------- | 40,959 | [4] 34,885 | 1,265 | 2,074 | [4] 5,480 | [4] 39.58 | -------- |
| 1900 | 29,875 | 41.9 | -------- | [4] 5,740 | [4] 841,202 | [4] 37.0 | [4] 147 | [5] 283,218 | -------- | 20,365 | [4] 16,614 | 750 | 1,991 | [4] 2,895 | [4] 19.82 | -------- |
| 1890 | 24,771 | 42.3 | -------- | 4,565 | 623,219 | 32.7 | 137 | [5] 219,706 | -------- | 16,439 | 13,279 | 494 | 1,346 | 2,909 | 21.31 | -------- |
| 1880 | 21,973 | 43.8 | -------- | 4,009 | 536,082 | 28.2 | 134 | [5] 166,187 | -------- | 12,404 | 10,197 | 406 | 1,064 | 2,544 | 19.02 | -------- |
| 1870 | -------- | -------- | -------- | 2,660 | 407,735 | 21.4 | 153 | -------- | -------- | 9,412 | 7,444 | 271 | 968 | 2,799 | 18.26 | -------- |
| 1860 | -------- | -------- | -------- | 2,044 | 407,213 | 21.4 | 199 | -------- | -------- | 7,980 | 6,645 | 246 | -------- | 3,251 | 16.32 | -------- |
| 1850 | -------- | -------- | -------- | 1,449 | 293,561 | 15.6 | 203 | -------- | -------- | 3,967 | 3,272 | 152 | -------- | 2,258 | 11.14 | -------- |

1. Intercensal estimates derived from straight-line interpolation. Excludes District of Columbia.
2. Census years as of date of enumeration. All other years as of March 1. Excludes District of Columbia.
3. Estimates as of January 1.
4. Includes Alaska and Hawaii.
5. Cropland harvested only.

Source: Series K 1–16, p. 457.

TABLE A.2. Farms, by race and tenure of operator, and acreage and value, by tenure of operator, 1880–1910

| Race and tenure of operator | 1910 | 1900 | 1890 | 1880 |
|---|---|---|---|---|
| NUMBER OF FARMS | | | | |
| **U.S.** | **6,365,822** | **5,739,657** | **4,564,641** | **4,008,907** |
| Full owner | 3,355,731 | 3,202,643 | | |
| Part owner | 593,954 | 451,515 | }3,269,728 | 2,984,306 |
| Manager | 58,353 | 59,213 | | |
| Tenant | 2,357,784 | 2,026,286 | 1,294,913 | 1,024,601 |
| White | 5,441,372 | 4,970,129 | ---------- | ---------- |
| Full owner | 3,159,560 | 3,026,214 | ---------- | ---------- |
| Part owner | 548,457 | 420,916 | ---------- | ---------- |
| Manager | 56,679 | 57,353 | ---------- | ---------- |
| Tenant | 1,676,676 | 1,465,646 | ---------- | ---------- |
| Negro and other | 924,450 | 769,528 | ---------- | ---------- |
| Full owner | 196,171 | 176,429 | ---------- | ---------- |
| Part owner | 45,497 | 30,599 | ---------- | ---------- |
| Manager | 1,674 | 1,860 | ---------- | ---------- |
| Tenant | 681,108 | 560,640 | ---------- | ---------- |
| **South** | **3,097,547** | **2,620,391** | **1,836,372** | **1,531,077** |
| Full owner | 1,329,390 | 1,237,114 | | |
| Part owner | 215,121 | 133,368 | }1,130,029 | 977,229 |
| Manager | 16,284 | 18,765 | | |
| Tenant | 1,536,752 | 1,231,144 | 706,343 | 553,848 |
| Croppers | ---------- | ---------- | ---------- | ---------- |
| White | 2,207,406 | 1,879,721 | ---------- | ---------- |
| Full owner | 1,154,100 | 1,078,635 | ---------- | ---------- |
| Part owner | 171,944 | 105,171 | ---------- | ---------- |
| Manager | 15,084 | 17,172 | ---------- | ---------- |
| Tenant | 866,278 | 678,743 | ---------- | ---------- |
| Croppers | ---------- | ---------- | ---------- | ---------- |
| Negro and other | 890,141 | 740,670 | ---------- | ---------- |
| Full owner | 175,290 | 158,479 | ---------- | ---------- |
| Part owner | 43,177 | 28,197 | ---------- | ---------- |
| Manager | 1,200 | 1,593 | ---------- | ---------- |
| Tenant | 670,474 | 552,401 | ---------- | ---------- |
| Croppers | ---------- | ---------- | ---------- | ---------- |

| Tenure of operator | 1910 | 1900 |
|---|---|---|
| LAND IN FARMS (1,000 acres) | | |
| **All farms** | **878,798** | **838,592** |
| Full owner | 464,923 | 431,261 |
| Part owner | 133,631 | 124,779 |
| Manager | 53,731 | 87,518 |
| Tenant | 226,513 | 195,034 |
| Croppers (South) | ---------- | ---------- |
| AVERAGE VALUE PER FARM OF FARMLAND AND BUILDINGS (dollars) | | |
| **All farms** | **5,471** | **2,896** |
| Full owner | 5,160 | 2,851 |
| Part owner | 8,515 | 4,347 |
| Manager | 25,075 | 13,114 |
| Tenant | 4,662 | 2,345 |
| Croppers (South) | ---------- | ---------- |

Source: Series K 109–153, pp. 465–66.

TABLE A.3. Farms and land in farms, by size of farm, 1880–1910 [farms in thousands; land in farms in thousands of acres]

| Year | Total | Under 10 acres: Total | Under 10 acres: Under 3 acres | Under 10 acres: 3–9 acres | 10–29 acres | 30–49 acres | 50–99 acres | 100–179 acres | 180–259 acres | 260–499 acres | 500–999 acres | 1,000 acres and over |
|---|---|---|---|---|---|---|---|---|---|---|---|---|
| Number of farms | | | | | | | | | | | | |
| 1910 [1] | 6,362 | 335 | 18 | 317 | 1,918 | | 1,438 | 2,051 | | 444 | 125 | 50 |
| 1900 | 5,740 | 268 | 42 | 227 | 1,665 | | 1,366 | | 2,291 | | 103 | 47 |
| 1890 | 4,565 | 150 | (NA) | (NA) | 1,168 | | 1,122 | | 2,009 | | 84 | 32 |
| 1880 | 4,009 | 139 | 4 | 135 | 1,036 | | 1,033 | | 1,696 | | 76 | 29 |
| Land in farms | | | | | | | | | | | | |
| 1910 | [2] 881,431 | (NA) | (NA) | (NA) | (NA) | | 103,121 | | 470,770 | | 83,653 | 167,082 |
| 1900 | 841,202 | 1,482 | 80 | 1,402 | 47,253 | | 98,600 | 295,978 | | 129,686 | 67,878 | 200,324 |

NA = not available.
Source: Series K 162–173, p. 467.

[1]Excludes Alaska and Hawaii.
[2]Total includes Alaska and Hawaii.

TABLE A.4. Wage rates for farmwork, 1866–1909

| Year | Per month [3]: With board and room | Per month [3]: With house | Per day [3]: With board and room | Per day [3]: Without board or room | Year | Per month [3]: With board and room | Per month [3]: With house | Per day [3]: With board and room | Per day [3]: Without board or room |
|---|---|---|---|---|---|---|---|---|---|
| | Dollars | Dollars | Dollars | Dollars | | Dollars | Dollars | Dollars | Dollars |
| 1909 | 22.00 | 28.00 | 1.00 | 1.25 | 1887 or 1888 | 13.50 | 19.50 | 0.70 | 1.00 |
| 1906 | 18.50 | 26.00 | 1.05 | 1.30 | 1884 or 1885 | 13.00 | 19.00 | .70 | .95 |
| 1902 | 15.50 | 22.00 | .85 | 1.10 | 1881 or 1882 | 13.00 | 19.00 | .70 | .95 |
| 1899 | 14.00 | 20.00 | .75 | 1.00 | | | | | |
| 1898 | 13.50 | 19.00 | .70 | .95 | 1880 or 1881 | 12.50 | 18.50 | .65 | .90 |
| | | | | | 1879 or 1880 | 11.50 | 17.50 | .65 | .90 |
| 1895 | 12.50 | 18.50 | .65 | .85 | 1877 [2] | 11.00 | 17.00 | .60 | .85 |
| 1894 | 12.50 | 18.50 | .65 | .85 | | | | | |
| 1893 | 14.00 | 20.00 | .70 | .90 | 1874 or 1875 | 11.00 | 17.00 | .70 | .95 |
| 1891 or 1892 | 13.50 | 20.00 | .75 | 1.00 | 1869 | 10.00 | 15.50 | .65 | .85 |
| 1889 or 1890 | 13.50 | 19.50 | .70 | .95 | 1866 [3] | 10.00 | 15.50 | .65 | .90 |

1. Annual averages are weighted averages of wage rates as reported quarterly by crop reporters.
2. 1877 or 1878, 1878 or 1879 (combined).
3. For years 1866 to 1878, paid in gold.
Source: Series K 174–183, p. 468.

TABLE A.5. Man-hours per unit and yield per unit of production of selected crops and livestock, 1800–1919 [figures for 5-year periods are annual averages]

| Year | Wheat | | | | | Corn for grain | | | | | Cotton | | | | |
|---|---|---|---|---|---|---|---|---|---|---|---|---|---|---|---|
| | Man-hours per acre | | | Yield per acre [1] (bu.) | Man-hours per 100 bushels | Man-hours per acre | | | Yield per acre [1] (bu.) | Man-hours per 100 bushels | Man-hours per acre | | | Yield of limit per acre [1] (pounds) | Man-hours per bale [2] |
| | Total | Before harvest | Harvest | | | Total | Before harvest | Harvest | | | Total | Before harvest | Harvest | | |
| 1910–1914 | 15.2 | 7.0 | 8.2 | 14.4 | 106 | 35.2 | 20.4 | 14.8 | 26.0 | 135 | 116 | 64 | 52 | 201 | 276 |
| 1900 | 15.0 | 7.0 | 8.0 | 13.9 | 108 | 38.0 | 22.0 | 16.0 | 25.9 | 147 | 112 | 62 | 50 | 189 | 284 |
| 1880 | 20.0 | 8.0 | 12.0 | 13.2 | 152 | 46.0 | 28.0 | 18.0 | 25.6 | 180 | 119 | 67 | 52 | 188 | 303 |
| 1840 | 35.0 | 12.0 | 23.0 | 15.0 | 233 | 69.0 | 44.0 | 25.0 | 25.0 | 276 | 135 | 90 | 45 | 147 | 438 |
| 1800 | 56.0 | 16.0 | 40.0 | 15.0 | 373 | 86.0 | 56.0 | 30.0 | 25.0 | 344 | 185 | 135 | 50 | 147 | 601 |

| Year | Hay | | | Potatoes | | | | | Tobacco | | | | |
|---|---|---|---|---|---|---|---|---|---|---|---|---|---|
| | Man-hours per acre harvested | Yield per acre (tons) | Man-hours per ton | Man-hours per acre | | | Yield per acre (cwt.) | Man-hours per ton | Man-hours per acre | | | Yield per acre (pounds) | Man-hours per 100 pounds |
| | | | | Total | Before harvest | Harvest | | | Total | Before harvest | Harvest | | |
| 1915–1919 | 13.0 | 1.25 | 10.4 | 73.8 | ---------- | ---------- | 56.9 | 26 | 353 | ---------- | ---------- | 803 | 44 |
| 1910–1914 | 11.9 | 1.15 | 10.3 | 76.0 | ---------- | ---------- | 59.8 | 25 | 356 | ---------- | ---------- | 816 | 44 |

| Year | Milk cows | | | Cattle, man-hours per cwt. of beef produced [3,4] | Hogs, man-hours per cwt. produced [4] | Chickens (laying flocks and eggs) | | | Chickens (farm raised) | | Chickens (broilers) | | Turkeys, man-hours per cwt. produced [4] |
|---|---|---|---|---|---|---|---|---|---|---|---|---|---|
| | Man-hours per cow | Milk per cow (pounds) | Man-hours per cwt. of milk | | | Man-hours per 100 layers | Rate of lay per year | Man-hours per 100 eggs produced | Man-hours per 100 birds | Man-hours per cwt. produced [4] | Man-hours per 100 birds | Man-hours per cwt. produced [4] | |
| 1915–1919 | 141.0 | 3,790.0 | 3.7 | 4.5 | 3.6 | ---------- | ---------- | ---------- | 33.0 | 9.4 | ---------- | ---------- | 31.1 |
| 1910–1914 | 146.0 | 3,842.0 | 3.8 | 4.6 | 3.6 | ---------- | ---------- | ---------- | 33.0 | 9.5 | ---------- | ---------- | 31.4 |

1. Data for 1800 and 1840 are estimates. Data for 1880 and 1900 are 5-year averages of published data, centered on year shown.

2. For statistical purposes, the bale of cotton is 500 pounds net weight of lint. Running bales reported prior to 1899 have been converted to bales of 478 pounds net weight. Actual bale weights vary considerably.

3. Production includes beef produced as a byproduct of the milk cow enterprise.

4. Live-weight production.

Source: Series K 445–485, pp. 500–501.

TABLE A.6. Value of farm gross output and product, in current and constant (1910–1914) dollars, 1800–1900 [in millions of dollars]

| Year | Gross output | | | | | | Farm gross product | Intermediate products consumed | Farm gross product, including improvements and manufactures | | | |
|---|---|---|---|---|---|---|---|---|---|---|---|---|
| | Total | Sales and home consumption of farm products | | | Livestock inventory changes | Gross rent from farm dwellings | | | Total | Farm gross product | Improvements to farms | Value of home manufactures |
| | | Total | Livestock | Crops | | | | | | | | |
| CURRENT DOLLARS | | | | | | | | | | | | |
| 1900 | 4,298 | 3,912 | 2,047 | 1,865 | 79 | 307 | 3,799 | 499 | 3,857 | 3,799 | 55 | 3 |
| 1890 | 3,397 | 3,106 | 1,515 | 1,591 | 44 | 247 | 3,035 | 362 | 3,107 | 3,035 | 67 | 5 |
| 1880 | 3,263 | 3,021 | 1,498 | 1,523 | 39 | 203 | 2,967 | 296 | 3,045 | 2,967 | 68 | 10 |
| 1870 | 2,774 | 2,553 | 1,393 | 1,160 | 52 | 169 | 2,542 | 232 | 2,631 | 2,542 | 67 | 22 |
| 1860 | 1,579 | 1,469 | 700 | 769 | 21 | 89 | 1,484 | 95 | 1,556 | 1,484 | 47 | 25 |
| 1850 | 904 | 837 | 414 | 423 | 10 | 57 | 851 | 53 | 914 | 851 | 34 | 29 |
| 1840 | 757 | 699 | 431 | 268 | 14 | 44 | 720 | 37 | 769 | 720 | 22 | 27 |
| 1830 | 466 | 427 | 251 | 176 | 15 | 24 | 445 | 21 | 491 | 445 | 17 | 29 |
| 1820 | 338 | 308 | 178 | 130 | 10 | 20 | 323 | 15 | 364 | 323 | 12 | 29 |
| 1810 | 336 | 311 | 186 | 125 | 9 | 16 | 324 | 12 | 363 | 324 | 9 | 30 |
| 1800 | 236 | 220 | 127 | 93 | 6 | 10 | 230 | 6 | 255 | 230 | 7 | 18 |
| CONSTANT (1910–14) DOLLARS | | | | | | | | | | | | |
| 1900 | 6,409 | 5,903 | 3,100 | 2,803 | 109 | 397 | 5,740 | 669 | 5,837 | 5,740 | 94 | 3 |
| 1890 | 4,990 | 4,604 | 2,612 | 1,992 | 70 | 316 | 4,527 | 463 | 4,638 | 4,527 | 106 | 5 |
| 1880 | 4,129 | 3,784 | 2,006 | 1,778 | 68 | 277 | 3,770 | 359 | 3,906 | 3,770 | 128 | 8 |
| 1870 | 2,694 | 2,436 | 1,436 | 1,000 | 74 | 184 | 2,479 | 215 | 2,597 | 2,479 | 106 | 12 |
| 1860 | 2,186 | 1,985 | 1,088 | 897 | 60 | 141 | 2,059 | 127 | 2,156 | 2,059 | 76 | 21 |
| 1850 | 1,521 | 1,379 | 826 | 553 | 42 | 100 | 1,442 | 79 | 1,536 | 1,442 | 69 | 25 |
| 1840 | 1,212 | 1,103 | 651 | 452 | 33 | 76 | 1,156 | 56 | 1,222 | 1,156 | 47 | 19 |
| 1830 | 854 | 764 | 462 | 302 | 34 | 56 | 819 | 35 | 879 | 819 | 44 | 16 |
| 1820 | 619 | 555 | 345 | 210 | 22 | 42 | 595 | 24 | 642 | 595 | 33 | 14 |
| 1810 | 463 | 415 | 260 | 155 | 16 | 32 | 448 | 15 | 485 | 448 | 26 | 11 |
| 1800 | 343 | 307 | 194 | 113 | 13 | 23 | 333 | 10 | 362 | 333 | 21 | 8 |

Source: Series K 240–250, p. 482.

TABLE A.7. Expenditures for, and consumption of, fertilizer and lime, 1850–1920

| Year | Farmers' expenditures for fertilizer and lime | Commercial fertilizer consumed in U.S. [1] | Lime consumed on farms [2] | Year | Commercial fertilizer consumed in U.S. [1] |
|---|---|---|---|---|---|
| | 192 | 193 | 194 | | 193 |
| | *Mil. dol.* | *1,000 short tons* | *1,000 short tons* | | *1,000 short tons* |
| 1920 | 390 | 7,176 | 2,653 | 1895 | 1,578 |
| 1919 | 358 | 6,751 | 2,476 | 1894 | 1,773 |
| 1918 | 311 | 6,580 | 2,306 | 1893 | 1,715 |
| 1917 | 232 | 6,087 | 2,136 | 1892 | 1,504 |
| 1916 | 193 | 5,214 | 1,966 | 1891 | 1,584 |
| 1915 | 165 | 5,418 | 1,796 | 1890 | 1,390 |
| 1914 | 195 | 7,194 | 1,626 | 1880 | 753 |
| 1913 | 175 | 6,416 | 1,456 | 1870 | 321 |
| 1912 | 161 | 5,852 | 1,286 | 1860 | 164 |
| 1911 | 168 | 6,108 | 1,116 | 1850 | 53 |
| 1910 | 152 | 5,547 | 946 | | |
| 1909 | 120 | 4,821 | ---------- | | |
| 1908 | ---------- | 4,449 | ---------- | | |
| 1907 | ---------- | 4,307 | ---------- | | |
| 1906 | ---------- | 4,249 | ---------- | | |
| 1905 | ---------- | 3,913 | ---------- | | |
| 1904 | ---------- | 3,704 | ---------- | | |
| 1903 | ---------- | 3,382 | ---------- | | |
| 1902 | ---------- | 3,084 | ---------- | | |
| 1901 | ---------- | 3,044 | ---------- | | |
| 1900 | ---------- | 2,730 | ---------- | | |
| 1899 | ---------- | 2,603 | ---------- | | |
| 1898 | ---------- | 2,333 | ---------- | | |
| 1897 | ---------- | 2,131 | ---------- | | |
| 1896 | ---------- | 1,888 | ---------- | | |

[1]Includes Puerto Rico, Hawaii, and Alaska for all years.
[2]Excludes Alaska and Hawaii for all years.
Source: Series K 192–194, p. 469.

TABLE A.8. Livestock – number, value per head, production, and price, 1867–1908 [Census figures in *italics*. All figures are as of January 1 except for 1870, 1880, 1890, 1900 (June 1); 1910 (April 15)]

| Year | Number on farms and value per head | | | | | | | | | |
|---|---|---|---|---|---|---|---|---|---|---|
| | All cattle | | Hogs | | Stock sheep | | Horses [1] | | Mules [1] | |
| | Number | Value per head | Number | Value per head | Number | Value per head | Number | Value per head | Number | Value per head |
| | *1,000 head* | *Dollars* | *1,000 head* | *Dollars* | *1,000 head* | *Dollars* | *1,000 head* | *Dollars* | *1,000 head* | *Dollars* |
| *1910* | *61,804* | ----- | *58,186* | ----- | *52,448* | ----- | *19,833* | ----- | *4,101* | ----- |
| 1910 | 58,993 | 24.54 | 48,072 | 9.05 | 46,939 | 4.06 | 19,972 | 107.70 | 4,239 | 119.98 |
| 1909 | 60,774 | 21.99 | 52,508 | 6.45 | 47,098 | 3.42 | 19,731 | 95.13 | 4,085 | 108.20 |
| 1908 | 61,989 | 20.92 | 58,388 | 5.99 | 45,095 | 3.87 | 19,444 | 92.76 | 3,949 | 107.81 |
| 1907 | 63,754 | 20.91 | 56,543 | 7.54 | 43,460 | 3.81 | 19,090 | 92.85 | 3,814 | 111.46 |
| 1906 | 65,009 | 19.65 | 53,633 | 6.07 | 41,965 | 3.51 | 18,806 | 79.77 | 3,680 | 97.75 |
| 1905 | 66,111 | 18.39 | 53,176 | 5.89 | 40,410 | 2.77 | 18,491 | 69.73 | 3,586 | 87.06 |
| 1904 | 66,442 | 19.69 | 51,623 | 6.08 | 41,908 | 2.55 | 18,331 | 67.59 | 3,465 | 78.02 |
| 1903 | 66,004 | 21.55 | 48,100 | 7.69 | 44,436 | 2.62 | 18,121 | 62.27 | 3,353 | 71.73 |
| 1902 | 64,418 | 21.48 | 47,858 | 6.95 | 46,196 | 2.62 | 17,968 | 58.52 | 3,264 | 67.23 |
| 1901 | 62,576 | 22.68 | 50,681 | 6.08 | 46,126 | 2.96 | 17,955 | 53.03 | 3,190 | 53.47 |
| *1900* [9] | *67,719* | ----- | *62,868* | ----- | *61,504* | ----- | *16,965* | ----- | *3,039* | ----- |
| 1900 | 59,739 | 26.50 | 51,055 | 5.36 | 45,065 | 2.97 | 17,856 | 43.56 | 3,139 | 51.46 |
| 1899 | 55,927 | 24.53 | 51,558 | 4.67 | 42,688 | 2.80 | 17,728 | 36.61 | 3,012 | 43.52 |
| 1898 | 52,868 | 22.79 | 53,282 | 4.70 | 40,097 | 2.51 | 17,698 | 33.35 | 2,918 | 42.31 |
| 1897 | 50,447 | 18.62 | 51,232 | 4.36 | 38,891 | 1.84 | 17,803 | 30.92 | 2,836 | 40.49 |
| 1896 | 49,205 | 17.86 | 49,154 | 4.50 | 39,609 | 1.71 | 17,876 | 32.34 | 2,782 | 44.08 |
| 1895 | 49,510 | 16.56 | 47,628 | 5.09 | 41,827 | 1.57 | 17,849 | 35.57 | 2,708 | 47.23 |
| 1894 | 51,713 | 16.84 | 46,522 | 6.06 | 43,414 | 1.97 | 17,709 | 46.63 | 2,632 | 60.65 |
| 1893 | 55,119 | 17.00 | 43,652 | 6.37 | 44,567 | 2.64 | 17,289 | 60.72 | 2,550 | 69.18 |
| 1892 | 58,126 | 16.81 | 45,165 | 4.65 | 44,628 | 2.60 | 16,846 | 64.56 | 2,459 | 74.31 |
| 1891 | 59,968 | 16.49 | 47,435 | 4.24 | 43,882 | 2.51 | 16,329 | 67.19 | 2,377 | 76.93 |
| *1890* | *57,649* | ----- | *57,427* | ----- | *40,876* | ----- | *15,266* | ----- | *2,252* | ----- |
| 1890 | 60,014 | 16.95 | 48,130 | 4.80 | 42,693 | 2.29 | 15,732 | 69.27 | 2,322 | 77.61 |
| 1889 | 59,178 | 18.77 | 44,508 | 5.80 | 42,365 | 2.14 | 15,064 | 72.39 | 2,295 | 78.95 |
| 1888 | 58,599 | 19.39 | 42,134 | 5.12 | 43,011 | 2.06 | 14,490 | 72.03 | 2,260 | 79.06 |
| 1887 | 56,602 | 21.18 | 42,563 | 4.60 | 44,217 | 2.05 | 13,821 | 71.59 | 2,213 | 78.39 |
| 1886 | 54,868 | 22.20 | 45,457 | 4.30 | 46,654 | 1.95 | 13,276 | 70.62 | 2,162 | 78.96 |
| 1885 | 52,463 | 24.40 | 47,330 | 5.06 | 49,620 | 2.19 | 12,700 | 72.94 | 2,102 | 81.88 |
| 1884 | 49,804 | 25.26 | 45,961 | 5.64 | 51,101 | 2.40 | 12,215 | 73.80 | 2,047 | 83.53 |
| 1883 | 47,387 | 23.87 | 43,440 | 6.74 | 50,935 | 2.53 | 11,794 | 69.92 | 1,975 | 77.79 |
| 1882 | 45,738 | 20.93 | 42,566 | 6.00 | 48,883 | 2.35 | 11,444 | 58.75 | 1,928 | 71.69 |
| 1881 | 44,501 | 18.67 | 43,076 | 4.80 | 47,371 | 2.35 | 11,187 | 57.91 | 1,912 | 68.84 |
| *1880* | *39,676* | ----- | *49,773* | ----- | *42,192* | ----- | *10,357* | ----- | *1,813* | ----- |
| 1880 | 43,347 | 17.80 | 44,327 | 4.40 | 44,867 | 2.18 | 10,903 | 53.74 | 1,878 | 61.74 |
| 1879 | 41,420 | 16.96 | 43,767 | 3.15 | 41,678 | 2.01 | 10,574 | 51.55 | 1,816 | 57.08 |
| 1878 | 39,396 | 19.05 | 43,375 | 4.89 | 38,942 | 2.12 | 10,230 | 55.38 | 1,746 | 62.61 |
| 1877 | 37,333 | 18.38 | 39,333 | 5.68 | 38,147 | 2.03 | 9,910 | 55.11 | 1,674 | 63.16 |
| 1876 | 36,140 | 18.76 | 35,715 | 5.97 | 37,477 | 2.20 | 9,606 | 56.48 | 1,608 | 65.51 |
| 1875 | 35,361 | 18.96 | 35,834 | 4.65 | 37,237 | 2.39 | 9,333 | 60.10 | 1,548 | 71.64 |
| 1874 | 34,821 | 19.51 | 38,377 | 3.93 | 36,234 | 2.33 | 9,055 | 64.12 | 1,485 | 80.26 |
| 1873 | 33,830 | 20.50 | 39,794 | 3.60 | 35,782 | 2.60 | 8,767 | 65.52 | 1,419 | 83.49 |
| 1872 | 33,078 | 21.64 | 39,296 | 3.96 | 34,312 | 2.51 | 8,441 | 66.54 | 1,360 | 86.02 |
| 1871 | 32,107 | 24.71 | 36,688 | 5.48 | 34,063 | 2.10 | 8,054 | 70.02 | 1,305 | 91.47 |
| *1870* | *23,821* | ----- | *25,135* | ----- | *28,478* | ----- | *7,145* | ----- | *1,125* | ----- |
| 1870 | 31,082 | 22.84 | 33,781 | 5.64 | 36,449 | 1.87 | 7,633 | 66.99 | 1,245 | 89.71 |
| 1869 | 30,060 | 20.74 | 32,570 | 4.60 | 39,802 | 1.65 | 7,304 | 60.48 | 1,130 | 78.57 |
| 1868 | 29,238 | 18.30 | 33,304 | 3.23 | 43,808 | 1.83 | 7,051 | 52.54 | 1,057 | 56.70 |
| 1867 | 28,636 | 19.13 | 34,489 | 3.95 | 44,997 | 2.40 | 6,820 | 57.56 | 1,000 | 67.73 |

1. Excludes spring-born colts.
Source: Series K 564–582, p. 520.

TABLE A.9. Cows kept for milk on farms, milk produced, manufactured dairy products, prices received by farmers, and wholesale prices of cheese and butter, 1830–1910 [Census figures in *italics*]

| Year | Cows and heifers 2 years old and over kept for milk, Jan. 1 | | Milk produced on farms during year | Production of dairy products | | | | Milk equivalent of manufactured dairy products [4] | Prices received by farmers | | | | Wholesale prices | |
|---|---|---|---|---|---|---|---|---|---|---|---|---|---|---|
| | Number | Value per head | | Butter [1] | Cheese [2] | Evaporated and condensed milk [3] | Ice cream | | Butter, per pound | Milkfat in cream, per pound | Whole milk: Wholesale, per 100 pounds | Whole milk: Retail, per quart | Cheese, American twins, per pound [5] | Butter at New York, per pound [6] |
| | *1,000 head* | *Dollars* | *Million pounds* | *Million pounds* | *Million pounds* | *Million pounds* | *Million gallons* | *Million pounds* | *Cents* | *Cents* | *Dollars* | *Cents* | *Cents* | *Cents* |
| *1910* | *20,625* | | | | | | | | | | | | | |
| 1910 | 19,450 | 35.40 | | 1,706 | 355 | 556 | | 41,132 | 25.5 | 26.4 | 1.58 | 6.6 | 14.7 | 31.1 |
| 1909 | 19,201 | 32.09 | 64,211 | 1,622 | 313 | 495 | 30 | 38,715 | 24.0 | 25.5 | | 6.4 | 14.6 | 29.9 |
| 1908 | 18,992 | 30.48 | | 1,763 | 313 | 450 | | 41,439 | | | | | 12.2 | 27.6 |
| 1907 | 18,629 | 30.63 | | 1,537 | 286 | 410 | | 36,290 | | | | | 13.4 | 28.1 |
| 1906 | 18,230 | 29.34 | | 1,545 | 292 | 373 | | 36,403 | | | | | 11.8 | 24.6 |
| 1905 | 17,823 | 27.19 | | 1,667 | 327 | 339 | | 39,210 | | | | | 11.7 | 24.6 |
| 1904 | 17,485 | 29.00 | | 1,540 | 331 | 308 | 12 | 36,468 | | | | | 9.3 | 21.7 |
| 1903 | 17,217 | 30.06 | | 1,485 | 323 | 279 | | 35,159 | | | | | 11.1 | 23.4 |
| 1902 | 16,992 | 29.08 | | 1,401 | 318 | 252 | | 33,248 | | | | | 11.2 | 24.7 |
| 1901 | 16,708 | 29.88 | | 1,575 | 362 | 228 | | 37,280 | | | | | 9.8 | 21.4 |
| *1900* | *17,136* | | | | | | | | | | | | | |
| 1900 | 16,544 | 31.30 | | 1,540 | 324 | 207 | | 36,106 | | | | | 10.0 | 22.2 |
| 1899 | 16,094 | 29.46 | 62,486 | 1,493 | 299 | 187 | 5 | 34,806 | | | | | [8] 10.6 | 21.3 |
| 1898 | 15,641 | 27.34 | | 1,473 | 281 | | | 34,145 | | | | | [8] 7.6 | 19.6 |
| 1897 | 15,382 | 23.08 | | 1,533 | 311 | | | 35,640 | | | | | [8] 8.5 | 19.0 |
| 1896 | 15,266 | 22.53 | | 1,604 | 240 | | | 36,385 | | | | | [8] 7.7 | 18.5 |
| 1895 | 15,230 | 22.11 | | 1,297 | 234 | | | 29,828 | | | | | [8] 7.4 | 21.2 |
| 1894 | 15,237 | 21.86 | | 1,063 | 257 | | | 25,113 | | | | | [8] 10.2 | 23.0 |
| 1893 | 15,164 | 21.90 | | 1,047 | 254 | | | 24,718 | | | | | [8] 9.6 | 27.1 |
| 1892 | 15,177 | 21.53 | | 1,058 | 318 | | | 25,561 | | | | | [8] 9.3 | 26.3 |
| 1891 | 15,133 | 21.73 | | 1,091 | 293 | | | 25,990 | | | | | [8] 8.9 | 26.2 |
| *1890* | *16,512* | | | | | | | | | | | | | |
| 1890 | 15,000 | 22.30 | | 1,171 | 318 | | | 27,906 | | | | | [8] 9.0 | 23.7 |
| 1889 | 14,706 | 24.03 | 44,807 | 1,292 | 301 | 45 | 1 | 30,260 | | | | | [8] 8.7 | 24.4 |
| 1888 | 14,350 | 24.82 | | 978 | 286 | | | 23,494 | | | | | [8] 8.1 | 27.5 |
| 1887 | 13,888 | 26.23 | | 978 | 268 | | | 23,301 | | | | | [8] 10.8 | 26.7 |
| 1886 | 13,478 | 27.52 | | 989 | 244 | | | 23,283 | | | | | [8] 9.6 | 26.8 |
| 1885 | 13,213 | 29.88 | | 933 | 260 | | | 22,258 | | | | | [8] 8.7 | 26.6 |
| 1884 | 12,883 | 31.58 | | 869 | 275 | | | 21,061 | | | | | [8] 11.1 | 30.3 |
| 1883 | 12,571 | 30.47 | | 844 | 281 | | | 20,584 | | | | | [8] 11.0 | 31.2 |
| 1882 | 12,234 | 26.12 | | 743 | 261 | | | 18,248 | | | | | [8] 11.9 | 35.6 |
| 1881 | 11,977 | 23.82 | | 803 | 304 | | | 19,934 | | | | | [8] 12.4 | 31.8 |

See footnotes at end of table.

TABLE A.9 (*continued*)

| Year | Cows and heifers 2 years old and over kept for milk, Jan. 1 | | Milk produced on farms during year | Production of dairy products | | | | Milk equivalent of manufactured dairy products [4] | Prices received by farmers | | | | Wholesale prices | |
|---|---|---|---|---|---|---|---|---|---|---|---|---|---|---|
| | Number | Value per head | | Butter [1] | Cheese [2] | Evaporated and condensed milk [3] | Ice cream | | Butter, per pound | Milkfat in cream, per pound | Whole milk: Wholesale, per 100 pounds | Whole milk: Retail, per quart | Cheese, American twins, per pound [5] | Butter at New York, per pound [6] |
| | *1,000 head* | *Dollars* | *Million pounds* | *Million pounds* | *Million pounds* | *Million pounds* | *Million gallons* | *Million pounds* | *Cents* | *Cents* | *Dollars* | *Cents* | *Cents* | *Cents* |
| *1880* | *12,443* | | | | | | | | | | | | | |
| 1880 | 11,754 | 23.31 | | 816 | 270 | | | 19,861 | | | | | [8] 12.5 | 30.5 |
| 1879 | 11,486 | 21.55 | | 807 | 243 | 13 | (Z) | 19,402 | | | | | [8] 8.0 | 24.2 |
| 1878 | 11,222 | 25.70 | | 726 | 303 | | | 18,307 | | | | | | 27.3 |
| 1877 | 11,004 | 25.14 | | 696 | 235 | | | 16,995 | | | | | | 28.5 |
| 1876 | 10,821 | 25.20 | | 677 | 214 | | | 16,390 | | | | | | 31.3 |
| 1875 | 10,714 | 25.29 | | 556 | 233 | | | 14,029 | | | | | | 32.8 |
| 1874 | 10,562 | 25.20 | | 585 | 206 | | | 14,347 | | | | | | 36.2 |
| 1873 | 10,348 | 26.32 | | 566 | 212 | | | 14,029 | | | | | | 35.4 |
| 1872 | 10,191 | 29.18 | | 434 | 187 | | | 10,997 | | | | | | 32.0 |
| 1871 | 9,941 | 33.62 | | 470 | 164 | | | 11,527 | | | | | | 33.6 |
| *1870* | *8,935* | | | | | | | | | | | | | |
| 1870 | 9,672 | 31.89 | | 412 | 181 | | | 10,472 | | | | | | 38.1 |
| 1869 | 9,205 | 28.86 | | 514 | 163 | 4 | (Z) | 12,434 | | | | | | 43.3 |
| 1868 | 8,705 | 26.96 | | | | | | | | | | | | 44.7 |
| 1867 | 8,263 | 29.40 | | | | | | | | | | | | 34.8 |
| 1866 | | | | | | | | | | | | | | 42.7 |
| 1865 | | | | | | | | | | | | | | 39.8 |
| 1864 | | | | | | | | | | | | | | 43.7 |
| 1863 | | | | | | | | | | | | | | 28.2 |
| 1862 | | | | | | | | | | | | | | 20.9 |
| 1861 | | | | | | | | | | | | | | 19.4 |
| *1860* | *8,586* | | | | | | | | | | | | | |
| 1860 | | | | | | | | | | | | | | 21.9 |
| 1859 | | | | 460 | 104 | | (Z) | 10,690 | | | | | | 23.9 |
| 1858 | | | | | | | | | | | | | | 23.8 |
| 1857 | | | | | | | | | | | | | | 25.7 |
| 1856 | | | | | | | | | | | | | | 25.8 |
| 1855 | | | | | | | | | | | | | | 26.4 |
| 1854 | | | | | | | | | | | | | | 23.0 |
| 1853 | | | | | | | | | | | | | | 23.0 |
| 1852 | | | | | | | | | | | | | | 23.6 |
| 1851 | | | | | | | | | | | | | | 18.4 |

| Year | Cows and heifers 2 years old and over kept for milk, Jan. 1, number | Production of dairy products: Butter [1] | Production of dairy products: Cheese [2] | Milk equivalent of manufactured dairy products [4] | Wholesale price, butter at New York, per pound [6] | Year | Wholesale price, butter at New York, per pound [6] | Year | Wholesale price, butter at New York, per pound [6] | Year | Wholesale price, butter at New York, per pound [6] |
|---|---|---|---|---|---|---|---|---|---|---|---|
| | *1,000 head* | *Million pounds* | *Million pounds* | *Million pounds* | *Cents* | | *Cents* | | *Cents* | | *Cents* |
| *1850* | *6,385* | | | | | 1845 | 17.7 | 1840 | 17.4 | 1835 | 19.2 |
| 1850 | | | | | 19.6 | 1844 | 15.2 | 1839 | 22.9 | 1834 | 14.4 |
| 1849 | | 313 | 106 | 7,636 | 18.9 | 1843 | 13.3 | 1838 | 23.4 | 1833 | 15.8 |
| 1848 | | | | | 20.1 | 1842 | 16.5 | 1837 | 21.6 | 1832 | 15.2 |
| 1847 | | | | | 20.7 | 1841 | 18.6 | 1836 | 23.9 | 1831 | 14.9 |
| 1846 | | | | | 16.7 | | | | | 1830 | 13.9 |

NA = not available.

Z Less than 500,000 gallons.

1. Farm and factory production combined.

2. Includes all types of cheese except cottage, pot, and bakers' cheese; full-skim American cheese excluded since 1908.

3. Includes total production of all condensed and evaporated milk as interpolated from census enumerations.

4. For 1849–1923, computed from data on estimated production of manufactured dairy products, using average milk equivalent factors; 1924–1970, as published by Agricultural Marketing Service. Data include farm butter.

5. On Wisconsin cheese exchange, based on weekly prices established on Friday each week.

6. Annual averages of monthly figures from sources and for grades as follows: 1830–1879, average of high and low for 2 days each week, high grade, New York shipping and commercial list; 1880–1895 average of monthly range, creamery extras, annual reports of New York Chamber of Commerce; 1896–1920, average daily quotations for extra fresh, specials, extras and firsts, or fresh extras, *New York Produce Review and American Creamery,* published by Urner-Barry Company.

Source: Series K 595–608, pp. 523–24.

TABLE A.10. Corn, wheat, oats, and barley—acreage, production, price, and stocks, 1839–1910 [Census figures in *italics*]

| Year | Corn for all purposes | | | All wheat for grain | | | Oats for grain | | | Barley for grain | | |
|---|---|---|---|---|---|---|---|---|---|---|---|---|
| | Acreage harvested | Production | Price per bushel [1] | Acreage harvested | Production | Price per bushel [1] | Acreage harvested | Production | Price per bushel [1] | Acreage harvested | Production | Price per bushel [1] |
| | *1,000 acres* | *Million bushels* | *Dollars* | *1,000 acres* | *Million bushels* | *Dollars* | *1,000 acres* | *Million bushels* | *Dollars* | *1,000 acres* | *Million bushels* | *Dollars* |
| 1910 | 102,267 | 2,853 | .52 | 45,793 | 625 | .91 | 36,844 | 1,106 | .36 | 7,546 | 142 | .61 |
| *1909* | [2] *98,386* | [2] *2,552* | *.56* | *44,263* | *683* | *.96* | *35,159* | *1,007* | *.41* | *7,699* | *173* | *.53* |
| 1909 | 100,200 | 2,611 | .62 | 44,262 | 684 | .99 | 35,062 | 1,014 | .43 | 7,697 | 173 | .56 |
| 1908 | 95,285 | 2,567 | .65 | 45,102 | 643 | .97 | 34,310 | 829 | .49 | 7,409 | 171 | .57 |
| 1907 | 96,094 | 2,614 | .51 | 44,139 | 629 | .87 | 34,439 | 801 | .44 | 6,854 | 151 | .67 |
| 1906 | 95,624 | 3,033 | .39 | 46,230 | 741 | .66 | 33,688 | 1,023 | .32 | 6,744 | 179 | .42 |
| 1905 | 95,746 | 2,954 | 41 | 46,306 | 706 | .75 | 33,426 | 1,104 | .29 | 6,658 | 172 | .39 |
| 1904 | 95,228 | 2,687 | 44 | 43,155 | 556 | .93 | 32,749 | 1,012 | .31 | 6,579 | 166 | .41 |
| 1903 | 93,555 | 2,515 | .42 | 48,456 | 663 | .69 | 32,187 | 885 | .34 | 6,231 | 149 | .45 |
| 1902 | 97,177 | 2,774 | .40 | 46,244 | 687 | .63 | 31,358 | 1,077 | .31 | 5,474 | 146 | .45 |
| 1901 | 94,422 | 1,716 | .60 | 50,847 | 763 | .63 | 30,891 | 800 | .40 | 4,963 | 124 | .45 |
| 1900 | 94,852 | 2,662 | 0.35 | 49,203 | 599 | 0.62 | 31,049 | 945 | 0.25 | 4,703 | 97 | 0.41 |
| *1899* | [2] *94,917* | [2] *2,666* | *.31* | *52,589* | *659* | *.56* | *29,540* | *943* | *.23* | *4,470* | *120* | *.35* |
| 1899 | 94,591 | 2,646 | .30 | 52,342 | 655 | .59 | 29,254 | 937 | .26 | 4,472 | 118 | .39 |
| 1898 | 87,784 | 2,351 | .29 | 50,506 | 768 | .58 | 29,327 | 842 | .25 | 4,113 | 98 | .39 |
| 1897 | 89,965 | 2,288 | .26 | 43,413 | 606 | .81 | 28,829 | 830 | .21 | 4,120 | 103 | .34 |
| 1896 | 89,074 | 2,671 | .21 | 40,828 | 523 | .72 | 30,248 | 775 | .18 | 4,131 | 97 | .30 |
| 1895 | 90,479 | 2,535 | .25 | 38,998 | 542 | .51 | 30,905 | 925 | .19 | 4,185 | 104 | .33 |
| 1894 | 80,069 | 1,615 | .45 | 40,167 | 542 | .49 | 29,556 | 750 | .32 | 3,639 | 74 | .44 |
| 1893 | 79,832 | 1,900 | .36 | 40,790 | 506 | .53 | 29,266 | 707 | .29 | 3,689 | 87 | .40 |
| 1892 | 76,914 | 1,897 | .39 | 42,979 | 612 | .62 | 28,168 | 722 | .32 | 3,857 | 95 | .47 |
| 1891 | 78,855 | 2,336 | .40 | 41,090 | 678 | .83 | 27,756 | 837 | .31 | 3,590 | 94 | .52 |
| 1890 | 74,785 | 1,650 | .50 | 36,686 | 449 | .84 | 28,275 | 609 | .42 | 3,250 | 70 | .62 |
| *1889* | [2] *72,088* | [2] *2,122* | | *33,580* | *468* | | *28,321* | *809* | | *3,221* | *78* | |
| 1889 | 77,656 | 2,294 | .28 | 36,098 | 504 | .70 | 28,697 | 831 | .22 | 3,352 | 81 | .42 |
| 1888 | 77,474 | 2,251 | .33 | 34,969 | 424 | .93 | 27,807 | 773 | .27 | 3,283 | 76 | .59 |
| 1887 | 73,296 | 1,605 | .43 | 36,873 | 491 | .68 | 26,272 | 696 | .30 | 3,258 | 72 | .52 |
| 1886 | 73,911 | 1,783 | .36 | 36,312 | 514 | .69 | 24,426 | 682 | .29 | 3,027 | 74 | .53 |

| | | | | | | | | | | | | | |
|---|---|---|---|---|---|---|---|---|---|---|---|---|---|
| 1885 | 71,854 | 2,058 | .32 | 35,095 | 400 | .77 | 23,351 | 674 | .28 | 2,862 | 64 | .56 |
| 1884 | 68,834 | 1,948 | .35 | 38,485 | 571 | .65 | 21,974 | 641 | .27 | 2,694 | 68 | .48 |
| 1883 | 68,168 | 1,652 | .42 | 35,587 | 439 | .91 | 20,621 | 606 | .32 | 2,474 | 57 | .59 |
| 1882 | 66,157 | 1,755 | .48 | 36,496 | 552 | .89 | 19,075 | 540 | .37 | 2,434 | 60 | .63 |
| 1881 | 63,026 | 1,245 | .63 | 36,795 | 406 | 1.20 | 16,916 | 446 | .46 | 2,201 | 49 | .82 |
| 1880 | 62,545 | 1,707 | .39 | 38,096 | 502 | .95 | 16,414 | 418 | .35 | 1,990 | 45 | .66 |
| *1879* | [2] *62,369* | [2] *1,755* | ----- | *35,430* | *459* | ----- | *16,145* | *408* | ----- | *1,998* | *44* | ----- |
| 1879 | 62,229 | 1,752 | .36 | 35,347 | 459 | 1.11 | 15,955 | 415 | .33 | 1,926 | 42 | .60 |
| 1878 | 59,659 | 1,565 | .31 | 33,379 | 449 | .77 | 15,830 | 443 | .24 | 1,848 | 37 | .58 |
| 1877 | 58,799 | 1,516 | .36 | 27,963 | 396 | 1.08 | 14,816 | 435 | .29 | 1,962 | 39 | .63 |
| 1876 | 55,277 | 1,478 | .36 | 28,283 | 309 | 1.04 | 14,589 | 327 | .35 | 1,973 | 41 | .69 |
| 1875 | 52,446 | 1,450 | .42 | 28,382 | 314 | 1.01 | 13,616 | 365 | .37 | 1,702 | 33 | .86 |
| 1874 | 47,640 | 1,059 | .64 | 27,310 | 356 | .95 | 12,775 | 273 | .52 | 1,628 | 36 | .96 |
| 1873 | 44,084 | 1,008 | .48 | 24,866 | 322 | 1.17 | 12,010 | 307 | .37 | 1,473 | 31 | .96 |
| 1872 | 43,584 | 1,279 | .38 | 22,962 | 271 | 1.24 | 11,789 | 327 | .32 | 1,421 | 32 | .74 |
| 1871 | 42,002 | 1,142 | .46 | 22,230 | 272 | 1.25 | 11,061 | 306 | .39 | 1,348 | 28 | .77 |
| 1870 | 38,388 | 1,125 | .52 | 20,945 | 254 | 1.04 | 10,348 | 268 | .43 | 1,331 | 29 | .85 |
| *1869* | ----- | [2] *761* | ----- | ----- | *288* | ----- | ----- | *282* | ----- | ----- | *30* | ----- |
| 1869 | 35,833 | 782 | .73 | 21,194 | 290 | .92 | 9,555 | 284 | .46 | 1,238 | 29 | .87 |
| 1868 | 35,116 | 920 | .62 | 19,140 | 246 | 1.46 | 8,897 | 230 | .54 | 1,064 | 23 | 1.49 |
| 1867 | 32,116 | 794 | .78 | 16,738 | 211 | 2.01 | 8,176 | 223 | .59 | 1,058 | 24 | 1.22 |
| 1866 | 30,017 | 731 | .66 | 15,408 | 170 | 2.06 | 7,935 | 232 | .47 | 754 | 18 | .95 |
| *1859* | ----- | [2] *839* | ----- | ----- | *173* | ----- | ----- | *173* | ----- | ----- | *16* | ----- |
| *1849* | ----- | [2] *592* | ----- | ----- | *100* | ----- | ----- | *147* | ----- | ----- | *5* | ----- |
| *1839* | ----- | [2] *378* | ----- | ----- | *85* | ----- | ----- | *123* | ----- | ----- | *4* | ----- |

1. December 1 price received by farmers prior to 1908; season average price thereafter.
2. Corn harvested for grain only.
Source: Series K 502–516, pp. 511–12.

TABLE A.11. Hay, cotton, cottonseed, shorn wool, and tobacco—acreage, production, and price, 1790–1910 [Census figures in *italics*]

| Year | Hay [1] | | | Cotton | | | Cottonseed | | Shorn wool | | Tobacco | | |
|---|---|---|---|---|---|---|---|---|---|---|---|---|---|
| | Acreage harvested | Production | Price per ton [2] | Acreage harvested | Production [3] | Price per pound [2] | Production | Price per ton [2] | Production [4] | Price per pound [5] | Acreage harvested | Production | Price per pound [6] |
| | *1,000 acres* | *Million tons* | *Dollars* | *1,000 acres* | *1,000 bales* | *Cents* | *1,000 tons* | *Dollars* | *Million pounds* | *Cents* | *1,000 acres* | *Million pounds* | *Cents* |
| 1910 | 68,332 | 75 | 11.66 | 31,508 | 11,609 | 13.96 | 5,156 | 26.11 | 306 | 21.7 | 1,398 | 1,142 | 9.3 |
| *1909* | *68,227* | *87* | *8.90* | *32,044* | *10,649* | | *5,325* | *22.74* | | | | | |
| 1909 | 68,703 | 87 | 10.20 | 30,555 | 10,005 | 13.52 | 4,442 | 24.35 | 310 | 22.2 | 1,292 | 1,054 | 10.1 |
| 1908 | 51,487 | 72 | 9.08 | 31,091 | 13,241 | 9.01 | 5,883 | -------- | 270 | 16.3 | 1,009 | 836 | 10.2 |
| 1907 | 49,833 | 66 | 11.60 | 30,729 | 11,106 | 10.36 | 4,931 | -------- | 256 | 20.5 | 1,042 | 886 | 10.0 |
| 1906 | 48,650 | 60 | 10.40 | 31,404 | 13,274 | 9.58 | 5,898 | -------- | 257 | 23.1 | 1,123 | 973 | 9.6 |
| 1905 | 48,333 | 67 | 8.49 | 27,753 | 10,576 | 10.78 | 4,700 | -------- | 253 | 22.2 | 1,103 | 939 | 8.2 |
| 1904 | 47,480 | 66 | 8.82 | 30,077 | 13,438 | 8.98 | 5,967 | -------- | 250 | 16.3 | 1,026 | 857 | 7.8 |
| 1903 | 46,650 | 64 | 9.18 | 27,762 | 9,851 | 10.49 | 4,379 | -------- | 245 | 15.4 | 1,212 | 976 | 6.7 |
| 1902 | 44,716 | 59 | 9.05 | 27,561 | 10,630 | 7.60 | 4,729 | -------- | 274 | 13.7 | 1,189 | 960 | 6.9 |
| 1901 | 43,555 | 53 | 9.88 | 27,050 | 9,508 | 7.03 | 4,226 | -------- | 266 | 13.7 | 1,098 | 886 | 7.2 |
| 1900 | 42,488 | 50 | 9.78 | 24,886 | 10,124 | 9.15 | 4,500 | -------- | 260 | 13.7 | 1,086 | 852 | 6.7 |
| *1899* | *61,691* | *79* | *6.11* | *24,275* | *9,535* | | *4,767* | *9.85* | | | | | |
| 1899 | 43,395 | 54 | 8.20 | 24,163 | 9,346 | 6.98 | 4,152 | -------- | 272 | 14.5 | 1,102 | 870 | 7.1 |
| 1898 | 43,083 | 60 | 6.52 | 24,715 | 11,278 | 5.73 | 5,120 | -------- | 267 | 13.7 | 1,116 | 909 | 6.1 |
| 1897 | 42,396 | 56 | 7.21 | 25,131 | 10,899 | 6.68 | 4,878 | -------- | 259 | 11.1 | 978 | 703 | 7.4 |
| 1896 | 40,971 | 51 | 7.60 | 23,230 | 8,533 | 6.66 | 3,778 | -------- | 272 | 10.3 | 1,038 | 760 | 5.5 |
| 1895 | 41,153 | 40 | 9.63 | 19,839 | 7,162 | 7.62 | 3,174 | -------- | 294 | 10.3 | 1,006 | 745 | 6.8 |
| 1894 | 41,864 | 48 | 8.98 | 21,886 | 9,091 | 4.59 | 4,448 | -------- | 325 | 11.1 | 993 | 767 | 6.6 |
| 1893 | 42,083 | 53 | 9.48 | 20,256 | 7,493 | 7.00 | 3,297 | -------- | 349 | 14.5 | 1,096 | 767 | 7.9 |
| 1892 | 41,328 | 53 | 8.78 | 18,869 | 6,700 | 8.34 | 2,956 | -------- | 294 | 16.3 | 1,039 | 757 | 8.9 |
| 1891 | 40,350 | 51 | 8.65 | 21,503 | 9,035 | 7.24 | 3,967 | -------- | 285 | 16.3 | 955 | 747 | 8.2 |
| 1890 | 39,613 | 51 | 8.11 | 20,937 | 8,653 | 8.59 | 3,802 | -------- | 276 | 17.1 | 851 | 648 | 8.0 |
| *1889* | *52,949* | *67* | | *20,175* | *7,473* | | | -------- | | | | | |
| 1889 | 38,867 | 50 | 7.74 | 20,191 | 7,473 | 8.55 | 3,318 | -------- | 265 | 18.0 | 758 | 525 | 6.6 |
| 1888 | 37,411 | 47 | 9.24 | 19,520 | 6,938 | 8.50 | 3,074 | -------- | 269 | 17.1 | 891 | 661 | 7.9 |
| 1887 | 36,480 | 42 | 10.09 | 18,793 | 7,047 | 8.55 | 3,056 | -------- | 285 | 18.0 | 722 | 469 | 10.5 |
| 1886 | 35,771 | 45 | 8.72 | 18,370 | 6,505 | 8.06 | 2,802 | -------- | 302 | 16.3 | 848 | 609 | 7.2 |
| 1885 | 34,507 | 43 | 10.07 | 17,922 | 6,576 | 8.39 | 2,828 | -------- | 308 | 14.5 | 815 | 611 | 7.5 |
| 1884 | 33,448 | 43 | 8.99 | 16,849 | 5,682 | 9.19 | 2,427 | -------- | 300 | 14.5 | 754 | 580 | 8.1 |
| 1883 | 32,077 | 44 | 8.77 | 16,295 | 5,713 | 9.13 | 2,450 | -------- | 290 | 17.1 | 750 | 509 | 8.8 |
| 1882 | 30,373 | 39 | 9.99 | 15,638 | 6,949 | 9.12 | 3,033 | -------- | 272 | 20.5 | 744 | 579 | 8.5 |
| 1881 | 28,619 | 35 | 12.25 | 16,483 | 5,456 | 10.66 | 2,280 | -------- | 240 | 22.2 | 698 | 426 | 9.6 |
| 1880 | 27,011 | 33 | 11.82 | 15,921 | 6,606 | 9.83 | 2,822 | -------- | 233 | 23.1 | 650 | 469 | 8.1 |
| *1879* | *30,631* | *35* | | *14,480* | *5,755* | | | -------- | | | | | |
| 1879 | 26,641 | 31 | 9.63 | 14,474 | 5,756 | 10.28 | 2,425 | -------- | 211 | 18.0 | 633 | 472 | 6.1 |
| 1878 | 25,627 | 33 | 7.32 | 13,539 | 5,074 | 8.16 | 2,106 | -------- | 208 | 18.8 | 651 | 455 | 5.8 |
| 1877 | 24,749 | 30 | 8.72 | 12,606 | 4,773 | 8.53 | 1,994 | -------- | 200 | 21.4 | 789 | 621 | 5.4 |
| 1876 | 23,986 | 29 | 9.80 | 11,747 | 4,474 | 9.71 | 1,826 | -------- | 192 | 19.7 | 625 | 466 | 7.3 |

| Year | | | | | | | | | | | | | |
|---|---|---|---|---|---|---|---|---|---|---|---|---|---|
| 1875 | 22,662 | 26 | 12.75 | 11,348 | 4,631 | -- | 1,909 | -- | 181 | 25.7 | 746 | 609 | 7.7 |
| 1874 | 21,861 | 25 | 13.85 | 10,753 | 3,836 | -- | 1,567 | -- | 170 | 25.7 | 378 | 217 | 13.7 |
| 1873 | 21,597 | 24 | 14.40 | 10,998 | 4,168 | -- | 1,718 | -- | 158 | 26.5 | 513 | 382 | 8.6 |
| 1872 | 21,081 | 24 | 15.35 | 9,580 | 3,933 | -- | 1,621 | -- | 150 | 31.7 | 492 | 385 | 10.7 |
| 1871 | 20,270 | 22 | 16.57 | 8,285 | 2,974 | -- | 1,223 | -- | 160 | 27.4 | 420 | 327 | 10.2 |
| 1870 | 19,719 | 21 | 14.45 | 9,238 | 4,352 | -- | 1,786 | -- | 162 | 22.2 | 424 | 345 | 9.7 |
| *1869* | -- | *27* | -- | -- | *3,012* | -- | -- | -- | -- | -- | -- | -- | -- |
| 1869 | 19,310 | 25 | 12.76 | 7,751 | 3,011 | -- | 1,118 | -- | 162 | 22.2 | 395 | 264 | 10.9 |
| 1868 | 19,568 | 23 | 13.90 | 6,973 | 2,366 | -- | 974 | -- | -- | -- | 369 | 286 | 12.2 |
| 1867 | 18,641 | 23 | 14.30 | 7,864 | 2,520 | -- | 1,042 | -- | -- | -- | 370 | 260 | 11.6 |
| 1866 | 18,250 | 21 | 14.48 | 7,666 | 2,097 | -- | 864 | -- | -- | -- | 394 | 316 | 11.6 |

| Year | Hay, production | Cotton, production | Year | Hay, production | Cotton, production | Year | Cotton, production | Year | Cotton, production | Year | Cotton, production | Year | Cotton, production |
|---|---|---|---|---|---|---|---|---|---|---|---|---|---|
| | *Million tons* | *1,000 bales* | | *Million tons* | *1,000 bales* | | *1,000 bales* | | *1,000 bales* | | *1,000 bales* | | *1,000 bales* |
| 1865 | -- | 2,094 | 1850 | -- | 2,136 | 1835 | 1,062 | 1820 | 335 | 1805 | 146 | 1795 | 17 |
| 1864 | -- | 299 | *1849* | *14* | *2,469* | 1834 | 962 | 1819 | 349 | 1804 | 136 | 1794 | 17 |
| 1863 | -- | 449 | 1849 | -- | 2,066 | 1833 | 931 | 1818 | 262 | 1803 | 126 | 1793 | 10 |
| 1862 | -- | 1,597 | 1848 | -- | 2,615 | 1832 | 816 | 1817 | 272 | 1802 | 115 | 1792 | 6 |
| 1861 | -- | 4,491 | 1847 | -- | 2,128 | 1831 | 805 | 1816 | 259 | 1801 | 100 | 1791 | 4 |
| | | | 1846 | -- | 1,604 | | | | | | | | |
| 1860 | -- | 3,841 | | | | 1830 | 732 | 1815 | 209 | 1800 | 73 | 1790 | 3 |
| *1859* | *19* | *5,387* | 1845 | -- | 1,806 | 1829 | 764 | 1814 | 146 | 1799 | 42 | | |
| 1859 | -- | 4,508 | 1844 | -- | 2,079 | 1828 | 680 | 1813 | 157 | 1798 | 31 | | |
| 1858 | -- | 3,758 | 1843 | -- | 1,750 | 1827 | 565 | 1812 | 157 | 1797 | 23 | | |
| 1857 | -- | 3,012 | 1842 | -- | 2,035 | 1826 | 732 | 1811 | 167 | 1796 | 21 | | |
| 1856 | -- | 2,874 | 1841 | -- | 1,398 | | | | | | | | |
| | | | | | | 1825 | 533 | 1810 | 178 | | | | |
| 1855 | -- | 3,221 | 1840 | -- | 1,348 | 1824 | 450 | 1809 | 172 | | | | |
| 1854 | -- | 2,708 | *1839* | *10* | *1,976* | 1823 | 387 | 1808 | 157 | | | | |
| 1853 | -- | 2,766 | 1839 | -- | 1,654 | 1822 | 439 | 1807 | 167 | | | | |
| 1852 | -- | 3,130 | 1838 | -- | 1,093 | 1821 | 377 | 1806 | 167 | | | | |
| 1851 | -- | 2,799 | 1837 | -- | 1,428 | | | | | | | | |
| | | | 1836 | -- | 1,129 | | | | | | | | |

1. All hay, 1909–1910; tame hay prior to 1909.

2. December 1 average price prior to 1909; season average price thereafter. Loose hay price after 1909.

3. Annual production estimates prior to 1962 are shown in 500-pound gross-weight bales. Figures for census years are shown in running bales, and are not comparable with annual production estimates.

4. Includes shearing at commercial feeding yards.

5. December 1 price.

Source: Series K 550–563, pp. 517–18.

# INDEX